Computer Architecture Tutorial
Using an FPGA

ARM & Verilog Introductions

Robert Dunne

Computer Architecture Tutorial Using an FPGA: ARM & Verilog Introductions

Contents at a Glance

Contents

Introduction

When Bob Dylan sang his song "The Times They Are a-Changin' " in the mid 1960s, he was not referring to computer architecture, but to political and social change. From a programmer's perspective, today's popular computer hardware architectures have barely changed from those popular in the 1960s. By no means am I saying there have been no advances in electronics and fabrication. During the past six decades, processor speeds have increased by an incredible factor of about 4,000, and computer hardware prices and physical sizes have plummeted by even a greater factor. However, today's popular X86 and ARM Central Processing Units (CPU) have very similar architectures to those of the IBM 360 and Control Data 6600 mainframes of the 1960s.

Computer Architecture Tutorial Using an FPGA: ARM & Verilog Introductions presents computer "building blocks," and how they are connected to form a computing system. Registers, instruction sets, word size, memory configuration, addressing modes, and data types are factors contributing to a computer's architecture. The Verilog Hardware Description Language (HDL) is a tool that can be used to describe digital electronics components and test their interconnections as they are formed into a computing system.

By working through the examples in this book and experimenting with the "building blocks," the reader will receive a "hands on" introduction to the following:

1. Computer Architecture in general
2. The ARM CPU in particular
3. The Verilog Hardware Description Language
4. Field Programmable Gate Arrays
5. Digital circuits such as decoders and multiplexers
6. Assembly language programming

FPGA Use is On the Rise

Today, the *times* really are "a-Changin'" in both computer architecture and education. Field Programmable Gate Arrays (FPGA) and their "cousins," the Complex Programmable Logical Devices (CPLD), provide flexible digital electronics platforms that can be configured as a CPU, neural net, and almost any other digital circuit. They are the digital equivalent of the shape-shifter in science fiction. At one time, these devices were used only at national laboratories and in specialized embedded systems applications. Now, they are used by Internet search engines, the stock market, graphics applications, neural nets, every cell phone, and many other applications.

Hands On Training at Home

For over fifteen years, I've taught digital electronics and embedded systems in the traditional college environment of lectures accompanied by supervised "hands-on" labs. However, much of today's education is transitioning into on-line learning with students "working at home." This tutorial has been written for students and computer enthusiasts to obtain a working knowledge of computer architecture and get practical experience using FPGAs while studying at home.

The topics begin with a graphical description of the operation of individual logic gates. More complex structures like buses and decoders are then constructed and demonstrated. The Verilog coding begins with simple circuits and culminates with the assembly and execution of many ARM machine code programs. Over 150 illustrations accompany detailed descriptions for setting up the FPGA and walking through each of the nearly 100 Verilog examples.

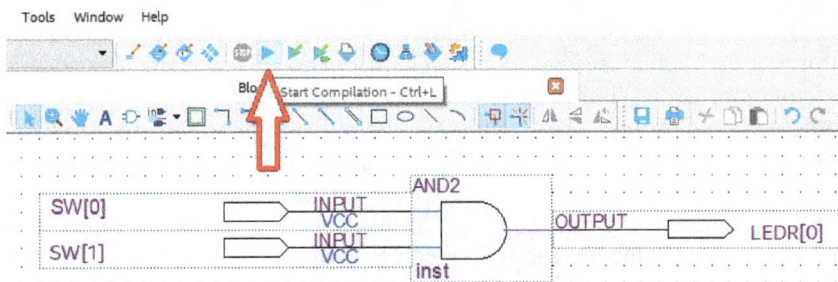

Copy of Figure 1.17: Examine operation of a single AND gate

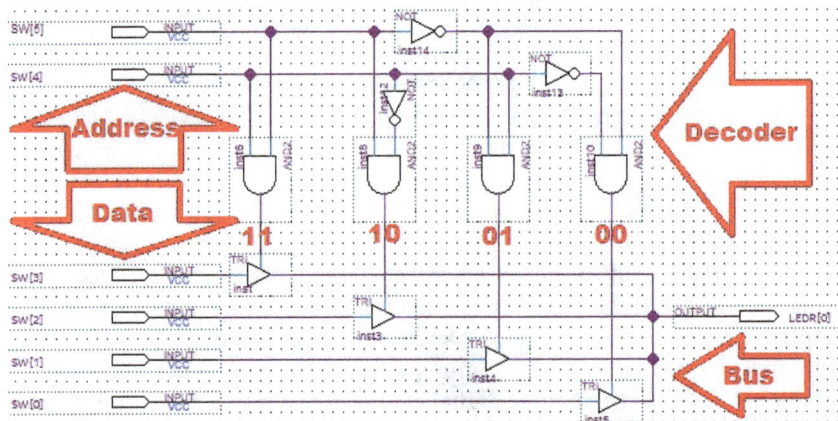

Copy of Figure 2.7: Graphical example demonstrating digital circuits

In this tutorial, a working model of a 32-bit ARM processor is gradually built from basic principles of computer architecture. The Verilog code not only compiles ARM assembly language into 100% ARM machine code, but also executes the programs that are written. The CPU instructions are described in detail, along with examples demonstrating their operation. This CPU imitation can be run at full speed, stepped through with break points, or paused within the fetch, decode, and execute cycle.

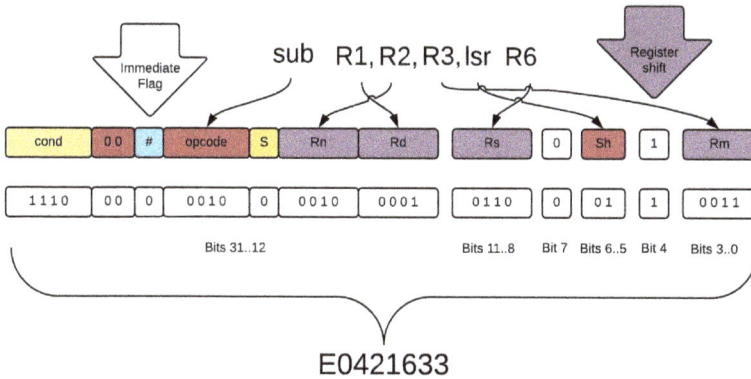

E0421633

Copy of Figure 12.11: Assembly language is presented beginning in Chapter 9

```
785. //
794. //          FactN: Function that calculates factorials, N! = FactN (N).
795. //          Input:      R2: Value of N. Range = [1, 12]
796. //                      LR: Return address
797. //                      SP: Stack must have at least 120 bytes available
798. //          Output:     R2: Calculated value of N. Range = [1, 479001600]
799. //                      R1: Used as scratch, i.e., not saved
800.
801. FactN    =          IP;          // Function FactN entry address
802.          `CMP       R2, 1        `_3 // 5: Test for special, 1! = 1
803.          `BX`EQ     LR           `_3 // 6: Return with factorial in R2
804.          `PUSH      R2           `_2 // 7: Save the input value of N
805.          `PUSH      LR           `_2 // 8: Save return address
806.          `SUB       R2, 1        `_3 // 9: Prepare argument of (N-1)
807.          `BL        FactN        `_2 // A: Get value for (N-1)! in R2
808.          `POP       LR           `_2 // B: Reload return address
809.          `POP       R1           `_2 // C: Restore input value of N into R1
810.          `MUL       R2, R1       `_3 // D: Calculate: N! = N * (N-1)! into R2
811.          `BX        LR           `_2 // E: Return N! in register R2
812.
```

From Listing 16.5: Recursive factorial function "assembled" using Verilog

Audience for This Book

This book is a tutorial having numerous hands-on examples written in the Verilog HDL specifically targeting the Terasic DE2-115 and DE10-Lite FPGA development boards. Either one is fine. Both are part of Intel's FPGA Academic Program. The DE10-Lite is available for purchase on-line for less than $100 at many Internet sites. The DE2-115 and its predecessor, the Altera® DE2, are present in numerous colleges and universities. All examples in this book use the Intel® Quartus® Prime Lite Edition software, which is available as a free download from Intel.

The goal of this book is to introduce students and computer enthusiasts to basic computer architecture, the increasingly popular FPGA technology, and the Verilog Hardware Description Language. The intended audience is the following:

- Someone desiring an introduction to CPU architecture in general and the ARM processor in particular
- Someone wanting to get "hands-on" experience using the Verilog Hardware Description Language
- Someone wanting to learn both at the same time, where Verilog provides a platform for "building" an ARM CPU, and the ARM instruction format provides a set of practical examples to learn Verilog

Expected prerequisites for someone reading this book and working through its examples:

- Access to the Intel® Quartus Prime software and an FPGA development board such as the Terasic DE10-Lite or DE2-115. Many students using this book will have access to computers already loaded with a professional or educational version of Intel® Quartus Prime. For those who don't, the "Lite Edition" is available as a free download from Intel. Basically, any FPGA or CPLD board with 10 slide switches, 10 LEDs, 6 seven-segment displays, 2 push-button switches, and an internal high-speed clock will be adequate for the examples in this book, but slight modifications will be needed from what is shown.
- The examples in this book have been "classroom tested" with students having very little, if any, previous programming or digital hardware design experience. The information is complete, allowing it to be used as an independent study.

Computer Architecture Tutorial Using an FPGA

Book Organization

The presentation begins with basic digital circuits and continues with a gradual construction of the ARM instruction set. The objective is to learn computer architecture by observing ARM machine code in action that has been constructed in Verilog. The concepts and demonstrations appear in the following order:

1. **Introduction to digital circuits and the FPGA boards:** In the first two chapters, the graphical schematic input of Quartus Prime is used to introduce the DE2-115 and DE10-Lite boards along with basic digital concepts and circuits.

2. **Introduction to the Verilog HDL:** Chapters 3 and 4 repeat many of the concepts and circuits, but use the Verilog HDL for implementation. Chapter 5 gives examples of the three main coding styles available in Verilog: Structural, Dataflow, and Behavioral. Verilog examples for implementing the computer building blocks of decoders, registers, and buses are presented in Chapter 6.

3. **Build a simple operational CPU in the FPGA:** In Chapter 7, the general concepts and structures of CPU "machine code" are developed using examples modeled as a calculator. A working model of a CPU using the op-codes of the ARM "data processing" instructions (ADD, SUB, AND, OR, ...) is developed and tested using Verilog in chapters 8 and 9.

4. **Build an ARM assembler and imitation:** In chapters 10 through 17, the 32-bit ARM instruction set is developed. An assembler is built using Verilog macros, tasks, and functions. An ARM machine code execution processor is built using Verilog behavioral style coding.

5. **Demonstrate software techniques:** Recursive function calls and data structures like stacks and arrays are demonstrated in ARM assembly language programs to explain the purpose for and the application of the machine code instructions.

6. **Encourage further development:** Questions and exercises at the end of each chapter reinforce the concepts and techniques presented. Further guidance and encouragement is provided in chapters 18 through 22 which cover block transfers, Thumb format, the NEON coprocessor, and floating point arithmetic.

7. **Background material:** Seven appendices provide supplementary information: Setting up the FPGA boards and the Quartus Prime® software, downloading instructions for over 90 source code files of Verilog demonstrations, binary arithmetic, ASCII, assembly language programming, and review of ARM instruction formats.

8. **Glossary, index, and linkages:** ARM instructions, Verilog commands, diagrams, and Verilog code examples are cross referenced.

Simulation, Emulation, Imitation

The major theme of this book is to learn computer architecture by observing ARM machine code in action. An imitation of the ARM CPU is built in Verilog code starting with a subset of the ARM instruction format and continuously adding more features. The final product can execute many ARM machine code programs, even those produced by commercial C compilers.

Sometimes the terms simulation, emulation, and imitation are confused. Emulation refers to a substitute or a fully operational replacement for a device. A simulation doesn't attempt to replace a product, but to study its operation in great detail, usually at a theoretical or mathematical level. An imitation replicates some functional aspects of a device, but does not provide a full working substitute.

The Verilog language has two major objectives:

- Simulation: Examine the theoretical design of a digital system in detail (such as its critical timing sequences).
- Synthesis: Generate a netlist that provides the physical description for building a working digital system.

In this book, synthesis is used to create hardware descriptions that are then downloaded into the FPGAs. When it comes to designing real production digital circuits, running a simulation is critical. It will provide details of timing faults that will not appear in simple "hands on" testing of the circuit. In the simulation, the functioning of internal "connections" that are inaccessible to a synthesized circuit can be examined and diagnosed.

Topics related to FPGA development or computer architecture that are not described in this book include the following:

- Study of FPGA internal structure and design
- Details of FPGA development boards other than the DE2-115 and DE10-Lite
- Verilog simulation (timings and waveforms)
- VHDL language
- Intel X86 architecture
- Digital design optimization and timings
- Computer board development and peripherals
- Full emulation of entire ARM and NEON architectures

Why the ARM Processor?

The ARM family of processors is one of the most popular in use today. The other very popular CPU family is the Intel X86. The ARM processors are categorized as Reduced Instruction Set Computers (RISC), while the Intel processors are described as Complex Instruction Set Computers (CISC). Since this book is an introduction which describes detailed internal instruction formats, the choice was obvious.

There are actually several variations of the ARM architecture. This book describes a 32-bit version that is similar to the Cortex-A7 and Cortex-A8 formats popular in "home" computers such as the Raspberry Pi and Beaglebone Black.

The ARM instruction format is eloquent and rather straight forward. However, it is not trivial. This tutorial consists of several Verilog modules that build an ARM assembler and execution processor. The modules are gradually developed and included as each new feature is added to the ARM imitation.

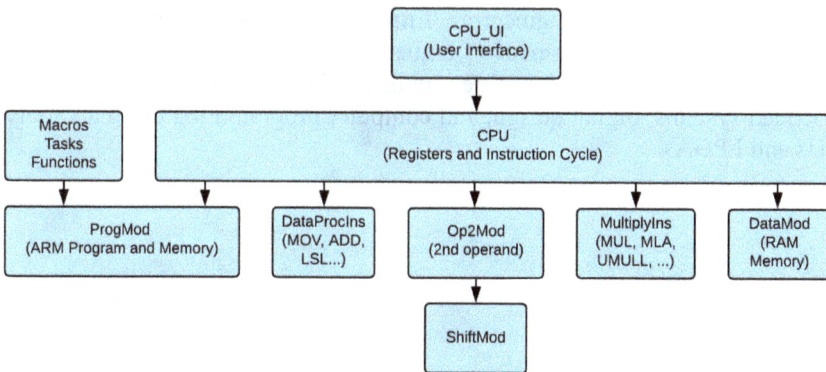

Copy of Figure F.1: Verilog modules gradually developed to describe ARM architecture

Why the Verilog Language?

I have been teaching both the VHDL and Verilog hardware description languages to digital electronics and embedded systems students for over ten years. I really like VHDL, which is both robust and rigorous. However, my new students have been much more successful implementing CPU architectures in Verilog due to its ease of use.

Why didn't I write a book about only the ARM or only Verilog, instead of both combined? I believe most students learn by "doing," and I needed "hands on" examples: Verilog is a great tool for expressing a computer architecture, and emulating the features of the ARM CPU provides a great set of practical examples for learning Verilog.

Let's Play Two

I remember hearing the Chicago Cubs great infielder Ernie Banks saying his famous line "Let's play two" on many fine days for baseball. I tell my electronics students that they are entering a very mature profession, and there is a lot to learn in as short a time as possible. Why not learn about the ARM CPU architecture while studying Verilog? Why not learn Verilog while studying the ARM CPU architecture? Time is precious, and there's a lot to learn. So, let's learn two.

About the Author

Robert Dunne has over 40 years of computer experience ranging from developing custom hardware interfaces for supercomputers to teaching technology courses in middle-school gifted-education programs. Starting out with degrees in physics and computer science, he was on staff at a national laboratory and a major engineering firm for ten years before becoming an entrepreneur in the development of embedded systems. During the past fifteen years, he has been teaching three undergraduate courses per semester in embedded systems, digital design, and computer programming using a variety of CPUs and FPGAs.

— 1 —
Gates & the FPGA

Logical "gates" are the building blocks forming a computer's architecture. Chapters 1 and 2 introduce gates and other basic electronic circuits such as flip flops, registers, and decoders using the graphical schematic design entry available in the Intel® Quartus® Prime Design Software. The same procedures demonstrated in Chapter 1 for setting up a project file, downloading to the FPGA, and testing the design are used in the remainder of this book, except that Verilog will be used for design entry.

The electronics and computer industries are littered with hundreds of abbreviations to save time and space. Lectures often sound like the random reciting of letters of the alphabet in seemingly random sequences. An acronym is an abbreviation that can be pronounced, such as LASER. By that definition, FPGA is certainly not an acronym. However, it's initials do describe what it is: an array of gates that can be programmed (i.e., linked and configured) in the field (i.e., not only at the factory).

Other examples of initialisms used in this book:

- **CPU**: Central Processing Unit
- **ALU**: Arithmetic Logic Unit
- **LED**: Light Emitting Diode
- **CPLD**: Complex Programmable Logical Device

Binary Signals

Digital devices are connected by electric signals that have two possible states: HIGH and LOW. These two (i.e., binary) states correspond to one and zero, true and false, etc.

"How high is up?" is a rhetorical question for philosophers, but in digital electronics, we do have the questions: "How low is LOW?" and how high is HIGH?" It depends. Sometimes a high/low scale is 5 volts, sometimes 3.3 volts, sometimes 24 volts, sometimes 24 VAC, and sometimes 20 milliamps. Where does LOW end and HIGH begin, and what about the undefined range in between? These issues are covered in a basic digital electronics course.

Although it's good to be familiar with the analog voltage/current aspects of connecting one device to another, digital circuit designers primarily focus on the logic, not the electrical. There are also mature applications in "Fuzzy Logic" and emerging applications in "Quantum Computing" where there are many more states than just two, but today it's mostly binary logic.

Gates

A physical gate has two states: open or closed. It allows or forbids movement into or out of an area such as a fenced field. A logical gate also has two states: True or False, which can be represented numerically by One or Zero, and electronically by High or Low.

A logical operation, such as AND, is commonly represented either graphically as a symbol in a schematic drawing or as text characters in an equation. Figure 1.1 shows both approaches including the associated truth table that defines the AND operation.

AND

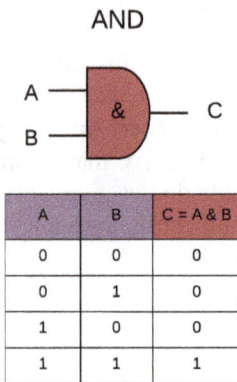

A	B	C = A & B
0	0	0
0	1	0
1	0	0
1	1	1

Figure 1.1: AND operation truth table

The truth table in Figure 1.1 gives the four possible outputs for an AND gate having two inputs.

1. If inputs A and B are both 0, the output will be 0.
2. If A is 0 and B is 1, the output will be 0.
3. If A is 1 and B is 0, the output will be 0.
4. Only if both A and B are 1 will the output be 1.

OR, XOR, and NOT are other fundamental gates having similar graphical and textual representation. They are also defined by their unique truth tables later in this chapter.

Netlist

Figure 1.2 shows a bowl of letters of the alphabet. Products are commercially available as a box of cereal, can of soup, or bag of pasta containing the letters of the alphabet. Although the bowl is composed of unconnected, randomly floating letters, sometimes words form. Sometimes children assemble words in their food. Sometimes the words are appropriate, and sometimes the parents don't approve.

Instead of letters, I would like to see a box of cereal or can of soup composed of the graphical symbols for gates. An FPGA would be like this bowl of gates that are floating around without any order or connections. We could "play" with our gate soup to move the gates into patterns to make digital circuits such as flip flops, registers, decoders, and multiplexers by the appropriate placement of the gates adjacent to each other.

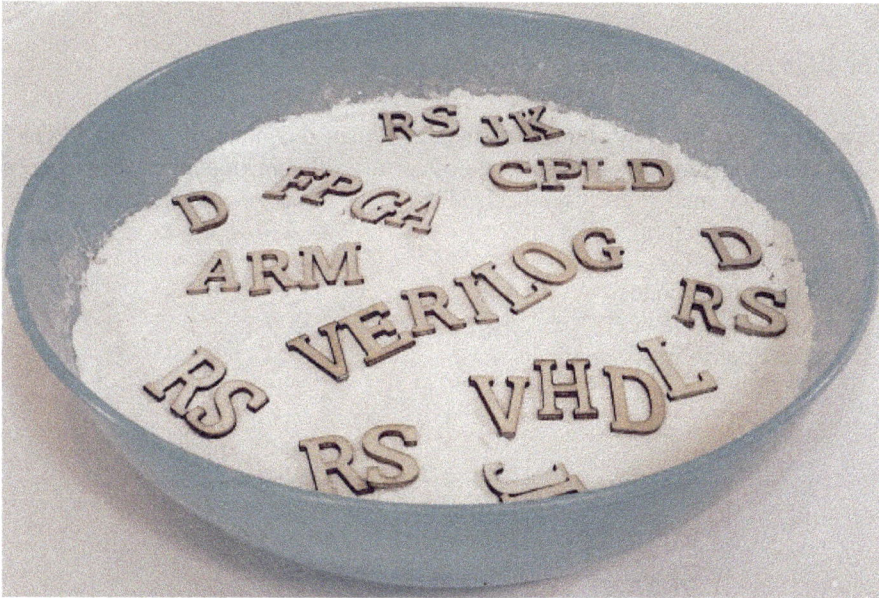

Figure 1.2: Random letters floating in soup or cereal

This gate soup is a metaphor, of course. The gates within an FPGA are not physically floating around, but these gates are not logically connected until the FPGA is "programmed" with a "netlist." A netlist is a pattern of connections forming a desired circuit, as illustrated in Figure 1.3.

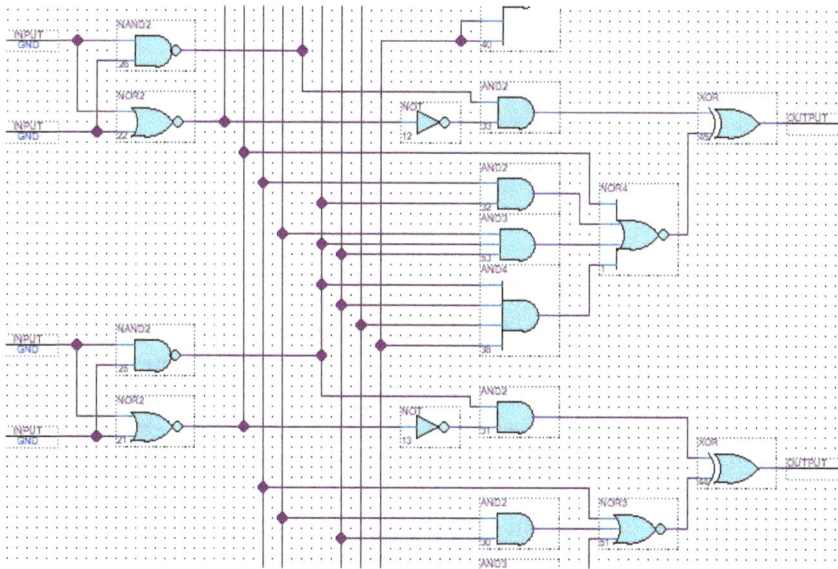

Figure 1.3: Gates "wired" to form netlist

The Digital Design Project

How do we describe and build a netlist, and then download it into an FPGA? Many disk files are involved, and the Quartus software organizes them into a "project." The netlist is "synthesized" by the Quartus software using the following three factors:

1. Source Code
2. Target Device
3. Board Pin Assignments

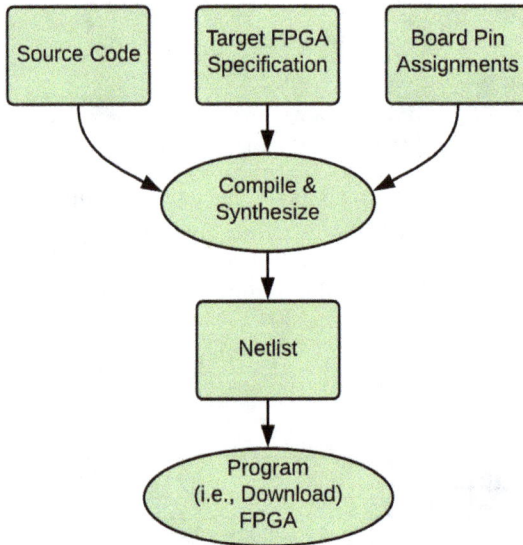

Figure 1.4: Compiler converts digital design's source code into netlist for download to FPGA device.

The first factor is the source code which describes the digital circuit to be built. This can be a graphical map showing the connection of gates or a textual description using a language such as Verilog.

The second factor identifies the type of device. There are hundreds of different FPGAs that can host a digital design. For a particular design, the format of the netlist must be compatible with the particular FPGA to be used.

The third factor involves the circuit board containing the FPGA. An FPGA has hundreds of pins and is mounted on a circuit board containing other integrated circuits as well as switches and displays.

Starting Quartus Prime

Many of you already have the Quartus software installed on your computers. For those who don't, Appendix A provides suggestions for downloading and installing the free "Lite" edition.

Once started, Quartus Prime presents the following opening page allowing the user to select either a recent project or create a new one.

Figure 1.5: Quartus Prime opening page

We are now going to generate a new project. The "New Project Wizard" can be invoked either from the opening page shown in Figure 1.5 or by left-clicking on the "File" pull-down menu as shown in Figure 1.6

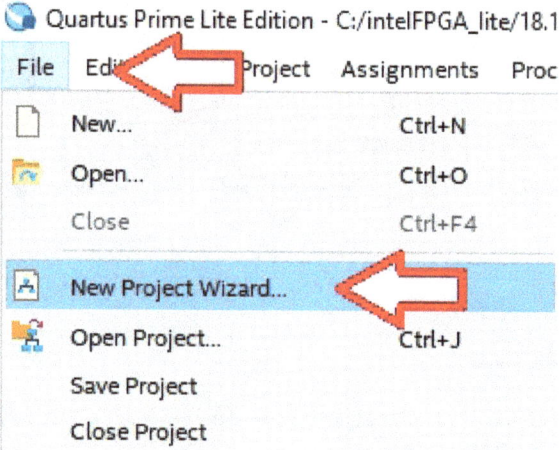

Figure 1.6: Use Project Wizard to specify target device and source language

Project Setup and FPGA Selection

Quartus Prime displays the following six pages in succession for specifying the project and FPGA model. The type of source code (Verilog, schematic, VHDL, etc.) and board pin assignments will be specified separately at a later time.

1. Directory, Name, Top-Level Entity
2. Project Type: Choose empty project or template
3. Add files
4. Family, Device, and Board Settings
5. EDA Tools Settings
6. Summary: Check project name, directory, and device

An "Introductions" page may appear before the "Directory, Name, Top-Level Entity," but it can be skipped. For this example, I will use a project name of "Schematic" and put the project into a directory named "BlockDesign" as shown in Figure 1.7. You, of course, can choose a directory appropriate for your own computer. The top-level design name is automatically copied from the project name. After clicking the "Next" box, a prompt to create a new directory must be answered "Yes."

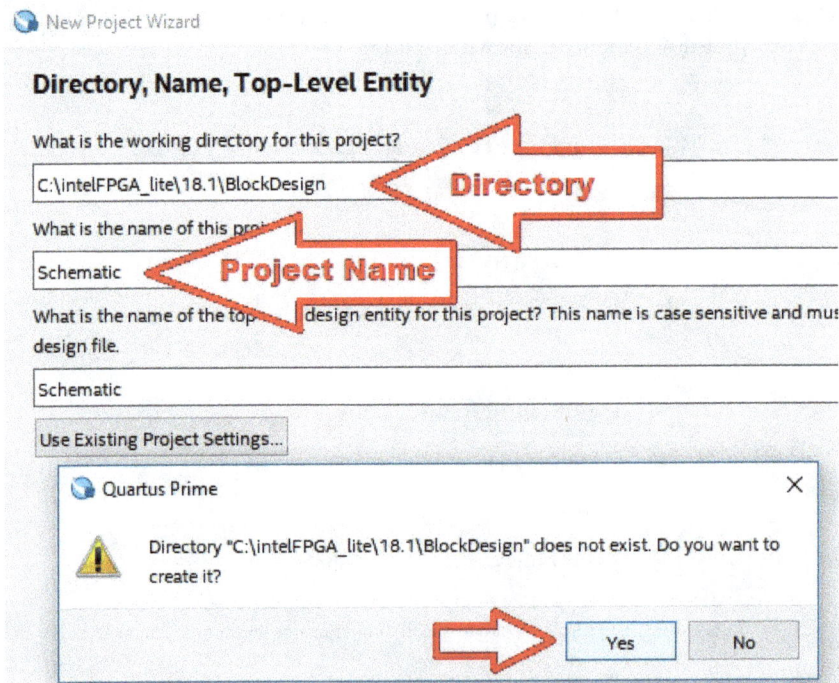

Figure 1.7: Identify the project

Computer Architecture Tutorial Using an FPGA

A "Project Type" page follows where "Empty Project" should be chosen unless you are at a site, such as a university, that has project templates set up corresponding to the FPGA equipment to be used. If you can, or are required to, use the "Project Template" option, then the following FPGA selection criteria pages will not be needed.

The "Add Files" page is next, but it will be skipped (i.e., click "Next") for now. In later chapters, we can use it to combine multiple design segments.

Families of FPGAs and CPLDs

FPGAs and CPLDs are very similar devices for implementing digital designs. Although they are both formed on silicon integrated circuits, their differences in internal structure lead to different operating properties, such as power consumption, speed, storage capacity, and memory volatility at power loss.

The examples in this book have been specifically targeted at two FPGAs: the Intel Cyclone IV E, resident on the Terasic DE2-115 development board, and the Intel MAX 10 DA, resident on the Terasic DE10-Lite development board. Other FPGAs and CPLDs will work with the same source code presented in this book, but the project setup will need to be altered.

Figure 1.8 shows the selections needed for the DE2-115. Note: You may have to widen the Name field in order to locate the "EP4CE115F29C7" part number. For DE10-Lite users, the Family is "MAX 10 DA" and the Name is "10M50DAF484C7G." If you cannot find the exact part number in the Name list, then your Quartus software requires an update. Please see Appendix A for download suggestions.

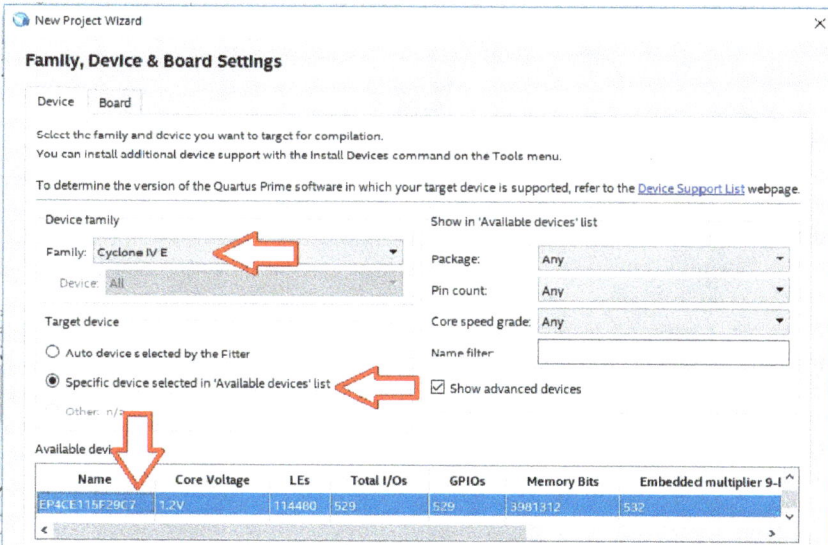

Figure 1.8: Target device

Terasic Board	FPGA Family	Intel Part Number
DE2-115	Cyclone IV E	EP4CE115F29C7
DE10-Lite	MAX 10 DA	10M50DAF484C7G

Table 1.1: FPGA families and part numbers on Terasic boards

"EDA Tools" is the next page in this sequence, and it should be skipped because it is not needed in this book. Finally, the Summary page appears. Please verify that the desired directory, project name, and device have been chosen. If everything looks correct as seen in Figure 1.9 for the DE2-115, then click finish, and the project setup has been completed.

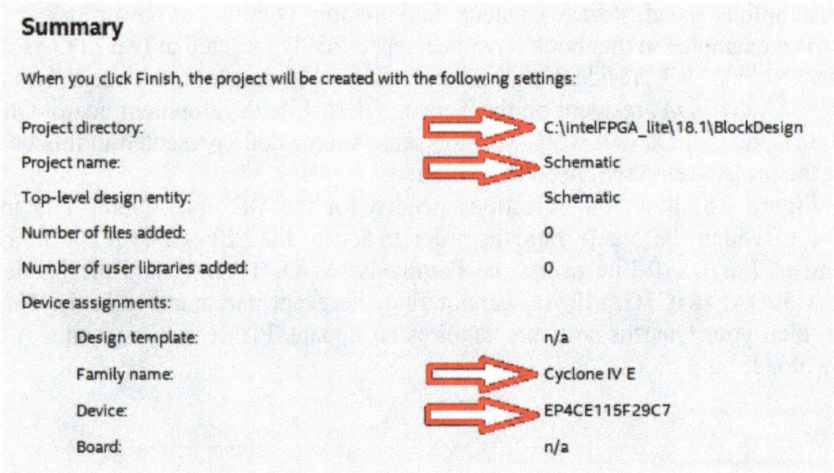

Summary

When you click Finish, the project will be created with the following settings:

Project directory: C:\intelFPGA_lite\18.1\BlockDesign

Project name: Schematic

Top-level design entity: Schematic

Number of files added: 0

Number of user libraries added: 0

Device assignments:

    Design template: n/a

    Family name: Cyclone IV E

    Device: EP4CE115F29C7

    Board: n/a

Figure 1.9: Verify project location and target device

Design File Setup and Schematic Selection

Now that the project has been set up and FPGA target device chosen, we must specify which "Design File" source code language and board pin assignments to use. The Design File is initialized by clicking "File" on the menu bar, then select "New" as seen in Figure 1.10.

Here in Chapters 1 and 2, elementary digital circuits will be described graphically in a schematic because a "picture" leads to a greater understanding for most people. In the remainder of the book, Verilog is used as the source language because at that level of complexity, the textual approach is actually simpler and leads to better organization.

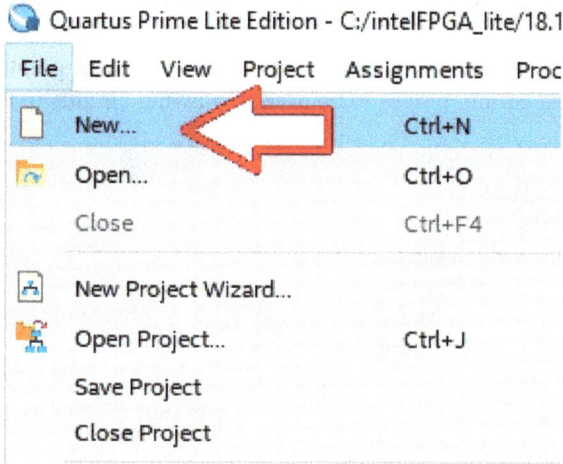

Figure 1.10: Specify source language

Figure 1.11 shows the nine possible source languages that are currently available with Quartus Prime. Choose "Block Diagram / Schematic File." Note: For most of this book, we will be using Verilog coding, so "Verilog HDL File" will be chosen in Chapter 3 and beyond.

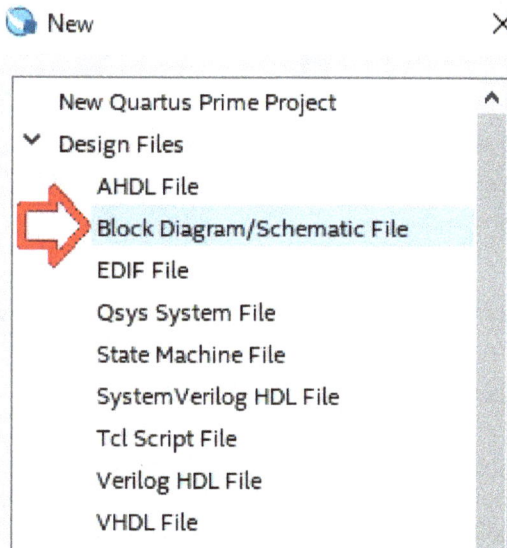

Figure 1.11: Specify schematic entry

A workspace now appears as shown in Figure 1.12. It looks similar to an electronics breadboard for mounting parts and wiring them together. A toolbar is present at the top of the breadboard. The following tools pointed to in Figure 1.12 are the most common ones to be used in drawing circuits in this chapter.

1. Selection Tool
2. Zoom Tool
3. Symbol Tool
4. Orthogonal Node Tool

The "Selection Tool" enables a left mouse click to choose a part for movement or deletion from the breadboard.

The "Zoom Tool" magnifies the breadboard with a left mouse click. This makes connecting the circuits elements much easier. The drawing scale can be reduced with a right mouse click.

The "Symbol Tool" provides symbols for gates, pins, and other basic components. Clicking on it will open a menu that can be "drilled down into" to find an extensive list of components.

The "Orthogonal Node Tool" provides the wires for connecting the gates and pins together. A wire is "drawn" by positioning the mouse at the starting point of a wire, then hold down the left mouse button while moving the mouse to the end of the wire, and then letting up on the button.

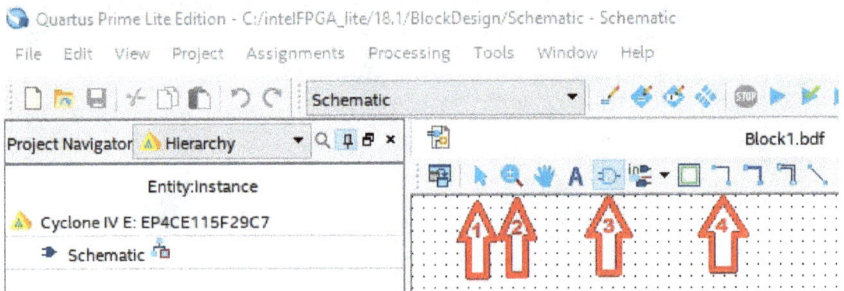

Figure 1.12: Four commonly used tools for building digital schematics

First Example: AND Gate

For our first example, let's simply demonstrate the truth table of an AND gate. Click on the Symbol Tool on the toolbar at the top of the breadboard. Then drill down to select "and2" as shown in Figure 1.13. After clicking OK, the AND gate may be placed one or more times on the breadboard by left mouse clicks. Place only one AND gate, and then push the ESC to keep from placing more copies of the AND symbol.

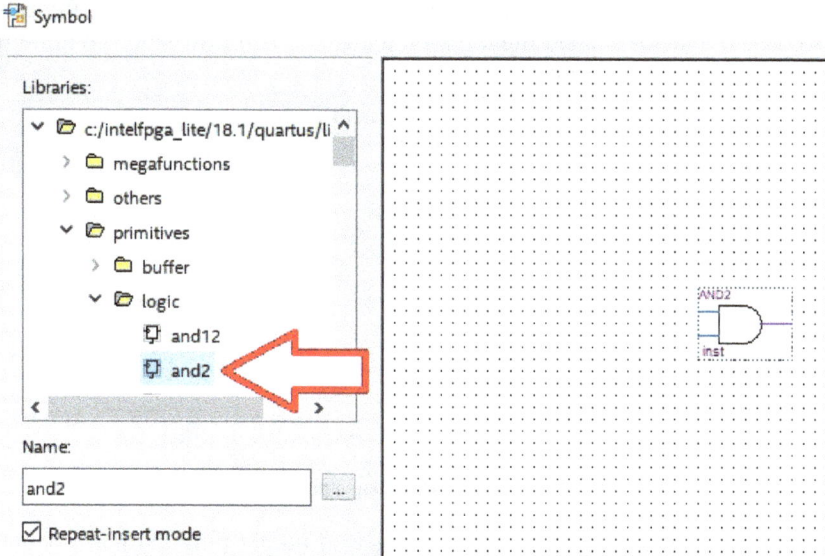

Figure 1.13: Select an AND gate having two inputs.

Now, we have to connect two input pins and one output pin to the AND gate. I recommend first using the Zoom Tool to enlarge the AND gate to make connecting its pins easier. Select "Zoom Tool" from the menu bar, and then left click a couple of times on the breadboard. Now go back to the Symbol Tool, and this time drill down to primitives/pin/input as shown in Figure 1.14.

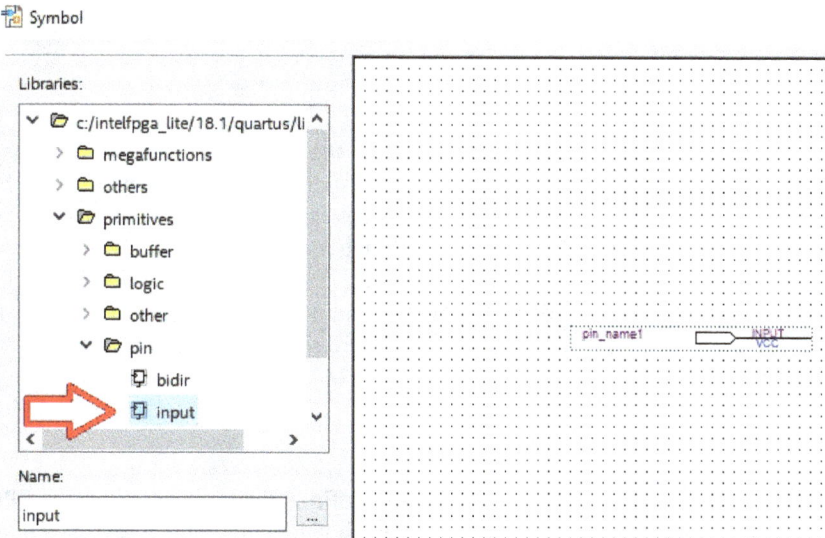

Figure 1.14: Select input pin

Using the mouse, position the first input pin as shown in Figure 1.15 and then left-click to lock it into position on the top input pin of the AND gate. Repeat the procedure to place a second input pin on the gate. Push ESC since no more than two input pins will be needed.

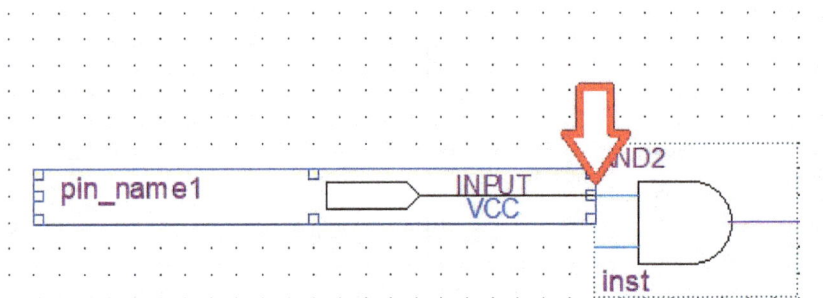

Figure 1.15: Connect input pin to gate

Go back again to the Symbol Tool, and get an output pin, and place it on the drawing as shown in Figure 1.16.

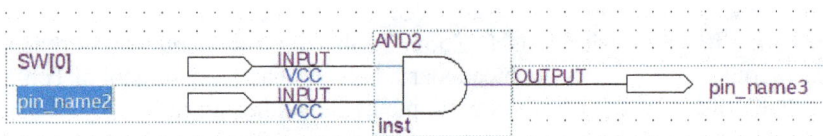

Figure 1.16: Add output pin and provide pin names

Pin Names

"Pins" are the interface from the FPGA to the outside world. If this was an automobile engine application, pin names such as "MAF Sensor" or "Fuel Injector 5" would be appropriate. In this book, common names associated with the DE2-115 and DE10-Lite will be used, such as SW, LEDR, KEY, and HEX.

Change the default names of "pin_name" in Figure 1.16 by double-clicking on each of the three pin_names. The two input names should be SW[0] and SW[1], and the output is LEDR[0]. Note: There are no blanks in the names.

Compilation

We have now prepared the source code and defined the target device. The FPGA board pin assignments are not yet defined, but let's first do a test compile to check for errors in the source code schematic.

A compiler converts source code (human language) to a format compatible with the target device. Here, we are converting a graphical expression of a simple circuit to a netlist to be downloaded into an FPGA. Figure 1.17 points to the triangle that must be clicked to start the compilation.

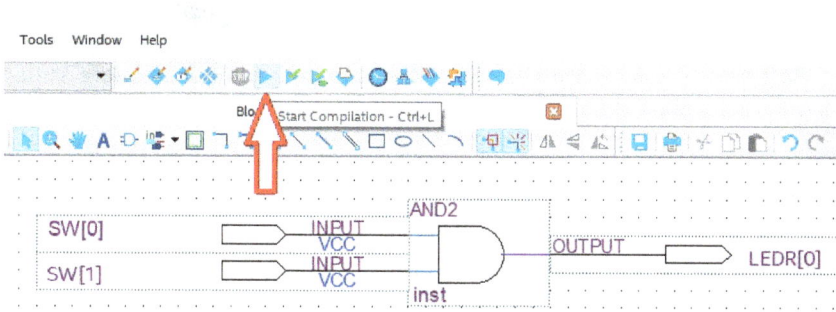

Figure 1.17: Start compilation

When Quartus Prime starts the compilation, it will ask if the source is be updated, so click the "Yes" button on the prompt as shown in Figure 1.18.

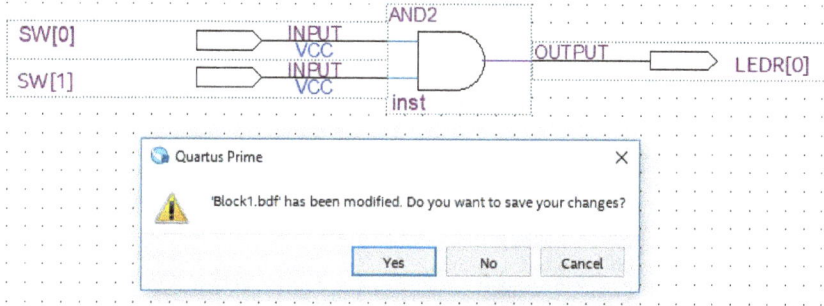

Figure 1.18: File must be updated

The compilation may take 20 seconds or more, which is much longer than programmers working with C, Java, or assembly language expect. Remember: We are generating a digital hardware netlist, not a program to be executed. When complete, messages such as the following will appear.

- Info (293000): Quartus Prime Full Compilation was successful. 0 errors, 13 warnings
- Warning (332068): No clocks defined in design.
- Critical Warning (169085): No exact pin location assignment(s) for 3 pins of 3 total pins.

Red, blue, and green messages can appear. Red messages are fatal, and no netlist is produced. For schematic input such as that performed above, the problem is often a broken connection or extra wire. Look for an "open" or an X on the schematic. Use the Selection Tool to move objects on the schematic to make corrections. If you wish to delete an object, click on it, and push the delete key. After making corrections, recompile, and check for errors again.

Blue messages are warnings that do not prevent a netlist from being generated. Sometimes, hundreds of warnings appear. For classroom exercises and testing, these can generally be ignored. However, for real production work, every warning should be examined.

The green messages are comments that generally do not interfere with the generation of a good netlist.

Pin Assignments

Earlier in this chapter, Figure 1.4 showed that a netlist is generated by compiling a project containing 1) source code, 2) the specifications of the target FPGA, and 3) the development board pin assignments. We have already named the pins as SW[0], SW[1], and LEDR[0], but we have not defined where these pins are located on the FPGA.

Using an Excel CSV file that provides a pin location for each pin name is a convenient method of providing this information. Although several different FPGAs can accept the same design having the same pin names, the CSV file associated with each board will be unique for each particular board.

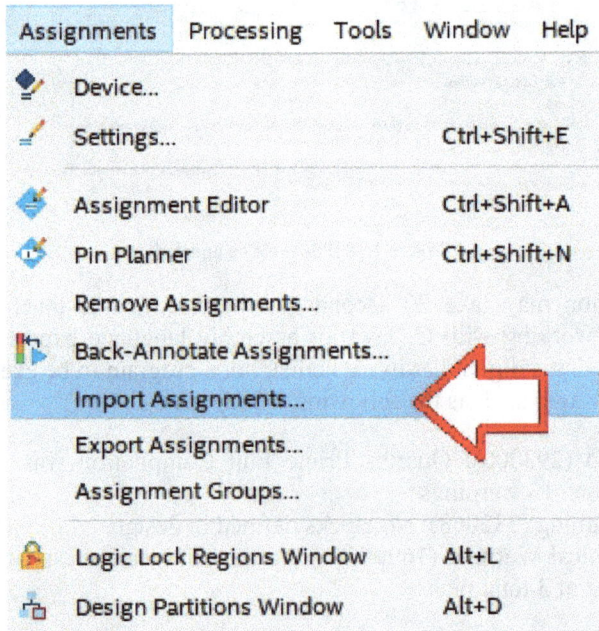

Figure 1.19: Import list of pin name to pin locations

Figure 1.19 shows how the pin assignment file can be included in the project: Select "Assignments" from the top menu bar, and then click on "Import Assignments." The screen in Figure 1.20 will pop up allowing a pin assignment file to be chosen. This CSV file is normally included with the DVD received with the development board and should be copied to a location on the computer's hard drive. It is not part of the Quartus Prime software.

For those who cannot locate the pin assignment CSV file, I have included three copies on GitHub: one for the DE10-Lite, one for the DE2-115, and one for the original Altera DE2. See Appendix A for details.

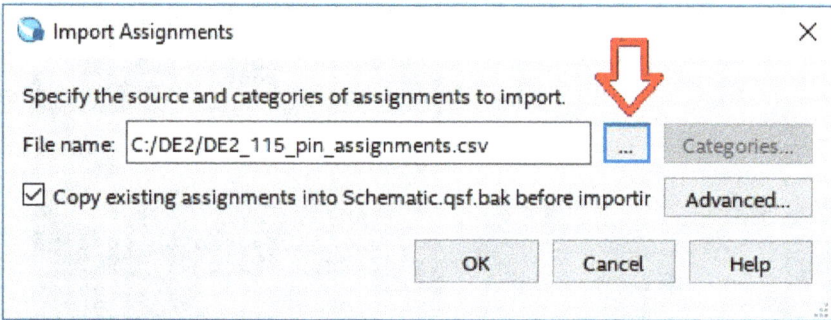

Figure 1.20: Select file containing the pin assignments

Name in Schematic	Direction	Location on Cyclone IV E
KEY[2]	Input	PIN_N21
KEY[1]	Input	PIN_M21
KEY[0]	Input	PIN_M23
LEDR[3]	Output	PIN_F21
LEDR[2]	Output	PIN_E19
LEDR[1]	Output	PIN_F19
LEDR[0]	Output	PIN_G19
SW[3]	Input	PIN_AD27
SW[2]	Input	PIN_AC27
SW[1]	Input	PIN_AC28
SW[0]	Input	PIN_AB28

Table 1.2: Sample of data in "DE2_115_pin_assignments.csv" file

I sometimes like to verify that the pins have been assigned. Start by selecting "Assignments" from the top menu bar, and then click on "Pin Planner." Figure 1.21 shows the pin configuration of the FPGA and the pin location associated with each Node Name (i.e., pin name). Pin locations could have been entered manually using this page, but the CSV file is much more convenient.

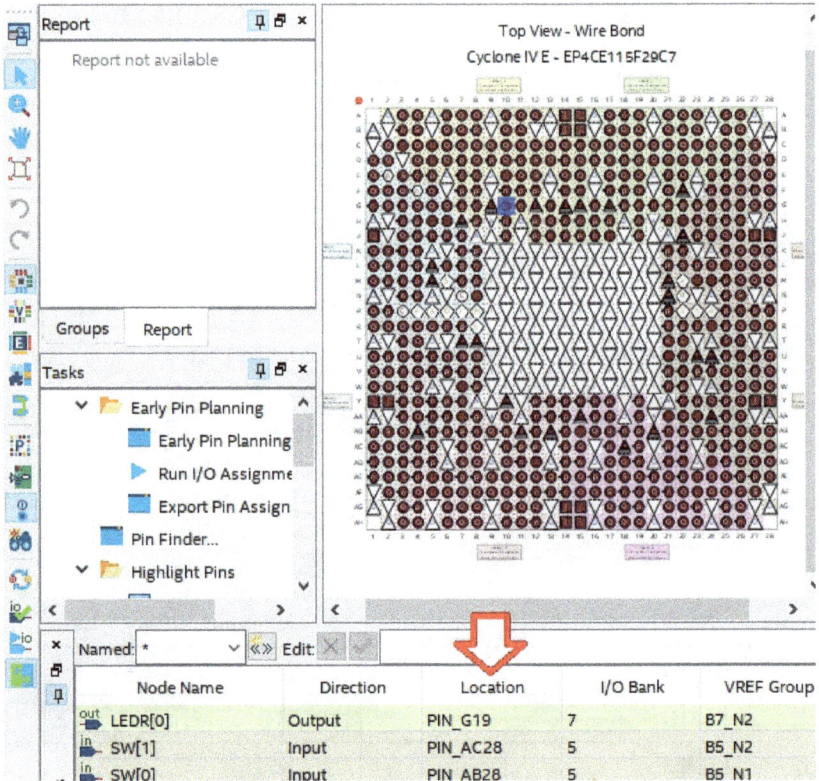

Figure 1.21: Pin location for each pin name

Now that the pins have been assigned, recompile the project to finish the netlist. In Figure 1.22, it is seen that the breadboard now has the pin locations. For the DE10-Lite, SW[0], SW[1], and LEDR[0] are PIN_C10, PIN_C11, and PIN_A8, respectively. For the original Altera DE2, they are PIN_N25, PIN_N26, and PIN_AE23.

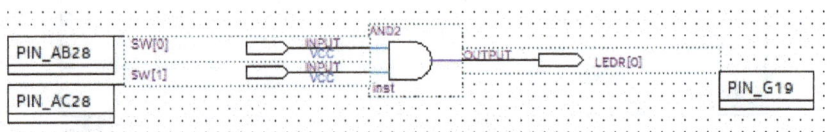

Figure 1.22: Switches and LED pin assignments on DE2-115 board

Download to the FPGA

The netlist is now finished and ready for downloading (i.e., programming) to the FPGA. The "Programmer" command is located six positions to the right of "Compile" command as shown in Figure 1.23. By hovering over each command icon, the Quartus software will identify its meaning.

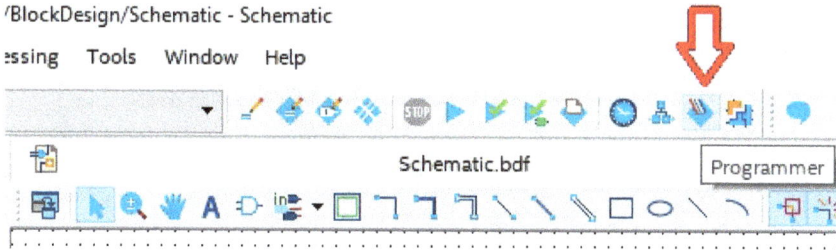

Figure 1.23: Programmer (download)

Before downloading the design into the FPGA, the development board should be powered on, and its USB cable connected. This gives the Windows operating system time to identify the type of device driver needed for communications.

Figure 1.24 shows that USB-Blaster has been selected. First click on the file name to highlight it, and then click "Start" to download the netlist into the FPGA. Typically, the download will take about three or four seconds.

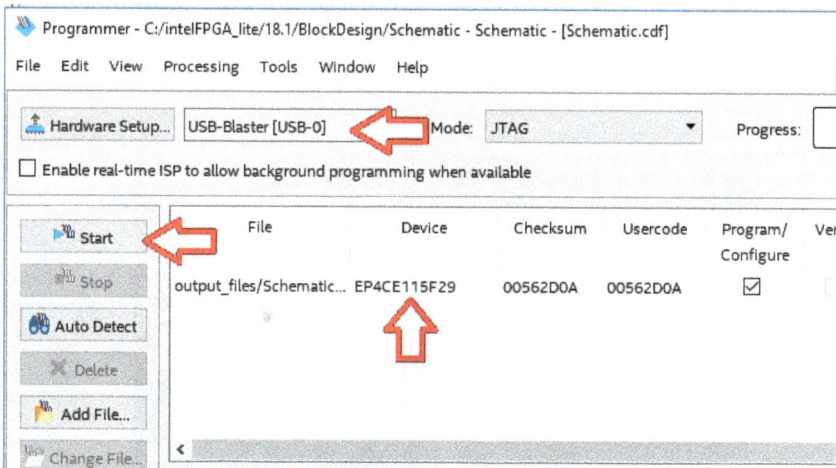

Figure 1.24: Click on the Schematic file name, then click "Start" to download

Sometimes, the Windows operating system has not identified a device driver, and the "Start" button is not ready. Go to "Hardware Settings" as shown in Figure 1.25 and double click on USB-Blaster. Then close the pop-up window if "USB-Blaster" appears in the "Currently selected hardware" line.

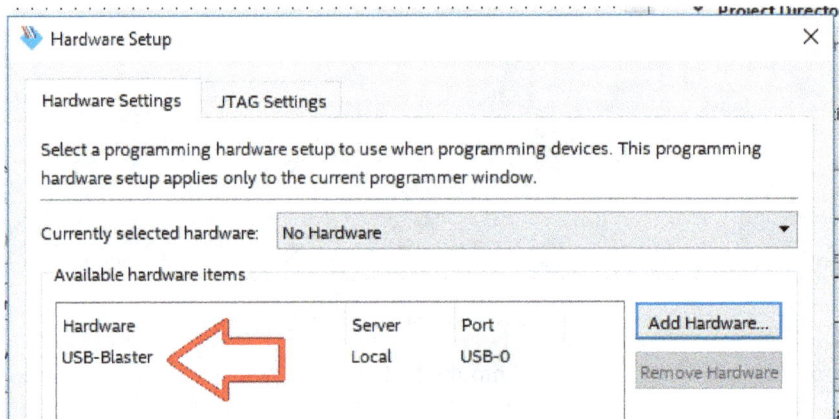

Figure 1.25: If no download driver or hardware

The DE10-Lite has only one USB connector, which also powers the board, but the DE2-115 has three USB connections. If USB-Blaster was not in the list of possible device drivers, please verify that the USB connector closest to the power plug is used (see Figure 1-26).

If the USB-Blaster device driver is still not found, it is probably not present on the computer that you are using. Please see Appendix A for suggestions. Do *not* select a different driver, such as the "Ethernet-Blaster."

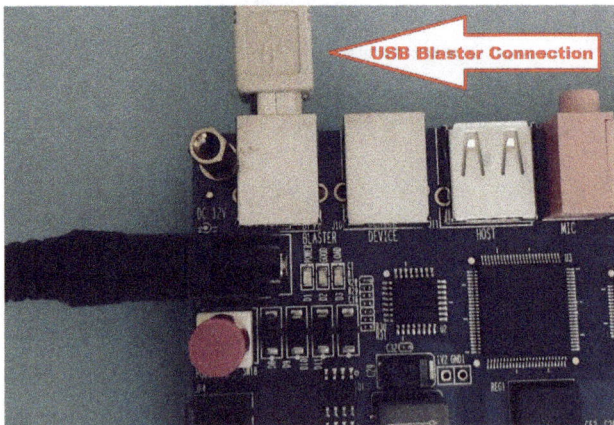

Figure 1.26: Upper left corner of DE2-115 with USB cable attached

Truth Table for the AND Gate

Now that the AND gate design has been downloaded, it is time to test it. With two input switches, there are four possible input configurations: both switches down, both switches up, left down with right up, and left up with right down. The red LED should only light for both switches being up.

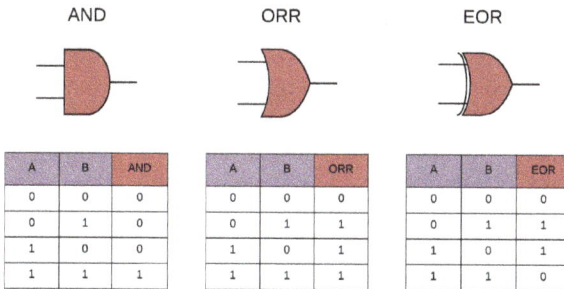

A	B	AND
0	0	0
0	1	0
1	0	0
1	1	1

A	B	ORR
0	0	0
0	1	1
1	0	1
1	1	1

A	B	EOR
0	0	0
0	1	1
1	0	1
1	1	0

Figure 1.27: Truth tables for the logical AND, OR, and exclusive OR operations

Figure 1.27 shows the truth tables for the AND logical operation and two versions of the OR. The inclusive OR is normally just referred to as "OR," but in the ARM instruction set, it is "ORR." Usually the exclusive OR operation is referred to as "XOR," but in the ARM, it is "EOR."

Let's run the truth tables for OR and XOR like that which was done for the AND operation. Go back to the breadboard, remove the AND, replace it with OR, recompile, re-download, and test the four possible switch settings.

1. Get the Selection Tool, and then click on the AND gate to highlight it in the schematic
2. Push "DEL" on keyboard to delete the AND from the circuit
3. Use the Symbol Tool to get the "or2" gate and carefully place it where the AND gate used to be
4. Recompile and download ("program" the FPGA)
5. Run all four possible switch permutations to verify the OR truth table

Perform the above process again, but use the XOR gate, and verify its truth table.

Highlights and Comparisons

Chapter 1 introduced the "gate." A netlist is a digital circuit composed of many gates that can be downloaded (a.k.a, programmed) into an FPGA or CPLD. The

Quartus Prime pull-down menus show many other gates in addition to the *and2*, *or2*, and *xor2* demonstrated in this chapter. I recommend trying some others like the *and3*, *or4*, *nand2*, and *band2* to see what they do.

Intel Corporation bought Altera in 2015 to become a major contributor in the FPGA marketplace. Xilinx, Inc. has been in the FPGA business since 1984 and is credited with inventing the FPGA. Its ISE software is similar to Intel's Quartus Prime, but of course, it generates netlists for its own large families of CPLDs and FPGAs.

Review Questions

1. What do the terms compile, synthesize, and program mean in the creation and testing of a digital hardware description?
2. What is the difference between an OR gate and an XOR gate?
3. *If a netlist is successfully downloaded, why might the FPGA not work properly as expected? In other words, why might the same source code work for one person, but not another (i.e., the LEDs do not light at all when expected)?
4. *Name two reasons why the USB Blaster device driver is not available and does not appear on the download (i.e., program) page?

Exercises

1. Check the Internet for a "brief" description of quantum computing and compare it to the "two state" Aristotelian logic used in digital electronics.
2. Check the Internet for a "brief" description of fuzzy logic and compare it to the "two state" Aristotelian logic used in digital electronics.
3. The designs in this book will also work with CPLDs. Check the Internet to compare CPLDs to FPGAs (power consumption, capacity, power off volatility, internal organization, etc.).
4. Check the Internet to see what makes a CPLD complex compared to the relatively simple SPLD?

— 2 —
Digital Circuits

The architect of a house not only uses bricks and lumber in the design, but also many compound components such as windows, doors, pumps, and lighting fixtures. Likewise a computer's architecture is not only specified as individual gates, but is organized using compound patterns of gates known as the following:

1. Bus
2. Decoder
3. Flip Flop
4. Register

The graphical approach for specifying a digital circuit is great for illustrating basic concepts. Chapter 2 will begin with two application-type examples using a combination of a few gates. It will then showcase "digital building blocks" used in computer architecture. These circuits will be recreated in Chapter 6 using Verilog.

Combinatorial Examples

Digital circuits where all the output signals are determined by a combination of the immediate input signals are referred to as "combinatorial circuits" (a.k.a., "combinational"). In order to get a little more practice with logical circuits, the following two simple combinatorial circuits will be built graphically here in Chapter 2 and repeated in Chapter 6 using Verilog.

1. Elevator Door Example
2. Build XOR from AND, OR, and NOT Gates

In Chapter 1, the truth table for the AND gate was demonstrated graphically using schematic entry. The characteristics of an OR gate were then examined by simply replacing the AND gate with an OR gate in the schematic, recompiling, downloading, and throwing the switches. In other words, we didn't have to create a new project and import the pin assignment file for each new test. Not only can one symbol be replaced, but the circuit can be partially or totally erased by highlighting it using the selection tool and pushing the delete key.

In the elevator door example, we will build a circuit implementing the logic for opening the door. Usually this involves pushing a floor selection button and waiting for the elevator to position itself properly, but there is also a "Firefighters" override.

Figure 2.1: Elevator floor buttons with "Fireman" key

The following four factors contribute to the decision to open the door:

1. Elevator is at correct floor
2. Elevator is stopped
3. Elevator is level with floor
4. Fireman key override

Basically, the door should be opened only when it is at the correct floor, it is stopped, AND it is level with the floor. However, firefighters have special override privileges to properly handle emergencies. Figure 2.2 shows the circuit that is built to implement this logic using a three-input AND gate along with a two-input OR gate.

Note that a wire must be "drawn" from the output of the AND gate to the input of the OR gate. Using the "Orthogonal Node Tool," position the mouse at the starting point of the wire, then hold down the left mouse button while moving the mouse to the end of the wire, and then let up on the button.

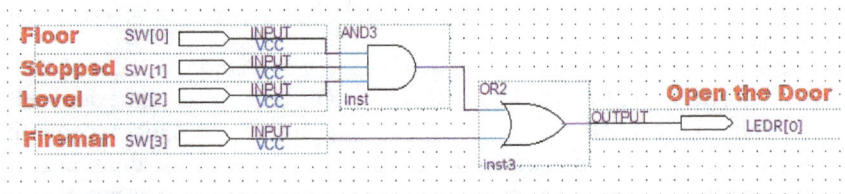

Figure 2.2: Elevator door-opening example

Take note of the "inst" and "inst3" names assigned to the AND and OR gates in Figure 2.2. Every instance (i.e., occurrence) of a gate must have a unique name. The names assigned automatically by Quartus Prime can be overwritten by double clicking on them. See question 2 at the end of this chapter.

Computer Architecture Tutorial Using an FPGA

XOR from AND, OR, and NOT

Logical gates can be built from other logical gates. For example, the XOR can be built from a combination of AND with OR, but we will also need another gate called the "inverter" (i.e., NOT gate).

A "bubble" attached to a gate indicates a signal inversion: A HIGH input results in a LOW output, and a LOW input results in a HIGH output. Figure 2.3 illustrates the following gates which are also available from the Symbol Tool:

- Buffer: From a logic perspective, a buffer is the same as a wire, but it does provide for an electrical power boost and slight timing delay.
- Inverter: A logical NOT is a "bubble" inline with a buffer.
- NAND: A "NOT AND" or NAND gate is an AND gate with its output inverted.
- NOR: A "NOT OR" or NOR gate is an OR gate with its output inverted.
- Bubble AND: This symbol indicates that the inputs are inverted before entering the AND gate. Although this is not a common name for this symbol, it does appear fairly frequently on digital schematics. Using DeMorgan's Law (algebraic technique), the BAND is the same as the NOR gate. Likewise, the BOR is the same as a NAND gate.

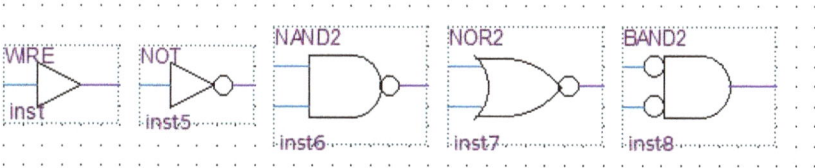

Figure 2.3: Some more gates popular in digital design

Figure 2.4 shows the XOR function being generated by two AND gates, two inverters, and an OR gate. As in the previous example, the "Orthogonal Node Tool" provides the wires for connecting the gates and pins together. Note the two big dots on the schematic. This shows that a wire can even be connected to another wire.

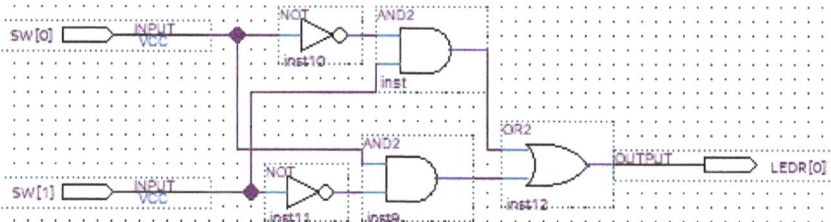

Figure 2.4: Build XOR from AND, OR, and NOT

The Bus

A bus is a common communications path among multiple digital devices. For example, a bus is the data path between the CPU and multiple memory and I/O devices. Although it can be implemented using a combination of AND/OR gates, it is commonly implemented on a computer's circuit board using a technique known as "tri-state" (also written as "3-state"). Digital logic has only two states: High and Low, One and Zero, True and False, etc. Right? Actually, some digital output lines can have a "high impedance" third state.

The schematic in Figure 2.5 illustrates a single wire bus having four inputs and one device to display the high/low status of the bus. Take note of the following:

1. The outputs from the four switches are wired together.
2. The buffer gates connected to each switch have a third wire coming in from the top.

If all four switches are High or all four are Low, the output value of the bus is obvious. But what if two switches are High and two are Low? Is the value going to be half way? Is there going to be a "fight" or possibly even some type of overload or short circuit? With tri-state, a bus master (i.e., the CPU) can select which one gate will output to the bus line. All of the outputs from the other gates will be inhibited.

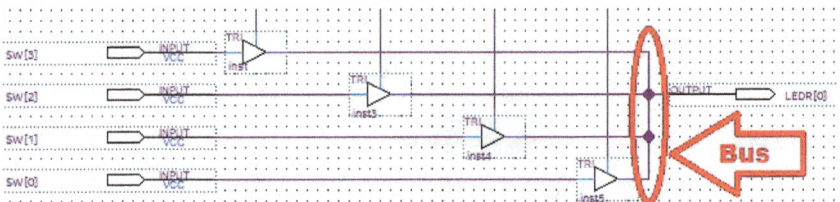

Figure 2.5: Bus: Four switches output to one LED

Not all digital electronics gates are tri-state, only those that output to a bus. The three possible outputs are the following:

1. **High:** The gate "drives" other gates with a voltage in the approximate range of 3 to 5 volts.
2. **Low:** The gate "sinks" the output bus line to near ground level (zero volts).
3. **High-Z:** The gate neither drives nor sinks the bus line. This high impedance state makes it look like the gate is not even connected to the bus line.

Go ahead and build the circuit in Figure 2.5. The tri-state buffer gates are available from the symbol tool as shown in Figure 2.6. Connect switches 4 through 7 to the control lines on the four buffer gates. When you run the example, make sure to only have one of the control switches (SW[5] through SW[7]) in the High state at a time. The other three must be Low, thereby inhibiting their outputs.

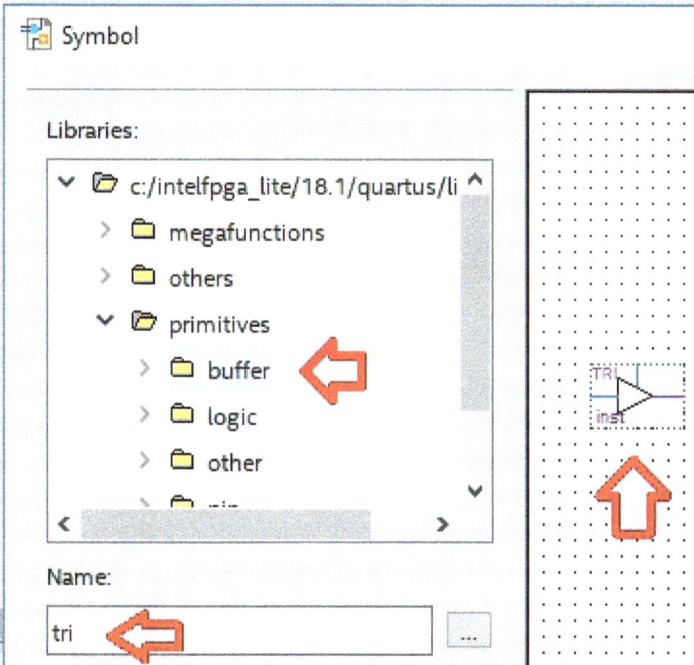

Figure 2.6: Select tri-state buffer gate

The bus is an important building block for computer systems:

1. In a typical computer architecture, there are three buses: data, address, and control.
2. Buses typically consist of multiple single bus lines. For example in a 32-bit processor, the data bus is actually 32 individual bus signals running in parallel.
3. Another technique popular for implementing a bus structure is "open collector," but it is normally applied to a network bus between computers.

Decoder

A decoder circuit basically converts a binary number into a set of signals. We will begin with a two-bit decoder which selects one of four possible choices. This is a follow-on to the previous "bus" example where one (and only one) tri-state buffer gate is to be selected. The other three must be inhibited.

The schematic in Figure 2.7 illustrates a single wire bus having four inputs and one output. This decoder is built with four AND gates and three NOT gates. Switches SW[4] and SW[5] will dictate which of the switches SW[0] through SW[3] drives the output line. This is similar to a memory bus within a computer where the address lines select which memory chip provides the data.

SW[5], SW[4]	Switch Signal Selected	Switch Signals Inhibited
0, 0	SW[0]	SW[3], SW[2], SW[1]
0, 1	SW[1]	SW[3], SW[2], SW[0]
1, 0	SW[2]	SW[3], SW[1], SW[0]
1, 1	SW[3]	SW[2], SW[1], SW[0]

Table 2.1: Decoder selects one of four inputs

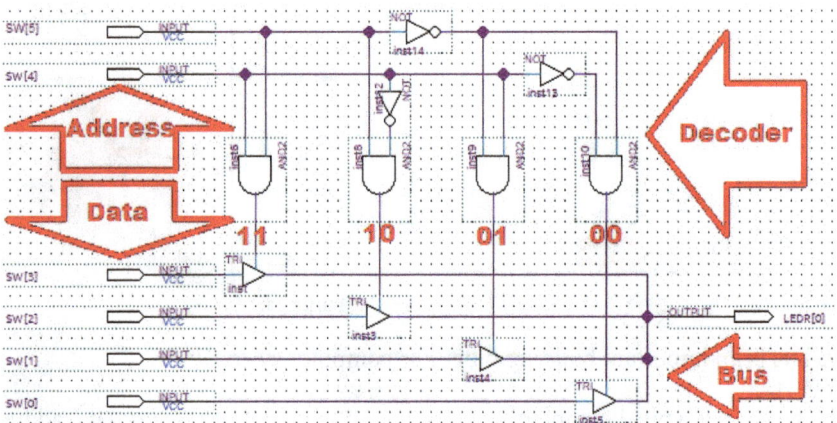

Figure 2.7: Decoder with tri-state

Likewise, a decoder with three inputs can select one of eight candidates. The ARM instruction format has several 4-bit fields that select one of sixteen possible values. For example, a 4-bit field indicates which of sixteen data processing instructions is to be performed. Another 4-bit field selects which of 16 general purpose registers is to be accessed.

Decoders are sometimes referred to as demultiplexers. They are a very common digital building block and are available as preassembled units. For example, they can be found under the "maxplus2" category using the Quartus Prime symbol tool as shown in Figure 2.8.

Go ahead and try the circuit in Figure 2.9 to get comfortable with how a 4-bit decoder selects one of sixteen possible data lines.

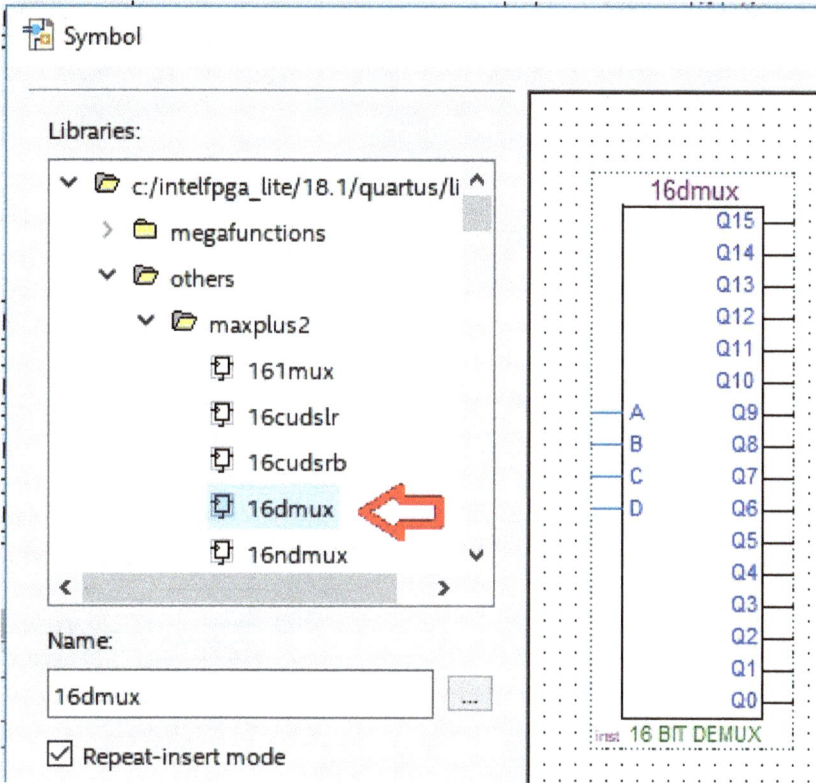

Figure 2.8: Select the 4-bit input, 16-bit output decoder (demultiplexer).

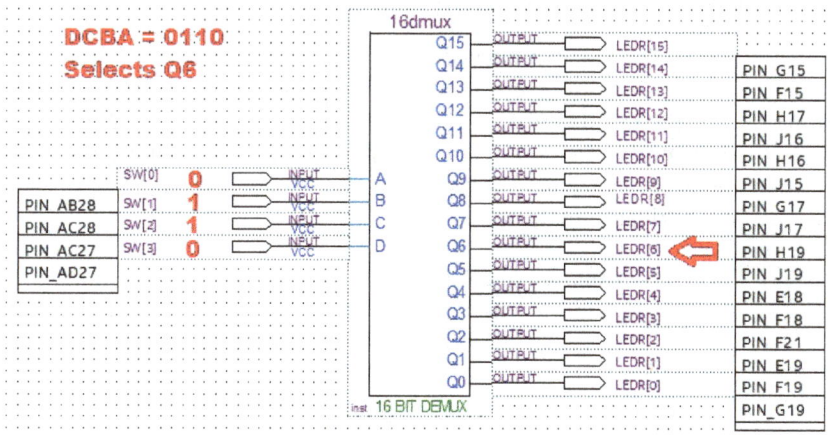

Figure 2.9: Decode 4-bit binary number to select one of sixteen lines.

2: Digital Circuits   49

The "16dmux" decoder selects its one output using a HIGH signal, while the other fifteen lines are held LOW. In most digital circuits, the decoder selects the desired line with a LOW signal with the other fifteen lines being HIGH. Use "16ndmux" to demonstrate this decoder as shown in Figure 2.10.

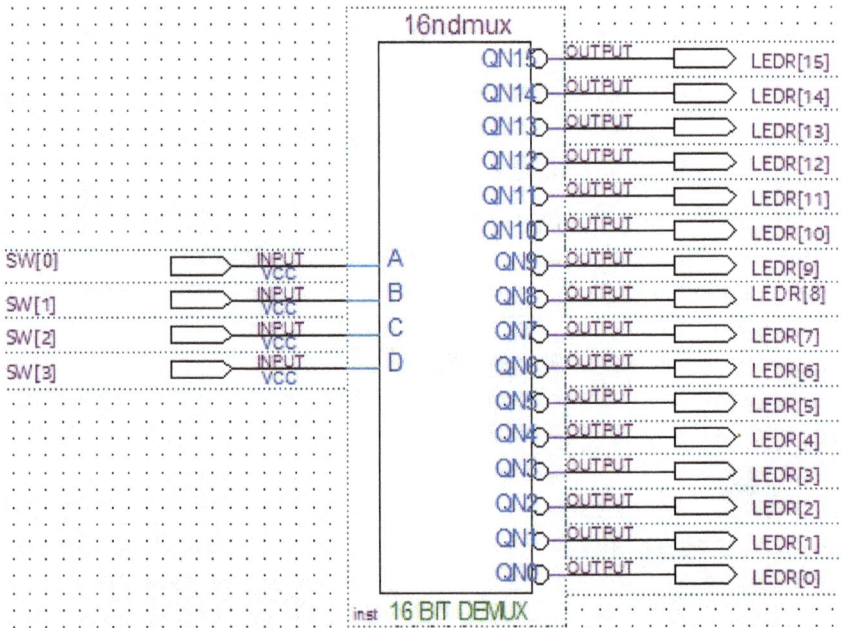

Figure 2.10: Fifteen outputs are HIGH and the selected line is LOW.

Sequential Circuit Examples

The bus and decoder are combinatorial digital circuits, where all their output signals are determined by their immediately present inputs. The other two building blocks, the flip flop and register, are categorized as "sequential circuits" because their outputs are determined not only by their immediate input values, but also by their previous input values. These circuits maintain an internal "state" which determines their outputs.

1. Flip Flop: A one-bit memory
2. Register: A multi-bit memory built from multiple flip flops

RS Flip Flop Circuit

A "flip flop" is a simple circuit that holds "state." In Figure 2.11, a momentary pushing of KEY[0] will result in LED[0] remaining lit. A momentary pushing of KEY[1] will result in LED[0] remaining off. In common electronics terms,

KEY[0] "Sets" the flip flop, and KEY[1] "Resets" it. This circuit is commonly referred to either as an SR or RS flip flop.

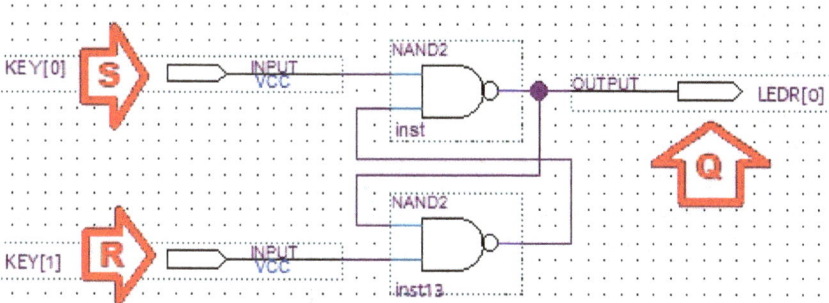

Figure 2.11: RS Flip Flop

The circuit in Figure 2.12 has more LEDs to provide more detail of what is happening in the RS flip flop. In both of these examples, the R and S lines are normally HIGH. A momentary LOW pulse on the S line "sets" the flip flop, and a momentary LOW pulse on the R line "resets" it. For a circuit where the R and S lines are normally LOW and the pulses are HIGH, then an RS flip flop built from two NOR gates would be appropriate.

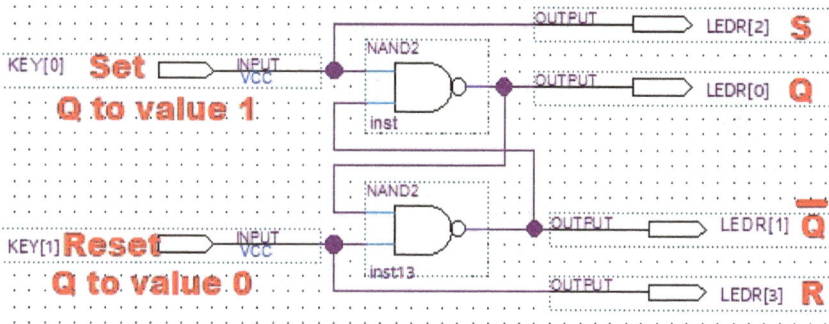

Figure 2.12: RS Flip Flop showing all input and output signals

D Flip Flop

A second type of flip flop is the "D" (for "Data") flip flop. It differs from the RS flip flop in that it has a single input "Data" line and one clock pulse input line. Basically, the D flip flop samples and holds the high/low state of the data line whenever the clock "ticks." The D flip flop is obtained from the symbol tool under the pull-down menu: primitives > storage > dff.

RS flip flops usually appear in schematics as a pair of NAND gates or a pair of NOR gates, but sometimes they appear as a rectangle with R and S inputs with Q and Q-bar outputs. The D flip flops almost always appear as a rectangle as

2: Digital Circuits   51 shown in Figure 2.13.

Go ahead and build, compile, download, and test the circuit in Figure 2.13. Switch[0] is the input data line that will be sampled when the clock on KEY[0] is pushed. Switches 1 and 2 are only for initializing the D flip flop, and for normal operation will be up (i.e., HIGH).

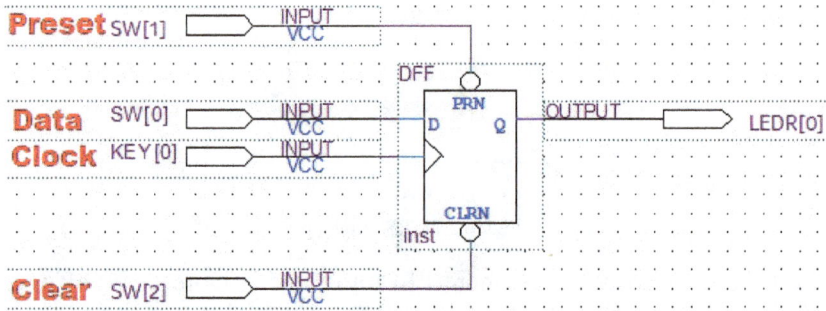

Figure 2.13: D Flip Flop

Registers

A register is a group of bits (i.e., flip flops) that holds a "number." The purpose of that number varies by application. The maximum size of that number depends on the how many bits are in the register. Computer CPUs contain between about five and one hundred registers. Some are assigned special purposes, while others are called general purpose that can be used both in arithmetic calculations and as pointers into memory.

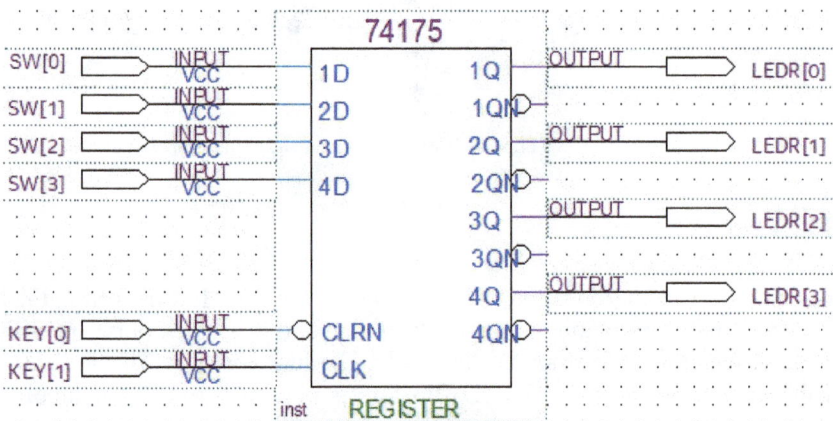

Figure 2.14: 74175 4-bit Register

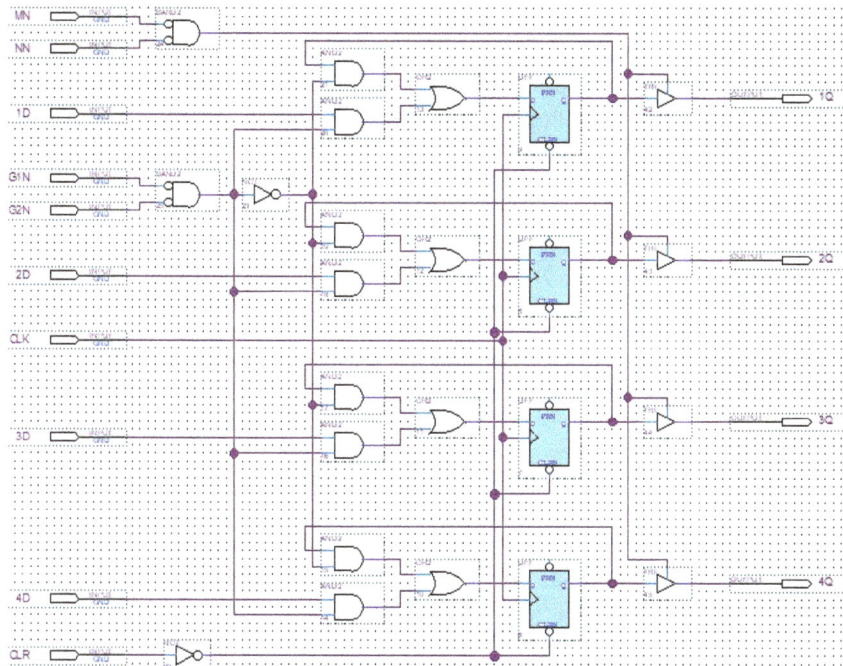
Figure 2.15: Register composed of flip flops

Using the same "maxplus2" symbol tool pull-down list used in a previous example (Figure 2.8), find the 74175 4-bit register and build it into the simple circuit in Figure 2.14. Notice that the data lines are only "strobed" into the memory when the clock (KEY[1]) ticks. Figure 2.15 shows a 4-bit register built from flip flops (indicated by the shaded rectangles).

Another chip to experiment with from the "maxplus2" list is the 74373 tri-state 8-bit register.

Moore and Mealy

Sequential circuits have a memory. That is, they internally hold state over time. In digital courses, we discus two general categories of "state machines": Moore and Mealy.

1. Moore machine: Outputs are determined only by the internal state.
2. Mealy machine: Outputs are determined by both the internal state as well as the immediate input signals present.

For example, the flip flop in Figure 2.13 is a Moore machine. If we modify that circuit so that its Q output was ANDed with either the R or S input lines, then it would be a Mealy machine. Another distinction between the two machine types is how a circuit is described relative to system clocks.

Arithmetic?

It appears that I am ignoring the proverbial "elephant in the room." Isn't arithmetic computation the main purpose of a computer?

For completeness sake and to further demonstrate gate-level construction, I will present the classical "half adder" and "full adder" circuits. These will also be shown in Chapter 5 being implemented in Verilog. However, they will not be used in later chapters to "build" the ARM CPU because Verilog has a more eloquent and easier way to construct arithmetic circuits.

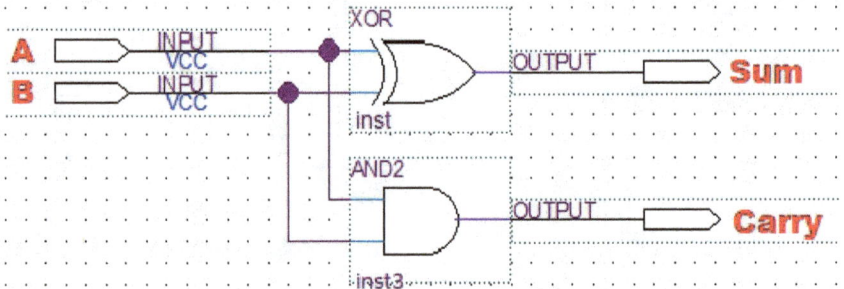

Figure 2.16: Half adder (adds two bits)

The half adder has two inputs with a sum and a carry output. Because registers have more than 1 bit and carries are common, three bits must be added in most cases. Build both circuits in figures 2.16 and 2.17 to see how they work. Use any three switches for the three inputs, and any two LEDs for the sum and carry outputs. These adders can even be replicated to add registers of any size.

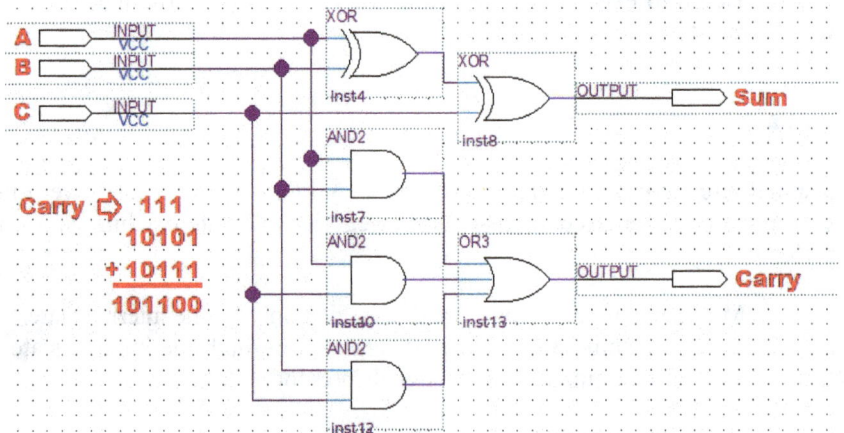

Figure 2.17: Full adder (adds three bits)

Highlights and Comparisons

Chapter 1 introduced gates: AND, OR, XOR, NOT (i.e., the fundamental construction material of digital circuits). Although a computer can be described and constructed entirely from individual gates, it is much easier to think of a computer as composed of larger structures such as decoders, flip flops, and registers, all of which are actually composed of gates themselves.

The graphical approach of describing digital circuits taken here in Chapter 2 is great for illustrating basic concepts. However, as the circuits get more complicated, the textual approach available in software like Verilog and VHDL makes hardware design much more manageable. I've wrestled with blueprints containing hundreds of gates. It's easy to get lost in that forest of logic and timings.

Review Questions

1. *How can a NAND gate be converted to an inverter?
2. *Why might you change the automated "instance" name (Figure 2.2) generated by the Quartus Prime compiler?
3. What are three buses typically found connected to a CPU?
4. *Why is it that the data bus requires tri-state logic while the other two buses do not?
5. Sometimes the terms decoder and demultiplexer are used interchangeably. Why do you think this happens?
6. Why are half adders and full adders categorized as combinatorial circuits rather than sequential circuits?
7. Is a half adder a Moore or a Mealy machine?
8. *How can an arch be a metaphor for describing a flip flop?
9. Why are the preset and clear input signal lines to a D flip flop called "asynchronous," while the data input line is called "synchronous"?
10. In an RS flipflop built from two NAND gates, what happens when both inputs are low?

Exercises

1. "Combinatorial" or "combinational" logic? Check the Internet to see which term you prefer.
2. There are two forms of DeMorgan's Theorem (BAND2 = NOR2, BOR2 = NAND2, using Quartus Prime schematic nomenclature). Build the circuits and prove DeMorgan's Theorem by running through all four cases for each circuit.
3. Using the decoder from Figure 2-9, route the signal from KEY[0] to any one of ten LEDs. Hint: Also use ten AND gates, and SW[3:0] will select which LED (SW[0] connects to the A input on the 16dmux,

SW[1] connects to the B input, etc.).

4. Build an OR gate from NAND gates (Hint: use DeMorgan's Theorem).

5. Build a 4-bit adder, where the two numbers are input on switches and the sum is displayed on the LEDs. The first number is in binary on switches SW[7:4], and the second number is on switches SW[3:0]. For example to add 5 + 7, binary 0101 + 0111, the switches would be XX01010111, where XX represents switches SW[9:8] that are not needed in this exercise. Display the output on LEDs 0 through 4. The circuit should combine a half adder for bit position 0 with three copies of a full adder for bit positions 1, 2, and 3. The half adder will have SW[0] and SW[4] as its inputs. Its sum will go to LEDR[0]. The first full adder will have three inputs: SW[1], SW[5], and the carry from the half adder. Its sum will go to LEDR[1]. The next two bit columns are similar, and the final carry goes directly to LEDR[4]. It's really not that much work because the Quartus Prime software allows you to copy and paste sections of the schematic.

6. Make an 8-bit register from two 74175 "chips" (Figure 2-14). Use SW[7:0] as the inputs to the register, LEDR[7:0] as the output of current register contents (the Qs), KEY[0] as the clock, and KEY[1] to clear all eight bits of the register.

— 3 —
Verilog Modules

Verilog is a Hardware Description Language (HDL). The architecture of a building is divided into floors and rooms. Likewise, Verilog coding divides a digital architecture into "modules." Chapter 3 provides basic examples of modules, and how modules can be embedded within each other to construct more complicated digital hardware devices.

Chapter 1 illustrated how to set up a digital design project using the Intel® Quartus® Prime design software in the block diagram schematic approach. Chapter 3 reviews that procedure, but does it using a Verilog approach for the circuit design specifications.

Introductions

In schematics, there are standard symbols for basic operations like the AND, OR, and XOR gates. Likewise in text-based languages, such as Verilog, there are several "key words" that have special meanings. Chapter 3 only introduces as many Verilog keywords as necessary to build very simple modules, embed them within each other, and test them on an FPGA development board containing switches and LEDs.

- **module, endmodule:** Marks the beginning and ending of a block of Verilog coding
- **input:** Defines input signals to a module
- **output:** Defines output signals from a module
- **assign:** Specifies continuous "dataflow" from one part of a circuit to another

Verilog HDL

Although the graphical approach for specifying a digital circuit is great for illustrating basic concepts, the textual approach of an HDL like Verilog is much more practical for describing complex devices such as computer CPUs. Verilog code looks like a computer program written in C or Java, but it's a hardware description, not a program. A program works in the time domain, while an HDL works primarily in the space domain (i.e, number of gates physically in an FPGA). For example, the following loop can appear in a C program or a Verilog hardware description, but what each produces is considerably different.

- for (m=0; m<4;m=m+1)
- u[m] = x[m] & y[m];

The above loop represents and is equivalent to the following four logical AND calculations.

1. u[0] = x[0] & y[0];
2. u[1] = x[1] & y[1];
3. u[2] = x[2] & y[2];
4. u[3] = x[3] & y[3];

In C, these four ANDs can be calculated on a single CPU, one at a time starting with the first. In Verilog, four separate sections of hardware are generated, allowing the four AND calculations to be performed simultaneously.

Increasing the loop from four to one thousand results in more time required to run the C program. Increasing the loop in Verilog uses more physical resources from the FPGA to build the one thousand AND gates that can operate in parallel (at the same time).

Source Code Download

This book contains about 90 examples of Verilog coding. I have made them available on the Internet so they can be easily downloaded using the GitHub website. GitHub "is a code hosting platform for version control and collaboration." It is composed of multiple public and private "repositories" holding text, image, and video files.

Enter the following command in your Internet browser to initiate the load of all the listings in this book.

https://github.com/robertdunne/FPGA-ARM

I recommend that you download and unpack the source code files because I will be using them in examples in the remainder of this book. If you are already familiar with and have experience with GitHub, then use a procedure with which you are most comfortable. Otherwise, please perform the following steps at the GitHub site:

1. Click on the green button labeled "Code" which will bring up a drop-down menu.
2. Select "Download Zip" from the drop-down menu which will download one file to your normal downloads directory.
3. You may now exit GitHub or close your browser since you will no longer need it.

From your downloads directory, perform the following to extract all the source code into C:\FPGA-ARM-master:

1. Right click on the FPGA-ARM-master.zip file just downloaded.
2. Select "Extract All..." from the pull-down menu.
3. In the "Select Destination and Extract" screen, change the file name to "C:\FPGA-ARM" or to a different directory name you chose for your work files.
4. Click on the "Extract" button.

The above procedure will generate all of the listing files as TXT files having file names corresponding to the captions under each listing in this book. Each will have to be copied and pasted as needed. There are also three CSV files and a README.md file in the downloaded directory.

Warning: The Verilog source code that appears in this book and is available for download is for learning computer architecture. No guarantee of its commercial utility is expressed or implied.

Simulation or Synthesis

In Chapter 1, an FPGA is described as a very large pool of individual disconnected gates. How these gates are to be logically connected to form a digital circuit is described graphically in a schematic in Chapter 1, but will be described in Verilog text coding here in Chapter 3. This pattern of gates, called a "netlist," is then downloaded into an FPGA to achieve the digital design goals.

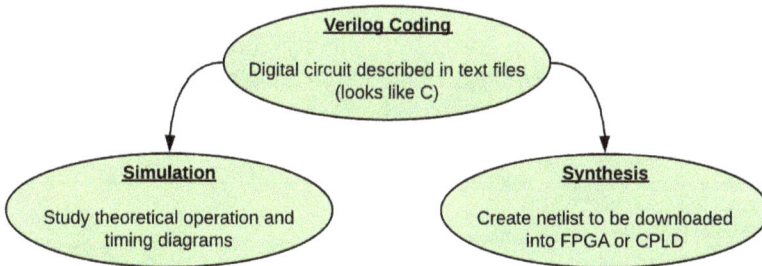

Figure 3.1: Compilation for Simulation and Synthesis

Originally, Hardware Description Languages (HDL), such as VHDL and Verilog, were used to document and test digital circuits, not create them. VHDL was originally developed in 1981, and became an Institute of Electrical and Electronic Engineers (IEEE) standard in 1987. Verilog became an IEEE standard in 1995 after having started in 1984 as a proprietary language. Both VHDL and Verilog have gone through multiple revisions, and each has the following capability:

- Simulation: Examine a working digital system in detail (such as its critical timing sequences).
- Synthesis: Generate a netlist that provides the physical description for building a working digital system.

When it comes to designing real production digital circuits, running a simulation is critical. It will provide details of timing faults that will not appear in simple "hands on" testing of the circuit. In the simulation, the functioning of internal "connections" that are inaccessible in a synthesized circuit can be examined and diagnosed.

Verilog has had major revisions in 2005 and 2009, with the latest revision generally referred to as SystemVerilog. This book uses Verilog features available in the free "Lite" edition of the Quartus Prime software, which includes many, but not all features of SystemVerilog.

First Verilog Example

The procedure for setting up a project for Verilog is almost identical to that which was done in Chapter 1 for the block diagram schematic approach. Actually, projects can be constructed using both Verilog and schematics and even include VHDL for that matter. We must provide the following to make a successful project:

1. Source Code
2. Target Device Specification
3. Board Pin Assignments

The Verilog source code in Listing 3.1 simply connects ten switches to ten LEDs. It will serve as a Verilog example for creating a netlist and testing it in an FPGA. All of the Verilog coding in this book can build upon this example.

```
module TopLevel (SW, LEDR);
   input [9:0] SW;
   output [9:0] LEDR;
   assign LEDR = SW;              // Connect all switches to LEDs
endmodule
```

Listing 3.1: Connect 10 switches to 10 LEDs

The "Target Device Specification" and "Board Pin Assignments" presented in the examples in Chapter 3 are for the Terasic DE10-Lite development board which contains an Intel MAX 10 DA FPGA. Its setup will be slightly different from that of the DE2-115 used in Chapter 1. With minor modifications, the examples will also work with almost any other FPGA or CPLD development board having at least ten slide switches, ten LEDs, and two push buttons.

Starting Quartus Prime

Start Quartus Prime, and then select "New Project Wizard" as was done in Chapter 1. The wizard will display the following six pages that were already "walked through" in Chapter 1. An "Introductions" page may appear before the "Directory, Name, Top-Level Entity" page, but it can be skipped.

1. Directory, Name, Top-Level Entity
2. Project Type: Choose empty project or template
3. Add files
4. Family, Device, and Board Settings
5. EDA Tools Settings
6. Summary: Check project name, directory, and device

Figure 3.2 shows the first page where three lines must be entered to identify the project. In the top line, please choose a working directory to contain your project. The middle line contains a project name which is automatically copied to the top-level design entry name.

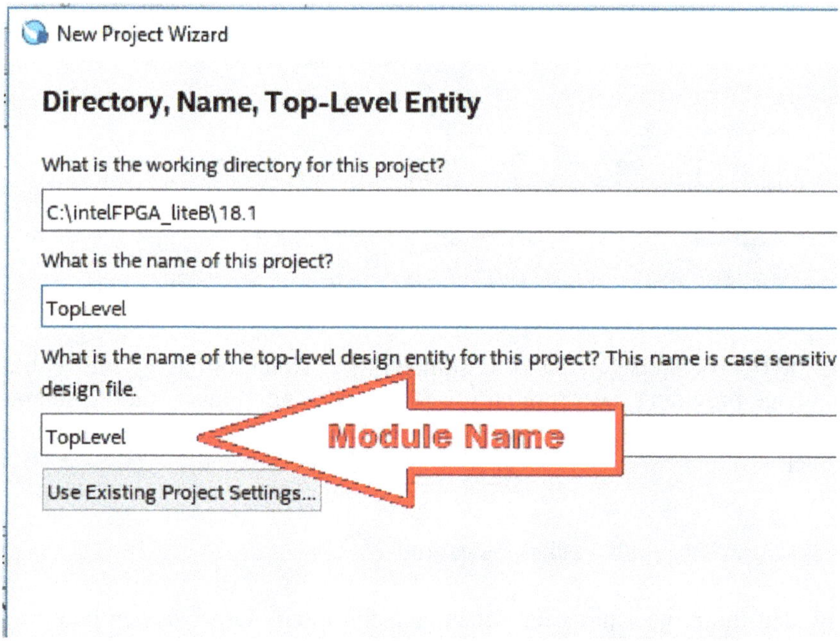

Figure 3.2: TopLevel module name

The directory, project name, and top-level design name can all be different and are of your choosing. However, the top-level design name must match the module name in the Verilog code. For example, I chose the project name of "TopLevel" which was automatically copied into the top-level design name field.

The name "TopLevel" matches the first text line of the module in Listing 3.1 Note: The Verilog module name and top-level design names are both case sensitive.

The third page is the "Add Files" page, and it should be skipped by clicking "Next." The fourth page is "Family, Device & Board Settings." Here in Chapter 3, I am showing the screens for the DE10-Lite, so the "MAX 10 FPGA" is selected in Figure 3.3. If you are using the DE2-115, please select the Cyclone IV E as was done in Chapter 1.

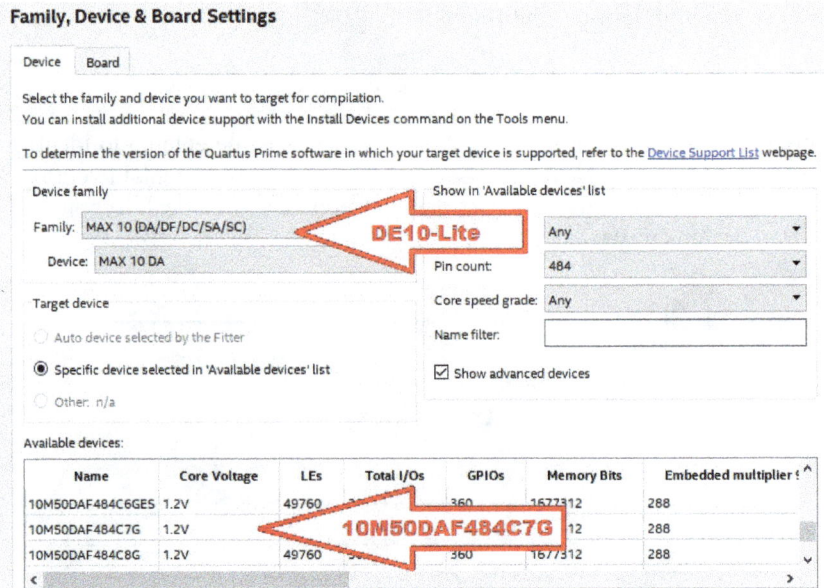

Figure 3.3: DE10-Lite target device

The "EDA Tools" page should be skipped next, which brings up the sixth and final page providing the "Summary." Before clicking "Finish" on the summary page, please check that the directory, device type, and top-level design name are correct.

Design File Setup and Verilog Selection

Now that the project has been set up and the FPGA target device chosen, we must specify the source code language and board pin assignments. The Design file is initialized by clicking "File" on the menu bar, and then select "New" as seen in the copy of Figure 1.10.

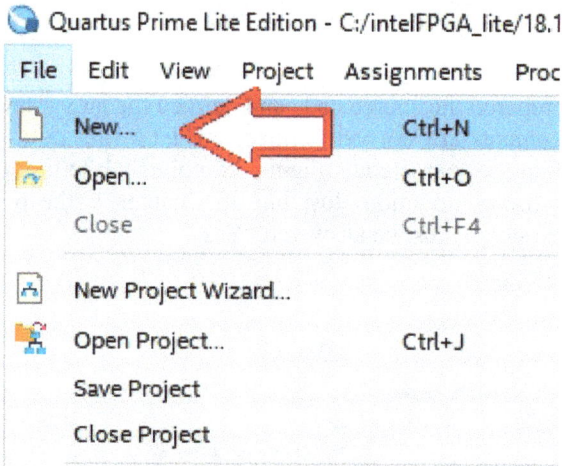

Copy of Figure 1.10: Specify source language

Figure 3.4 shows the nine possible source languages that are currently available with Quartus Prime. Here in Chapter 3, the "Verilog HDL File" is chosen.

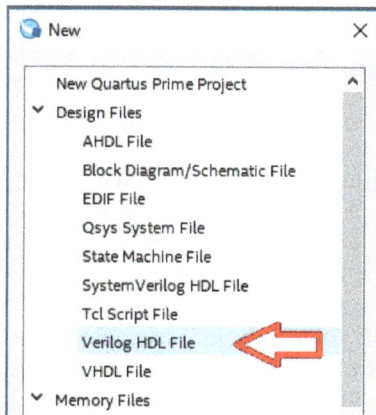

Figure 3.4: Specify that Verilog source code is to be used.

A text editor page appears next where the Verilog text can be entered directly from the keyboard or copied and pasted from Listing_3-1.txt (downloaded from GitHub). A few hints for Verilog text entry:

1. Indentation (i.e., leading blanks) is not required, but is a standard good practice and is very helpful for updates for maintenance.
2. Verilog keywords are case sensitive (almost all are lower case).
3. The Verilog text editor will display keywords in blue.
4. Most Verilog command lines end with a semicolon.

Compilation

We have now prepared the source code and defined the target device. The FPGA board pin assignments are not yet defined, but let's first do a test compile to check for errors in the source code. Figure 3.5 points to the triangle that must be clicked to start the compilation. Just like in Chapter 1, the pop up asking if changes are to be saved must be answered "Yes."

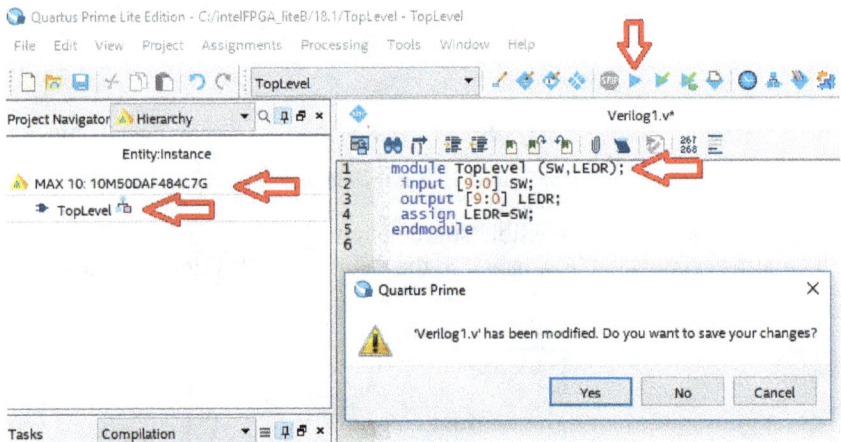

Figure 3.5: Verilog module to "wire" 10 switches to 10 LEDs

As in Chapter 1, red messages are fatal, blue messages are warnings, and green messages are comments. No netlist is produced if red error message appear. Blue messages do allow a netlist to be generated, but sometimes, hundreds of warnings appear. For classroom exercises and testing, these can generally be ignored. However, for real production work, every warning should be examined.

Most errors result from "key in" oversights or misspellings. First check for the obvious: 1) Keywords are blue if spelled correctly, 2) Most lines end with a semicolon, but some like "endmodule" do not. For other fatal errors, that cannot be easily found by a quick check of the text, scroll to the first fatal error message, and double click on it. The line where the problem was found will then be shown. Note: The actual problem may be on a previous line (such as a missing semicolon).

Pin Assignments

A working netlist is generated by compiling a project containing 1) source code, 2) the specifications of the target FPGA, and 3) the development board pin assignments. The netlist just created in Figure 3.5 can be downloaded into the FPGA, but it will not work because the pin assignments for the DE10-Lite board have not yet been made. We have already named the pins as SW[0] through SW[9] and LEDR[0] through LEDR[9], but we have not defined where these pins are located on the FPGA.

Using an Excel CSV file that assigns a pin location for each pin name is a convenient method for providing this information. Although several different FPGAs can accept the same design having the same pin names, the CSV file associated with each board will be unique.

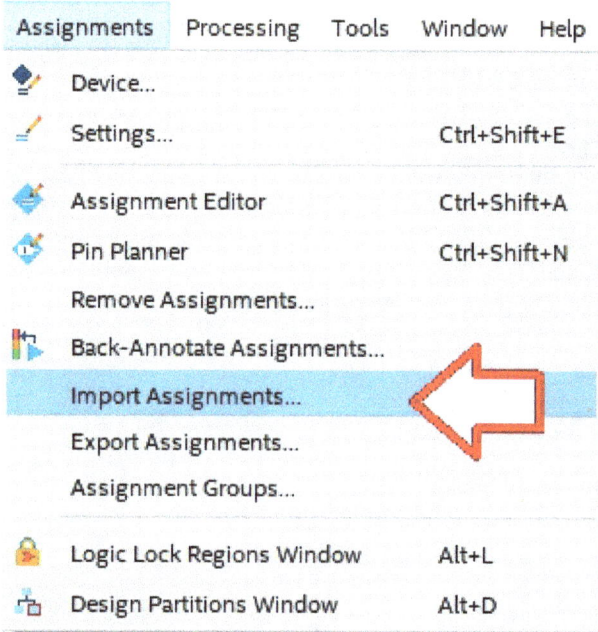

Copy of Figure 1.19: Import list of pin name to pin locations

The copy of Figure 1.19 shows how the pin assignment file can be included in the project: Select "Assignments" from the top menu bar, and then click on "Import Assignments." Then on the next page, browse (i.e., the . . . box) to the directory containing your pin assignment file. This CSV file is often included with the DVD received with the development board and should be copied to a location on the computer's hard drive. It is not included as part of the Quartus Prime software.

For those who cannot locate the pin assignment CSV file, I have included three copies on github: one for the DE10-Lite, one for the DE2-115, and one for the original Altera DE2. See Appendix F for details.

Signal Name	Direction	Location on MAX 10 DA
LEDR[0]	Output	PIN_A8
LEDR[1]	Output	PIN_A9
LEDR[2]	Output	PIN_A10
SW[0]	Input	PIN_C10
SW[1]	Input	PIN_C11
SW[2]	Input	PIN_D13

Table 3.1: Sample of data in "DE10_Lite_pin_assignments.csv" file

It's not required, but if you want to verify that the pins have been assigned, select "Assignments" from the top menu bar, and then click on "Pin Planner." Figure 3.6 shows the pin configuration of the FPGA and the pin location associated with each Node Name (i.e., pin name). Pin locations could have been entered manually using this page, but for a large project, the CSV file is much more convenient.

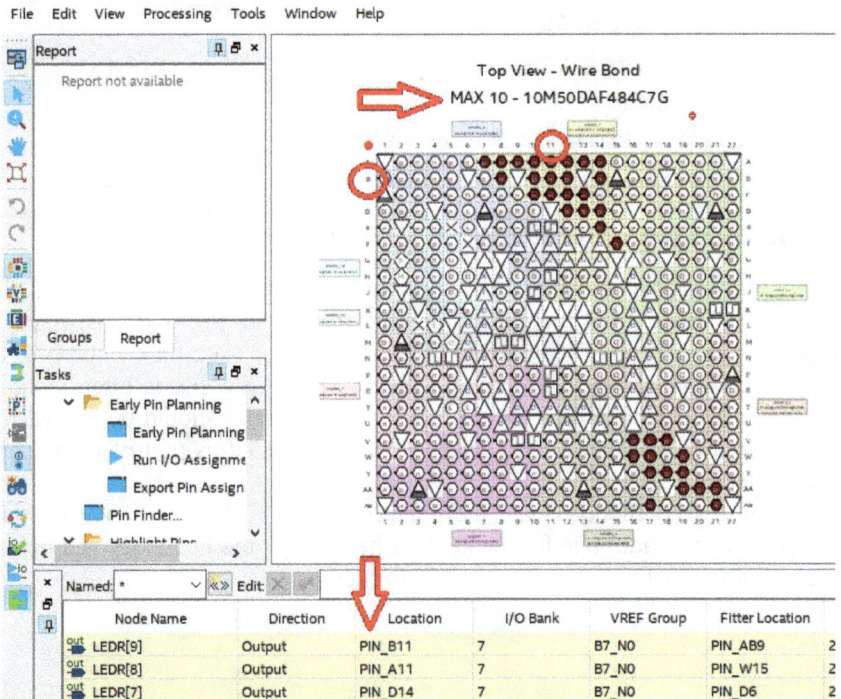

Figure 3.6: Pin assignments for DE10-Lite board

Computer Architecture Tutorial Using an FPGA

Download and Test the Design

Now that the pins have been assigned, recompile the project to finish the netlist, and then download to (i.e., program) the FPGA. The "Programmer" command is located six positions to the right of the "Compile" command as shown in the copy of Figure 1.23.

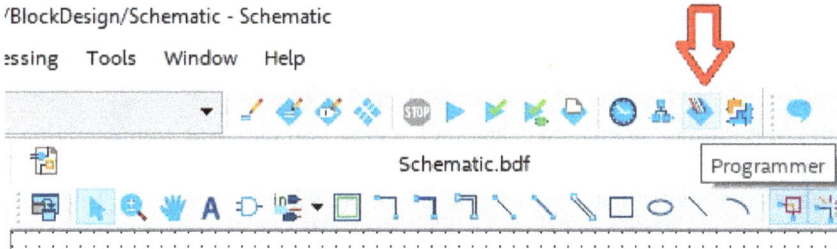

Copy of Figure 1.23: Programmer (download)

module, endmodule

The fundamental building block of a Verilog hardware description is the *module*. The first text line of a module contains its name, followed by a list of parameters enclosed within parenthesis that provides input and output signals to and from the module.

- module ModuleName (Parameter1, Parameter2, ...);

A module ends with the *endmodule* text line. Note: Verilog is a case-sensitive language: The lower case "module" is required rather than "Module" or "MODULE" or any other combination of upper and lower case letters. There is a semicolon after the parameter list on the module line, but not one after *endmodule*. Also, *endmodule* is one word, not two words with a blank between "end" and "module."

input, output

The parameters on the module command line provide names for the signals connecting the digital circuit within the module to the "outside world." These signals can be *input*, *output*, or *inout* (bidirectional such as a data bus). Each name can represent a single line or a vector (group of lines used collectively). The individual elements within a vector can be either identified by counting from the "big end" (such as input[0:9]) or the "little end" (such as input[9:0]).

assign

The Verilog *assign* command is a "dataflow" command that essentially builds wires between the elements on the right side of the equals sign to those on the left. For vectors, multiple wires will be produced connecting corresponding individual signals on each side of the equals. Basically, the following three formats are possible:

1. assign LEDR = SW;
2. assign LEDR[5] = SW[2];
3. assign LEDR[5:3] = SW[0:2];

In the first format, all switches that were defined on the input command line are connected to corresponding LEDs on the output command line. This could be as simple as one switch connected to one LED, or as in the example in Listing 3.1, ten switches are connected to ten LEDs. The order of the connections matches the order in the input and output statements. Change [9:0] to [0:9] on either the input or output command (but not both) in Listing 3.1. Recompile, download, and enjoy the test.

In the second line, a single element from the right is connected to the left. In the third line, three pairs of elements are connected: SW[0] with LEDR[5], SW[1] with LEDR[4], and SW[2] with LEDR[3].

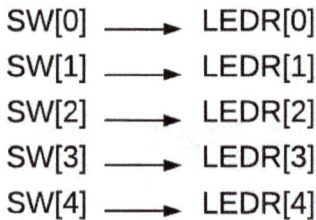

SW[0] ⟶ LEDR[0]

SW[1] ⟶ LEDR[1]

SW[2] ⟶ LEDR[2]

SW[3] ⟶ LEDR[3]

SW[4] ⟶ LEDR[4]

Figure 3.7: Assignment statement "wires" one device to another "continuously."

It is generally a good idea to make sure that the number of elements on the left side of the equals is the same as that on the right. However, Verilog is somewhat forgiving, and will synthesize a netlist even if the sizes don't match. It will also generate some "blue" warning messages. Basically, it truncates extra elements on the right side and zero fills extra elements on the left side.

This is only a brief introduction to the *assign* command. The details of the types of variables on each side of the equals, where in a Verilog hardware description the *assign* can be used, and that the right side can be a rather complex Boolean expression will be described in the next few chapters.

Computer Architecture Tutorial Using an FPGA

Sub-circuits

At a housing construction site, many components arrive preassembled such as windows and prehung doors. Likewise, in the architecture of a hardware description, we assemble modules (i.e., components) from smaller modules, each having a specific purpose. In Listing 3.2, the Wire5 module simply "wires" its inputs (named opcode) to its outputs (named display) using the assign command. The TopLevel main module uses the Wire5 module twice to "wire" ten input signals to ten outputs, five wires in each use of Wire5.

In programming languages, we have subroutines and macros to help organize the coding. A subroutine is one physical section of code that is called from various parts of the main program at various times. A macro actually generates a separate local copy of the code at each place it is "called." From a programmer's perspective, Verilog modules are more like macros than subroutines because each use generates a new circuit.

```
module TopLevel (SW, LEDR);
    input [9:0] SW;
    output [9:0] LEDR;
    Wire5 FirstSet (SW[4:0], LEDR[4 :0]);
    Wire5 SecondSet (SW[9:5], LEDR[9:5]);
endmodule

module Wire5 (opcode, display);
    input [4:0] opcode;
    output [4:0] display;
    assign display = opcode;
endmodule
```

Listing 3.2: Example of nested modules in Verilog

Module and Parameter Names

The *module* command contains the module name and a list of names of parameters. When each new instance of the module is created, the module name can be followed by an instance name and a list of arguments.

Well-chosen names, formally called identifiers, lead to good self documentation and more effective maintenance of Verilog hardware descriptions. The choice of names is very flexible and will be described more fully in Chapter 6. For now keep in mind the following restrictions on identifiers.

1. Cannot be a Verilog keyword (such as *module, input, assign,* ...)
2. Cannot contain a blank, semicolon, comma, or other punctuation.
3. Names are case sensitive (i.e., WIRE5 is not the same as wire5).
4. The module name of the top-level design entry must match the name provided to the project wizard (Figure 3.2).

Highlights and Comparisons

Verilog is a Hardware Description Language (HDL) and so is VHDL. Both can produce simulations as well as synthesize digital circuits. Both can describe a circuit directly by its structure and indirectly through its desired behavior. What are some differences? Verilog's syntax and coding looks like that of the C programming language. VHDL is modeled after the Ada programming language. VHDL coding is rather verbose compared to the somewhat abridged Verilog coding.

Review Questions

1. What type of message (severity level) is represented by each color of compilation messages?
2. *Although we will generally ignore "blue" messages in the classroom, why should they all be checked in real production work?
3. *Why do you think we do not make all ports *inout* rather than specifically *input* or *output*?

Exercises

1. Check the Internet to see the difference among the terms simulation, emulation, and imitation as related to computer products.
2. Modify Listing 3.1 and change the inputs from switches to keys. The DE10-Lite board only has KEY[0] and KEY[1], so the range for both the KEYs and LEDRs must be [1:0].
3. Modify Listing 3.1 and change the outputs from LEDs to the seven-segment hex display 2. The range for both the SW and HEX2 must be [6:0] on the DE2-115 and [7:0] on the DE10-Lite.

— 4 —
Verilog Functions

Binary numbers are awkward for us humans due to the large number of columns required. Who would prefer replacing the decimal representation of 7094, 1620, 6600, 3033, and 7800 with their binary equivalents 1101110110110, 11001010100, 1100111001000, 101111011001, and 1111001111000?

A Verilog *module* generates a digital circuit. A Verilog *function* generates a single value, such as a bit vector, that can be used within a module. Chapter 4 develops a hexadecimal display function that works with both the DE2-115 and DE10-Lite as well as any other board supporting 7-segment displays.

Introductions

Verilog Commands:

- **begin, end:** Combines a group of Verilog text lines
- **case, endcase:** Provide multiple subcircuits for various conditions
- **else:** Provide an alternate circuit from that of an *if* command
- **function, endfunction:** Generate one value (bit, multi-bit vector, or number)
- **if:** Define a subcircuit to process a specific condition

Seven-Segment Display

The 7-segment display has been an economical and very popular output device for many decades. What many people don't realize is that the seven segments are actually independent of each other, and the controlling software must light each segment individually. Listing 4.1 is a slight modification to the example from Chapter 3, but instead of lighting the row of red LEDs, the LEDs within a 7-segment display are lit.

```
module TopLevel (SW, HEX0);
   input [6:0] SW;
   output [6:0] HEX0;
   assign HEX0 = SW;          // Connect switches to HEX0 segments
endmodule
```

Listing 4.1: Connect 7 switches to the 7 segments of HEX0

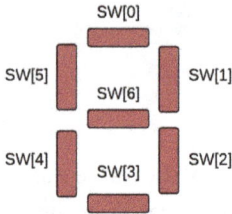

Figure 4.1: Order of LEDs in 7-
segment display

When Listing 4.1 is compiled, downloaded, and tested, the switch-to-segment mapping noted in Figure 4.1 is produced. Note that zero-bits turn on the LED segments, while one-bits turn the segments off. The pattern of switches to provide images of a zero and a one are the following:

- Image of 0: 7'b1000000
- Image of 1: 7'b1111001

"Play around" with the switches to see what digits you can display. With only SW[1] and SW[2] down (pattern of SW[6:0] = 1111001), a "1" is produced on the 7-segment display. Likewise, with only SW[4] and SW[5] up (pattern of SW[6:0] = 0110000), a "3" is displayed.

Is it really an 8-segment display? Yes, there is usually a decimal point present for each digit. On the DE2 boards, the decimal point is not connected to a pin on the FPGA, so it's not really an issue. However, on the DE10-Lite, each 7-segment display has eight signal lines connected to the FPGA, where the eighth line is the decimal point.

Function to Display Base 2

Let's display each switch position on both an LED and a 7-segment display. If a switch is up, then the corresponding LED will be lit and the corresponding 7-segment display will show a "1." When the switch is down, the LED will be off, and a "0" will be displayed.

A Verilog *function* is a feature which easily places copies of a digital hardware segment into multiple locations within a hardware description. Basically, it converts one or more input arguments to a single output value. It is used within a *module* and is similar to a macro in programming languages. It is a substitution convenience that operates when the code is being compiled. It is not like a function or method in a programming language that actually exists as a separate run-time component.

The "digit" function in Listing 4.2 returns binary 11000000 when its input argument is a 0 and returns binary 11111001 when its input argument is a 1. What is done with the returned value depends on the code that "calls" the function. In this code, the returned value is assigned to a vector of LED segments in a 7-segment display, which then shows up as the image of a zero or the image of a one.

Computer Architecture Tutorial Using an FPGA

```
module TopLevel (SW, LEDR, HEX0, HEX1, HEX2, HEX3, HEX4, HEX5);
   input [5:0] SW;
   output [5:0] LEDR;
   output [7:0] HEX0, HEX1, HEX2, HEX3, HEX4, HEX5;
   function automatic [7:0] digit;
      input num;
      begin
         if (num = = 0) digit = 8'b11000000;              // Display 0
         if (num = = 1) digit = 8'b11111001;              // Display 1
      end
   endfunction
   assign LEDR=SW;
   assign HEX0 = digit(SW[0]);
   assign HEX1 = digit(SW[1]);
   assign HEX2 = digit(SW[2]);
   assign HEX3 = digit(SW[3]);
   assign HEX4 = digit(SW[4]);
   assign HEX5 = digit(SW[5]);
endmodule
```

Listing 4.2: Binary on 7-segment displays

The TopLevel module text line has one input vector (SW) and seven output vectors (LEDR, HEX0, ..., HEX5). Compile, download, and test how the code in Listing 4.2 works. Take note of the following:

- The function declaration line indicates that function "digit" will produce an 8-bit output vector. The "automatic" keyword is optional here, but good practice if a simulation were to be run.
- The input to the function is only one bit, named "num."
- The *if* statement contains a double equals sign. This is a common syntax for programming languages like *C* to test for equal. A single equal sign is an assignment, not a test.
- The first *assign* statement "wires" each of the six switches to a corresponding LED.
- The remaining *assign* statements "wire" each of the six switches to a corresponding 7-segment HEX display. Actually, each switch is one bit that is translated by the "digit" function into an 8-bit vector that is "wired" to the eight segments of a HEX display.
- This code only uses six switches because there are only six 7-segment displays on the DE10-Lite. On the DE2-115, the code can be easily modified to include HEX6 and HEX7 if you like.
- The HEX 7-segment displays have eight signal lines. The high order (leftmost) bit is a one to turn off the decimal point on the 7-segments displays of the DE10-Lite. The DE2 boards only have seven signal lines so a "blue" warning message may appear.

```
function automatic [7:0] digit;
  input num;
  begin
    if (num = = 0) digit = 8'b11000000;        // Display 0
    else digit = 8'b11111001;                  // Display 1
  end
endfunction
```

Listing 4.3: Combine *if* with *else*.

Listing 4.3 shows another way to code the digit function. Since the input to the function is only one bit, then if it's not a 0, it must be a 1. Go ahead and change the one line in the digit function. Recompile, download, and test that it still works the same as before.

Base 4

Let's change the display to quaternary (i.e., base four). There are four digits in the set {0, 1, 2, 3} to represent any number. Quaternary is twice as compact as binary, but it really does not have much application in computer programming or hardware descriptions. Quaternary does have some relevance and application in representing DNA genetic codes.

Listing 4.2 can now be modified to provide four digits: 0, 1, 2, and 3. Because it is more compact, base 4 only requires 3 digits to display the contents of the six switches.

```
module TopLevel (SW, LEDR, HEX0, HEX1, HEX2);
  input [5:0] SW;
  output [5:0] LEDR;
  output [7:0] HEX0, HEX1, HEX2;
  function automatic [7:0] digit;
    input [1:0] num;
    begin
      if (num = = 0) digit = 8'b11000000;      // Display 0
      if (num = = 1) digit = 8'b11111001;      // Display 1
      if (num = = 2) digit = 8'b10100100;      // Display 2
      if (num = = 3) digit = 8'b10110000;      // Display 3
    end
  endfunction
  assign LEDR=SW;
  assign HEX0 = digit(SW[1:0]);
  assign HEX1 = digit(SW[3:2]);
  assign HEX2 = digit(SW[5:4]);
endmodule
```

Listing 4.4: Base 4 display

The TopLevel module in Listing 4.4 has one input vector and four output vectors. Compile, download, and test how this code works. Take note of the following:

- The input to the digit function is still named "num," but now has a size of two bits.
- Four *if* statements are present to handle the 2-bit binary numbers: 00, 01, 10, and 11. A better approach is in Listing 4.5 which uses a *case* statement
- The first *assign* statement "wires" each of the six switches to a corresponding LED.
- The other three *assign* statements "wire" each pair of switches to a corresponding 7-segment display.

```
function automatic [7:0] digit;
    input [1:0] num;
    case (num)
        0: digit = 8'b11000000;          // 0
        1: digit = 8'b11111001;          // 1
        2: digit = 8'b10100100;          // 2
        3: digit = 8'b10110000;          // 3
    endcase
endfunction
```

Listing 4.5: Base 4 display using *case* statement

The // character sequence in the above Verilog code examples indicates that the remainder of the text line is a comment. Like in the C programming language, a partial line or multiple lines can also marked as comments by a /* at the beginning and a */ at the end.

Base 8, Octal

Octal was a very popular display format during the mainframe and mini-computer days and even in the early days of microcomputers. It is about three times more compact than binary. Every group of three bits is represented by one octal digit.

Listing 4.6 presents a version of the "digit" function that converts a 3-bit pattern into the 8-bit pattern for displaying an octal digit. The six 7-segment displays on the DE10-Lite can now display an 18-bit number, so the code in Listing 4.6 has been expanded to accommodate all 10 switches on the DE10-Lite. DE2 users can expand this to display all 18 switches if they choose.

```
 1.  module TopLevel (SW, LEDR, HEX0, HEX1, HEX2, HEX3);
 2.    input [9:0] SW;
 3.    output [9:0] LEDR;
 4.    output [7:0] HEX0, HEX1, HEX2, HEX3;
 5.      function automatic [7:0] digit;
 6.        input [2:0] num
 7.        case (num)
 8.          0: digit = 8'b11000000;              // 0
 9.          1: digit = 8'b11111001;              // 1
10.          2: digit = 8'b10100100;              // 2
11.          3: digit = 8'b10110000;              // 3
12.          4: digit = 8'b10011001;              // 4
13.          5: digit = 8'b10010010;              // 5
14.          6: digit = 8'b10000010;              // 6
15.          7: digit = 8'b11111000;              // 7
16.        endcase
17.      endfunction
18.    assign LEDR=SW;
19.    assign HEX0 = digit(SW[2:0]);              // Low order octal digit
20.    assign HEX1 = digit(SW[5:3]);
21.    assign HEX2 = digit(SW[8:6]);              // High order octal digit
22.    assign HEX3 = digit(SW[9:9]);
23.  endmodule
```

Listing 4.6: Base 8 display

The TopLevel module text line has one input vector and four output vectors. Compile, download, and test how the code in Listing 4.6 works. Take note of the following:

- The input to the digit function is still named "num," but now has a size of three bits.
- Eight cases are present to handle the 3-bit binary numbers: 000, 001, 010, 011, 100, 101, 110, and 111.
- The first *assign* statement "wires" each of the ten switches to a corresponding LED.
- The other four *assign* statements "wire" each triplet of switches to a corresponding 7-segment display.

Base 16, Hexadecimal

Once again we will double the base and reduce the number of digits needed to display a particular binary number.

```
1.  module TopLevel (SW, LEDR, HEX0, HEX1, HEX2);
2.    input [9:0] SW;
3.    output [9:0] LEDR;
4.    output [7:0] HEX0, HEX1, HEX2;
5.      function automatic [7:0] digit;
6.        input [3:0] num
7.        case (num)
8.          0: digit = 8'b11000000;          // 0
9.          1: digit = 8'b11111001;          // 1
10.          2: digit = 8'b10100100;          // 2
11.          3: digit = 8'b10110000;          // 3
12.          4: digit = 8'b10011001;          // 4
13.          5: digit = 8'b10010010;          // 5
14.          6: digit = 8'b10000010;          // 6
15.          7: digit = 8'b11111000;          // 7
16.          8: digit = 8'b10000000;          // 8
17.          9: digit = 8'b10010000;          // 9
18.          10: digit = 8'b10001000;         // A
19.          11: digit = 8'b10000011;         // b
20.          12: digit = 8'b11000110;         // C
21.          13: digit = 8'b10100001;         // d
22.          14: digit = 8'b10000110;         // E
23.          15: digit = 8'b10001110;         // F
24.        endcase
25.      endfunction
26.    assign LEDR=SW;
27.    assign HEX0 = digit(SW[3:0]);          // Low order hex digit
28.    assign HEX1 = digit(SW[7:4]);
29.    assign HEX2 = digit(SW[9:8]);          // High order hex digit
30.  endmodule
```

Listing 4.7: Hexadecimal 7-segment display function

The TopLevel module text line has one input vector and four output vectors. Compile, download, and test how the code in Listing 4.7 works. Take note of the following:

- The input to the digit function is still named "num," but now has a size of four bits.
- Sixteen cases are present to handle the 4-bit binary numbers: 0000

through 1111.

- The first *assign* statement (line 26) "wires" each of the ten switches to a corresponding LED.
- The two *assign* statements on lines 27 and 28 "wire" groups of four switches to a corresponding 7-segment display.
- The last *assign* statement "wires" the remaining two switches to a corresponding 7-segment display. Verilog will fill the upper two bits of the "num" argument with zeros.

Does anyone really like mixing letters with digits as in hexadecimal? Who likes replacing binary 111000100 with the awkward-looking hexadecimal 1C4 instead of the comfortable octal 704? Yes, hexadecimal is more compact, but that's not the main reason for its current popularity.

In the 1960s and 1970s, most computer manufacturers used 6-bit bytes to represent their designs' 64 printable characters. These "bytes" were conveniently represented as two 3-bit octal digits. The rise of the ASCII and EBCDIC character codes along with 32-bit word sizes led to the popularity of 8-bit bytes. These 8-bit bytes can be presented as three octal digits (with a bit left over), but two 4-bit hex digits is ultimately better.

Binary Coded Decimal (BCD)

It looks like we jumped over our favorite base of 10. Ten is not an integer multiple of two, so there is no way to "slice up" a binary number into groups of bits to get decimal digits. Please see Appendix C if you want more detail.

There are many computer and industrial applications that store a decimal "number" as a sequence of BCD values 0 through 9, represented by binary 0000 through 1001, respectively. The digit function in Listing 4.7 will work fine for BCD. The six unassigned digits of A through F (1010 through 1111) will not occur in BCD.

There have been popular CPU architectures that can perform arithmetic using these "packed decimal" BCD values. The ARM is not one of them, and this capability is now mostly obsolete as it's being pushed out by opcodes needed for newer operations.

Parameters or Arguments

"Parameter" names are used within *module* and *function* definitions. When a *module* or *function* is instantiated (i.e., "called") within another module, the variables on the list are typically referred to as "arguments." In other words, the *function* named digit defined starting on line 5 of Listing 4.7 ("function automatic [7:0] digit;") has one parameter named "num" appearing on line 6. When it is instantiated on line 27 ("assign HEX0 = digit(SW[3:0]);"), SW[3:0] is referred to as an argument (to be assigned to parameter "num").

Verilog also has a command named *parameter* which will be introduced in Chapter 8. Its purpose is to provide a substitute name for commonly-used constants, such as PI for 3.1416. These parameters typically represent constants such as the number of switches or LEDs to be used in a hardware description, which might be changed when moving from one target device to another. These parameter names assist in documentation and maintenance.

Highlights and Comparisons

What's wrong with binary? Why use another base like hexadecimal (base 16)? The simple answer is hexadecimal is compact, and it is very easy for us humans to convert between binary and hexadecimal. Decimal is also more compact than binary, and we would prefer to use it. However, conversion between binary and decimal is difficult to do "in our heads." The difficulty stems from the fact that 10 is not an integer power of two, but base 16 is 2^4 thereby making it easy to convert every 4-bit pattern to a hexadecimal digit. Please see Appendix C for more information on binary and hexadecimal.

A Verilog *function* is a subcircuit defined within a module that can have multiple inputs, but only one output, such as a single bit or a bit vector. It has been used in this chapter to convert a binary number to a seven-segment display format. Functions will be used extensively in the remainder of this book for implementing the machine code for ARM instructions. A Verilog *task* command will be introduced in Chapter 10 that is similar to the *function* command, but it provides for multiple outputs.

Terminology in the electronics and computer industries can sometimes be confusing and overlapping. A Verilog *function* looks like functions in programming languages that are also sometimes called procedures or methods depending on the language. However, remember Verilog is describing hardware in a physical domain, not software in a time domain. Software functions physically create one instance of the code that is called multiple times. Hardware functions, whether in Verilog, VHDL, or another hardware description language, create a physical instance for each time the function is "called." From a software perspective, Verilog functions are more like software macros.

Verilog also has a macro capability that will be introduced in Chapter 9 with the *define* statement. It is basically like the "cut and paste" feature in editors and word processors. It will also be used extensively later in this book to implement ARM assembly language features.

Remember that whether hardware or software is being created, the objective is not to impress or confuse those maintaining the code in the future, but to be as clear and straightforward as possible. Dividing a large application into multiple subunits like modules, functions, and macros is very important, but it should not be overdone or create arbitrary new constructs for someone else to have to learn.

Review Questions

1. What is difference between the the 7-segment displays of the DE10-Lite and the DE2-115?

2. *The Terasic boards light up their LED segments when they receive zeros, and the segments are off when they receive ones. About half the seven-segment displays on the market work this way, and the other half work the opposite. Search the Internet to find the names associated with the two types of displays. Which kind is used in the DE2 and DE10-Lite boards?

3. *The seven segments of a seven-segment display are commonly assigned letter names. Search the Internet to find the names that would be assigned to segments used in this chapter.

4. *"By hand, without a calculator or computer," convert the following numbers expressed in decimal to binary format: 21, 63, 16, and 129. See Appendix C if you need some background in binary.

5. *"By hand, without a calculator or computer," convert the following numbers expressed in binary to decimal format: 1011, 1100101, 10110, 100001, and 1111011. See Appendix C if you need some background in binary.

Exercises

1. Logical operations such as a two-input AND gate can be described by a four-row truth table. Using a case statement as shown in Listing 4.5, create an AND2 *function* and test it with switches and an LED output. In addition to the AND2 function definition, you will need "assign LEDR[0] = AND2 (SW[1:0]);" to test it. The function definition line will be "function AND2;" because only a single bit will be generated instead a vector containing multiple bits. Likewise, implement the other gates: XOR2, OR2, NAND2, and NOR2.

2. Use an eight-row truth table and case statement to implement a three-input OR3 gate. First approach: Enter all eight lines where the first case is "0: OR3 = 0;" and the remaining seven cases are "1: OR3 = 1;", "2: OR3 = 1;", etc. The alternate approach has the same first case, but the remaining seven cases are replaced by the one line "default : OR3 = 1;".

3. Implement a half adder (see Figure 2.16) and full adder (see Figure 2.17) in a manner similar to that of Exercise 1. The function definitions will be "function [1:0] HALF;" and "function [1:0] FULL;" because the output will be a two-bit vector (the sum and carry). Note: This exercise will be expanded in the next chapter when concatenations, variables, and the nesting of functions are introduced.

Computer Architecture Tutorial Using an FPGA

— 5 —
Verilog Coding Styles

Some managers in the business world tell their people *what* is to be done, while others "micromanage" by telling them exactly *how* it is to be done. Likewise, Verilog can be told what "behavior" a circuit must have, or it can be told exactly which gates are to be used and exactly how to connect them. Both of these "coding styles" have advantages, and both are used extensively in digital design.

The digital circuit examples described graphically in chapters 1 and 2 will now be redone in the coding styles available in Verilog.

Introductions

Several Verilog keywords and commands are introduced in Chapter 5:

- **always:** Marks the beginning of a block of "behavioral" coding
- **and, or, xor, ...:** Logical gates in "structural" coding style
- **reg:** A circuit element, such as a register, used in "always" block.
- **wire:** "Hardwired" continuous connection between components
- **{ }:** Braces used to concatenate bits
- **<= :** Non-blocking assignment in always block

Verilog Coding Styles

Verilog provides four models for expressing a digital design. Each of the following can be used alone or together within a module.

1. **Switch Level:** The exact arrangement of transistors composing gates and memory can be specified.
2. **Structural:** The desired digital circuit is expressed as a combination of gates, the same as in the graphical approach used in Chapter 1, except words are used to select the gates and their connections. In this approach, keywords such as *and*, *or*, and *xor* are present.
3. **Dataflow:** The desired circuit is expressed in Boolean equations such as "Y = X & 1." In this approach, the *assign* keyword is present.
4. **Behavioral:** Rather than dictating the exact gates to be used and how to connect them, this approach specifies the desired digital outputs. Verilog determines the internal circuit details based on a circuit's behavior which is described in what looks like a C or Java program. In this approach, the *always* keyword is present.

The *switch level* is for special applications and not commonly used today. It will not be described in this book. I will first provide examples of the *structural* and *dataflow* models since they easily express basic combinatorial digital concepts. The *behavioral* model will be used in almost every ARM processor example in this book because of its ease of use, and it works well with synchronous sequential (i.e., clocked) circuits.

Verilog Syntax

An example using a single gate will introduce the basic features of the three popular Verilog coding styles. Each example will be coded, compiled, downloaded, and tested. It is assumed that you have already worked through the Verilog examples in chapters 3 and 4 that set up a new project including the pin assignments.

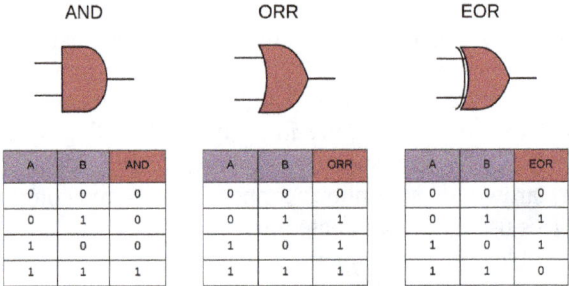

A	B	AND
0	0	0
0	1	0
1	0	0
1	1	1

A	B	ORR
0	0	0
0	1	1
1	0	1
1	1	1

A	B	EOR
0	0	0
0	1	1
1	0	1
1	1	0

Copy of Figure 1.27: Truth tables for the logical AND, OR, and exclusive OR operations

Structural (Gate Level) Modeling

Let's begin with a simple circuit where a single AND gate has inputs from switches SW[0] and SW[1] with an output on LEDR[0]. The general format for structural gates is listed below:

1. Gate type such as *and, or, xor, not, nand, nor, xnor, buf*
2. Instance name, which can be omitted
3. List of parameters.

Each *instance*, (i.e., occurrence) of a gate in a circuit must have a unique identifier. If this instance name is not provided, Verilog will make one up.

Multiple parameters are possible. For example, an *and* can have multiple inputs, but only one output. The *not* and *buf* gates can only have one input, so Verilog permits them to have multiple outputs in their parameter list.

Computer Architecture Tutorial Using an FPGA

Gate Name	Instance Name	Parameters
and, or, xor, nand, nor, xnor	Optional	(Output, Input, Input, ...)
not, buf	Optional	(Output, Output, ..., Input)

Table 5.1: Structural commands

```
module TopLevel (SW, LEDR);
    input [9:0] SW;
    output [9:0] LEDR;
    and OnlyGate (LEDR[0], SW[0], SW[1]);          // LEDR[0] = SW[0] & SW[1]
endmodule
```

Listing 5.1: Logical AND gate with two inputs

Go ahead and compile, download, and test the circuit in Listing 5.1. Verify that it is consistent with the AND truth table in Figure 1.27.

In addition to the *and*, test the other gates as well. Also try multiple inputs using multiple switches. Listing 5.2 demonstrates the inverter which can have multiple outputs. The instance name is optional. It was given the arbitrary name "OnlyGate" in Listing 5.1 and is omitted in Listing 5.2.

```
module TopLevel (SW, LEDR);
    input [9:0] SW;
    output [9:0] LEDR;
    not (LEDR[0], LEDR[1], LEDR[2], SW[0]);
endmodule
```

Listing 5.2: Logical NOT with multiple outputs, but only one input

Dataflow Modeling

The Dataflow Model describes the circuit to be built using Boolean equations. The Verilog *assign* keyword precedes each equation as shown in Listing 5.3. Try other Boolean operators: the caret (^) and vertical bar (|) for XOR and OR, respectively. Preceding a signal name with the tilde (~) will invert it. Multiple assigns for multiple output signals are possible, and the order does not matter.

```
module TopLevel (SW, LEDR);
    input [9:0] SW;
    output [9:0] LEDR;
    assign LEDR[0] = SW[0] & SW[1];                // Dataflow AND gate
endmodule
```

Listing 5.3: Logical AND gate

Behavioral Modeling

The Behavioral Model describes the desired digital outputs, and Verilog determines the internal circuit details. The *always* keyword identifies the beginning of a block of behavior modeling code. Typically, the behavior model would not be used for a circuit as simple as a single gate.

In structural and dataflow modeling, the circuit is "hard wired." Any change in the input signals is immediately propagated throughout the circuit. In the behavior approach, the circuit is only updated on specific conditions: either a clock pulse or a change in a specific set of signals. The "Sensitivity List" following the @ on the always command indicates when the circuit should be updated. In Listing 5.4, there will only be a change in the output when there is a change to either SW[0] or SW[1]. Only variables of type *reg* can be updated inside an always block (i.e., "net" type variables, like LEDR[0], cannot be directly altered inside an always block).

```
module TopLevel (SW, LEDR);
    input [9:0] SW;
    output [9:0] LEDR;
    reg result;
    assign LEDR[0] = result;
    always @ (SW[0], SW[1])
        result = SW[0] & SW[1];                    // Behavioral AND gate
endmodule
```

Listing 5.4: Logical AND gate

Try the following changes to Listing 5.4, and see what happens when you test the circuit.

1. Demonstrate each of the other logical operators (caret, vertical bar, tilde for XOR, OR, and NOT, respectively).
2. Put only SW[0] or only SW[1] in the sensitivity list.
3. Put only SW[2] in the sensitivity list instead of SW[0] and SW[1].
4. Replace "result" in the *always* block with "LEDR[0]" (and "enjoy" the error messages).

Multiple command lines can be placed in an *always* block, but they must appear within *begin* and *end* statements.

Combinatorial Examples

A digital circuit where all of the outputs are determined by an immediate combination of the input signals is referred to as a "combinatorial" (a.k.a., "combinational") circuit. The following circuits from chapters 1 and 2 will be rebuilt here using the *structural*, *dataflow*, and *behavioral* models.

1. Elevator Door Example
2. Build XOR from AND, OR, and NOT Gates
3. Half Adder
4. Full Adder

Table 5.2 compares the code among the three models. Variables A, B, C, and F represent input signals that will be assigned to switches. Open, eor, sum, and carry represent output signals that will be assigned to LEDs. ABC, AB, AC, BC, BA, notA, and notB represent internal connections between gates.

Example	Structural	Dataflow and Behavioral
Elevator Door	and (ABC,A,B,C); or (open, F, ABC);	open = (A & B & C) | F;
XOR from AND, OR, and NOT	not (notA, A); not (notB, B); and (BA, notA, B); and (AB, A, notB); or (eor, BA, AB);	eor = ~A & B | A & ~B;
Half Adder	xor (sum, A, B); and (carry, A, B);	sum = A ^ B; carry = A & B;
Full Adder	xor (sum, A, B, C); and (AB, A, B); and (AC, A, C); and (BC, B, C); or (carry, AB, AC, BC);	sum = A ^ B ^ C; carry = A & B | A & C | B & C;

Table 5.2: Compare Structural, Dataflow, and Behavioral commands

Elevator Door

In the elevator door example, we will build a circuit implementing the logic for opening the door. Usually this involves pushing a floor selection button and waiting for the elevator to position itself properly, but there is also a "Firefighters" override. The following four factors contribute to the decision to open the door:

1. The elevator is at correct floor
2. AND the elevator is stopped
3. AND the elevator is level with floor
4. OR the fireman key override

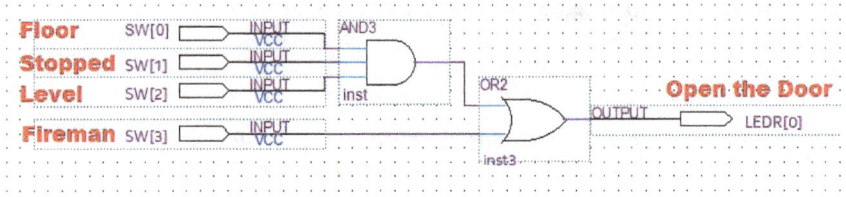

Copy of Figure 2.2: Elevator door-opening example

Note that this example has an internal signal connection known as a *wire* connecting the output from the *and* gate to one of the inputs of the *or* gate.

```
module TopLevel (SW, LEDR);
    input [9:0] SW;
    output [9:0] LEDR;
    wire abc;                              // Internal connection
    and (abc, SW[0], SW[1], SW[2]);        // "Safe" open condition
    or (LEDR[0], abc, SW[3]);              // "Safe" or fireman
endmodule
```

Listing 5.5: Structural model for elevator door example

In the Dataflow approach shown in Listing 5.6, the Boolean equation follows the *assign* keyword. We could have put parentheses around the first three switches, but they are not needed because in Boolean algebra: the *and* has higher precedence than the *or*.

```
module TopLevel (SW, LEDR);
    input [9:0] SW;
    output [9:0] LEDR;
    assign LEDR[0] = SW[0] & SW[1] & SW[2] | SW[3];
endmodule
```

Listing 5.6: Dataflow model for elevator door example

```
module TopLevel (SW, LEDR);
    input [9:0] SW;
    output [9:0] LEDR;
    reg openDoor;
    assign LEDR[0] = openDoor;
    always @ (SW[0], SW[1], SW[2], SW[3])
        openDoor = SW[0] & SW[1] & SW[2] | SW[3];
endmodule
```

Listing 5.7: Behavioral model for elevator door example

The Behavioral approach shown in Figure 5.7 uses the same Boolean equation,

Computer Architecture Tutorial Using an FPGA

but requires the *always* and *reg* keywords. Only *reg* variables can be assigned values within an always block.

Listing 5.7 contains all input signals used in the Boolean equation in the sensitivity list. Verilog knows which signals are used within the *always*, and it will include them if an asterisk appears in the sensitivity list in place of the list of signal names.

Most *always* blocks are longer than one statement, so a *begin-end* sequence is needed. The Behavioral code in Listing 5.8 builds the same netlist as Listing 5.7, but it is more descriptive and easier to maintain:

```
module TopLevel (SW, LEDR);
   input [9:0] SW;
   output [9:0] LEDR;
   reg openDoor;
   reg safeOpen;
   assign LEDR[0] = openDoor;
   always @ (*)
      begin
         safeOpen = SW[0] & SW[1] & SW[2];      // Normal condition to open
         openDoor = safeOpen | SW[3];           // Include fireman override
      end
endmodule
```

Listing 5.8: Behavioral model with better design

XOR from AND, OR, and NOT

Logical gates can be built from other gates. In this example, an exclusive OR is built from a combination of AND, OR, and NOT gates as is shown in the copy of Figure 2.4.

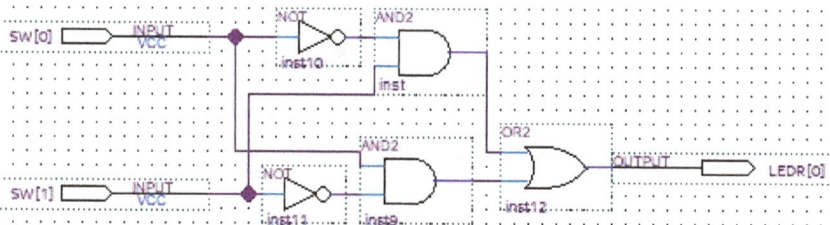

Copy of Figure 2.4: Build XOR from AND, OR, and NOT

Listing 5.9 implements the circuit from Figure 2.4 using the Verilog structural style of coding. Note that there are four *wires* (notA, notB, AB, and BA) internally connecting the gates. The code also generates the exclusive *or* directly for comparison.

```
module TopLevel (SW, LEDR);
   input [9:0] SW;
   output [9:0] LEDR;
   wire notA, notB, AB, BA;          // Internal connections between gates
   not (notA, SW[0]);                // ~sw[0]
   not (notB, SW[1]);                // ~sw[1]
   and (AB, notA, SW[1]);
   and (BA, notB, SW[0]);
   or (LEDR[0], AB, BA);             // Calculated XOR
   xor (LEDR[1], SW[0], SW[1]);      // Direct XOR
endmodule
```

Listing 5.9: Structural model for XOR from AND, OR, and NOT

The dataflow style is used in Listing 5.10 to produce the exclusive *or* by a combination of gates output to LEDR[0]. This output can be compared to the exclusive *or* directly output to LEDR[1]. Listing 5.11 shows the same two outputs using the behavioral approach.

```
module TopLevel (SW, LEDR);
   input [9:0] SW;
   output [9:0] LEDR;
   assign LEDR[0] = ~SW[0] & SW[1] | ~SW[1] & SW[0];   // XOR from other gates
   assign LEDR[1] = SW[0] ^ SW[1];                     // XOR directly
endmodule
```

Listing 5.10: Dataflow model for XOR from AND, OR, and NOT

```
module TopLevel (SW, LEDR);
   input [9:0] SW;
   output [9:0] LEDR;
   reg XOR_calc, XOR_dir;
   assign LEDR[0] = XOR_calc, LEDR[1] = XOR_dir;
   always @ (*)
     begin
       XOR_calc = ~SW[0] & SW[1] | ~SW[1] & SW[0];
       XOR_dir = SW[0] ^ SW[1];
     end
endmodule
```

Listing 5.11: Behavioral model for XOR from AND, OR, and NOT

Half and Full Adders

Figures 2.16 and 2.17 show two more combinatorial circuits previously implemented in the block diagram schematic format. The "full adder" produces the sum and carry for adding three bits, while the "half adder" only adds two.

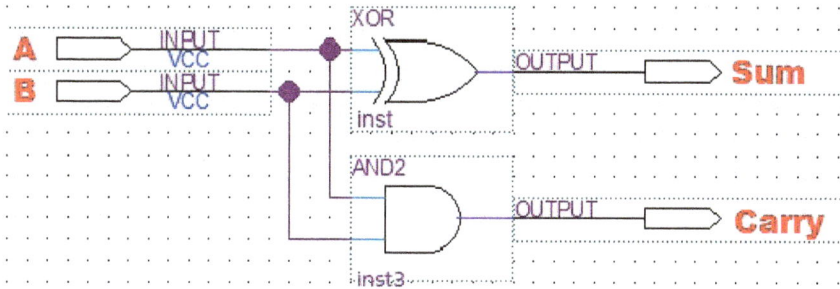

Copy of Figure 2.16: Half adder (two bits added)

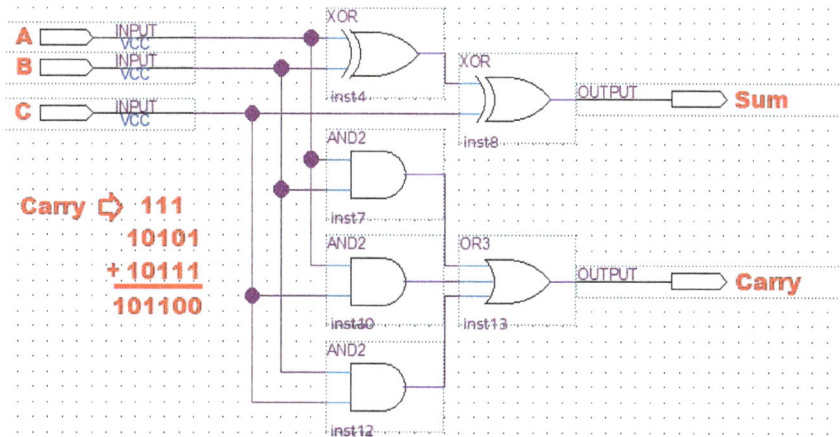

Copy of Figure 2.17: Full adder (three bits added)

The full adder reduces to the half adder when the C input is 0, so I'll only build the full adder using Verilog. Listing 5.12 provides a structural version in Verilog. The A, B, and C inputs are on switches 0, 1, and 2 respectively. LEDR[0] will indicate the sum, and LEDR[1] will be the carry.

```
module TopLevel (SW, LEDR);
  input [9:0] SW;
  output [9:0] LEDR;
  wire AB, AC, BC;                          // Internal connections for carry
  xor SumSig (LEDR[0], SW[0], SW[1], SW[2]);   // sum = A ^ B ^ C
  and (AB, SW[0], SW[1]);
  and (AC, SW[0], SW[2]);
  and (BC, SW[1], SW[2]);
  or CarrySig (LEDR[1], AB, AC, BC);        // carry = A & B | A & C | B & C
endmodule
```

Listing 5.12: Structural model for full adder

The dataflow style is used in Listing 5.13, and the behavior style is used in Listing 5.14, to implement the full adder circuit.

```
module TopLevel (SW, LEDR);
    input [9:0] SW;
    output [9:0] LEDR;
    assign LEDR[0] = SW[0] ^ SW[1] ^ SW[2];                         // Sum
    assign LEDR[1] = SW[0] & SW[1] | SW[0] & SW[2] | SW[1] & SW[2];   // Carry
endmodule
```

Listing 5.13: Dataflow model for full adder

```
module TopLevel (SW, LEDR);
    input [9:0] SW;
    output [9:0] LEDR;
    reg sum, carry;
    assign LEDR[0] = sum;
    assign LEDR[1] = carry;
    always @ (SW[0], SW[1], SW[2])
        begin
            sum = SW[0] ^ SW[1] ^ SW[2];
            carry = SW[0] & SW[1] | SW[0] & SW[2] | SW[1] & SW[2];
        end
endmodule
```

Listing 5.14: Behavioral model for full adder

Although Listing 5.14 uses the mechanics of the behavioral style, it is really not in the spirit of what can be done. Listing 5.15 shows an easier approach that also provides a level of self documentation.

```
module TopLevel (SW, LEDR);
    input [9:0] SW;
    output [9:0] LEDR;
    reg sum, carry;
    assign LEDR[0] = sum;
    assign LEDR[1] = carry;
    always @ (*)
        {carry, sum} = SW[0] + SW[1] + SW[2];
endmodule
```

Listing 5.15: Verilog figures out the gates for full adder

Note the use of the concatenation {carry, sum} to form a 2-bit value from two 1-bit values. The output from SW[0] + SW[1] + SW[2] is two bits, so the low-order bit (the sum) and the high-order bit (the carry) will be connected to their intended signal lines.

Multi-bit Adder

As a final combinatorial example, let's build a multi-bit adder. Since the DE10-Lite has only 10 switches, let's add the 5-bit binary number input on switches 0 to 4 to the 5-bit number on switches 5 to 9. The sum will be displayed on LEDs 0 to 5. Listing 5.16 shows a structural approach using nested modules. Listing 5.17 shows a behavioral approach where Verilog "figures out" the details of gates and configuration.

```
module TopLevel (SW, LEDR);
    input [9:0] SW;
    output [9:0] LEDR;
    wire carry0, carry1, carry2, carry3;        // Internal connections for carry
    FullAdder bit0 (SW[0], SW[5], 0, LEDR[0], carry0);
    FullAdder bit1 (SW[1], SW[6], carry0, LEDR[1], carry1);
    FullAdder bit2 (SW[2], SW[7], carry1, LEDR[2], carry2);
    FullAdder bit3 (SW[3], SW[8], carry2, LEDR[3], carry3);
    FullAdder bit4 (SW[4], SW[9], carry3, LEDR[4], LEDR[5]);
endmodule

module FullAdder (bitA, bitB, bitC, sum, carry);
    input bitA, bitB, bitC;
    output sum, carry;
    wire AB, AC, BC;                            // Internal connections for carry
    xor SumSig (sum, bitA, bitB, bitC);         // sum = A ^ B ^ C
    and (AB, bitA, bitB);
    and (AC, bitA, bitC);
    and (BC, bitB, bitC);
    or CarrySig (carry, AB, AC, BC);            // carry = A & B | A & C | B & C
endmodule
```

Listing 5.16: 4-bit adder using structural approach

```
module TopLevel (SW, LEDR);
    input [9:0] SW;
    output [9:0] LEDR;
    reg [4:0] sum;
    reg carry;
    assign LEDR[4:0] = sum;
    assign LEDR[5] = carry;
    always @ (*)
        {carry, sum} = SW[4:0] + SW[9:5];
endmodule
```

Listing 5.17: Verilog figures out 4-bit adder

In Listing 5.17, the concatenation {carry, sum} forms a 6-bit value. The output from SW[4:0] + SW[9:5] is six bits, so the low-order five bits (the sum) and the high-order bit (the carry) will be connected to their intended signal lines.

Sequential Examples

Digital circuits are generally divided into two categories: combinatorial and sequential. Examples of combinatorial circuits include the structural and dataflow coding styles, as well as all of the behavioral coding done thus far in this chapter. In every combinatorial circuit, all of the outputs are determined by the immediate values of the circuit's inputs.

Sequential digital circuits, on the other hand, have the following characteristics:

1. A sequential circuit holds an internal "state" (i.e., it has some sort of memory that is independent of the immediate input signals).
2. A sequential circuit's output is determined by its internal "state." Sometimes the output is also dependent on one or more input signals.

A sequential circuit is divided into two categories depending on its sensitivity to changes of its inputs:

1. Asynchronous: State changes whenever an input changes. Example: RS flip flop
2. Synchronous: State changes only when a clock "ticks" (rising or falling edge, but not both). Example: D flip flop

The copy of Figure 2.12 shows an RS flip flop which holds state (i.e., one bit of data), and is updated whenever its inputs (R and S) are updated. This circuit is coded in the structural style in Listing 5.18 and in the behavioral style in Listing 5.19.

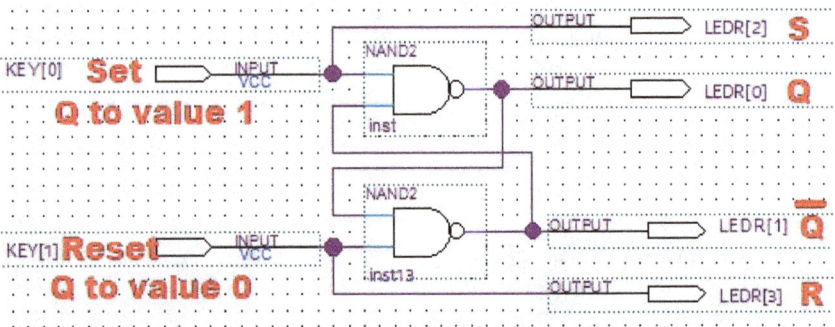

Copy of Figure 2.12: RS flip flop is sequential and updated asynchronously.

The D flip flip shown in Listing 5.20 "holds" the input data value present when the clock signal last transitioned from high to low (negative edge). Compile, download, and test the D flip flop, and notice how LED[0] displays the value that was on SW[0] when KEY[0] was last pushed.

```
module TopLevel (KEY, LEDR);
  input [1:0] KEY;
  output [3:0] LEDR;
  wire q,qBar;
  buf (LEDR[0],q);
  buf (LEDR[1],qbar);
  buf (LEDR[2],KEY[0]);
  buf (LEDR[3],KEY[1]);
  nand(q,KEY[0],qBar);                    // Set with KEY[0]
  nand(qBar,KEY[1],q);                    // Reset with KEY[1]
endmodule
```

Listing 5.18: RS flip flop (a.k.a., latch)

```
module TopLevel (KEY, LEDR);
  input [1:0] KEY;
  output [3:0] LEDR;
  reg q, qBar;
  assign LEDR[0] = q;
  assign LEDR[1] = qBar;
  assign LEDR[2] = KEY[0];
  assign LEDR[3] = KEY[1];
  always @ (*)
    begin
      q = ~(KEY[0] & qBar);               // Set with KEY[0]
      qBar = ~(KEY[1] & q);               // Reset with KEY[1]
    end
endmodule
```

Listing 5.19: RS flip flop in behavioral style of coding

```
module TopLevel (SW, KEY, LEDR);
  input [1:0] SW;
  input [1:0] KEY;
  output [1:0] LEDR;
  reg q;
  assign LEDR[0] = SW[0];
  assign LEDR[1] = q;
  always @ (negedge(KEY[0]))
    q <= SW[0];                           // "Clock in" the data bit
endmodule
```

Listing 5.20: D flip flop

Sensitivity List

The sensitivity list tells the Verilog compiler when the circuit within the always block should be updated. Combinatorial circuits are updated whenever any of the listed input variables change, while sequential circuits are only updated on either the positive or negative edges of one or more listed clock signals.

Suppose we have a circuit with five input variables: A, B, C, D, and E. Three sensitively lists are shown below. The "always @ (A, B, C)" example updates the circuit when any of the three variables A, B, or C changes (both positive and negative transitions). Variables, such as D or E which are not in the list, will indeed affect the circuit, but that effect will be delayed until either A, B, or C changes. The asterisk in the sensitivity list tells the Verilog compiler to update the circuit whenever any of the input variables to the circuit change.

The third example will only update the circuit when variable A moves from low to high (0 to 1).

- always @ (A, B, C)
- always @ (*)
- always @ (posedge(A))

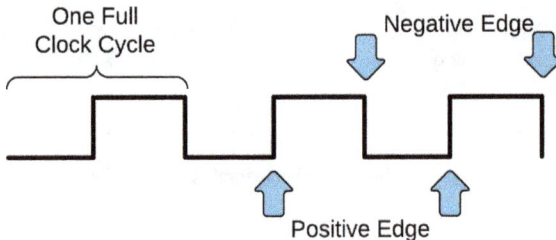

Figure 5.1: Update on positive or negative clock edge

Blocking and Nonblocking

Behavioral style coding within an always block looks like a computer program, but it's not. A computer program executes one statement after another in the order provided. Verilog statements generate digital subcircuits which can physically execute in parallel at exactly the same time, which means the statement order doesn't have to matter.

The Verilog compiler can build circuits such that one statement is "blocked" from starting until the previous one completes. As shown in Figure 5.2, statements containing only an equals sign (=) indicate the blocking approach, while statements using <= represent the nonblocking approach.

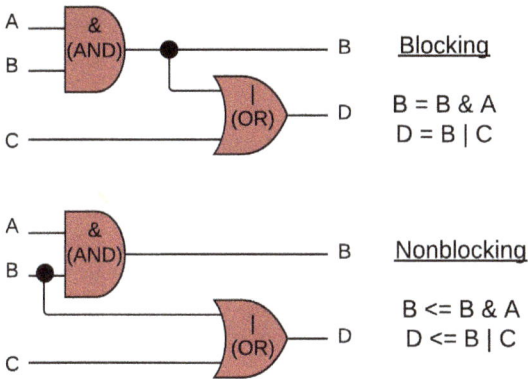

Figure 5.2: Blocking and nonblocking in always

Compile and download Listing 5.21 to demonstrate nonblocking statements containing an exclusive OR followed by an inclusive OR. With switches 0 and 1 down, push KEY[0] until the LEDs are off. Now raise SW[0] to have a value of 1, and then "clock in" new values for B and D by pushing KEY[0]. You will notice that only LEDR[0] lights because only B has a value of 1, not D which remains at zero. Push KEY[0] again, and LEDR[1] will also light. Note: Sometimes this appears to not work properly if the KEY[0] does not not provide one clean single pulse (i.e., debounced switch).

```verilog
module TopLevel (SW, KEY, LEDR);
   input [1:0] SW;
   input [1:0] KEY;
   output [1:0] LEDR;
   wire A, C;
   reg B, D;
   assign A = SW[0];
   assign C = SW[1];
   assign LEDR[0] = B;
   assign LEDR[1] = D;
   always @ (posedge(~KEY[0]))
      begin
         B <= B ^ A;                        // XOR
         D <= B | C;                        // OR
      end
endmodule
```

Listing 5.21: Nonblocking example

Edit the code to change to blocking statements (i.e., change <= to = on both lines). Recompile, download, and perform the same test. This time, you will notice that the first time KEY[0] is pushed, both LEDR[0] and LEDR[1] will light. In other words, for the one clock pulse, the second statement (D <= B | C) is blocked from starting until the first is completed (which updated the value of B from 0 to 1).

Table 5.3 summarizes some of the differences between the blocking and nonblocking approaches. Since computer circuits generally use clocks, almost all of the coding in the remainder of this book will use nonblocking statements within the always blocks.

Although the compiler might allow mixing of blocked and nonblocked statements, it is generally not recommended to do so. Also keep in mind, that even though multiple always blocks can be compiled within a single project, variables updated in different blocks do interact. Obviously, the same variable cannot be updated in two locations in the same block, but neither can it be updated in two different always blocks.

Example	Blocking	Non-blocking
Syntax	Dest = Source;	Dest <= Source;
Sensitivity List	* to get all variables	Positive edge or Negative edge of a clock
Digital Circuit Type	Combinatorial	Sequential
Order of statements	Important	Not important

Table 5.3: Always block statements

Synchronous, Sequence, and Sequential

Be careful with the terminology. Synchronous sequential circuits do *not* process their Verilog statements in sequence within the always block.

Sequential circuits are those that hold an "internal state." Their outputs depend upon a sequence of events that previously happened to set the values stored in their internal state. Combinatorial circuits, on the other hand, have no internal state, thereby resulting in their outputs solely determined by the immediate values appearing on their inputs.

An asynchronous sequential circuit updates its internal state whenever any signal listed in its sensitivity list changes value. Asynchronous sequential circuits typically use the blocking approach (i.e., their assignments within the always block use the = notation). The order of their assignments is very important meaning that they are processed in sequence.

Synchronous sequential circuits update their internal states only when a clock "ticks" (either the positive or negative edge of a designated clock signal). Synchronous sequential circuits generally use the nonblocking approach (i.e., their assignments within the always block use the <= notation). The order of their assignments is not important meaning they are not processed in sequence.

Highlights and Comparisons

Digital Systems can be outlined as the following:

1. Combinatorial
2. Sequential
   a. Asynchronous (* in sensitivity list)
   b. Synchronous (posedge, negedge in sensitivity list)

Verilog Codling Styles can be outlined as the following:

1. Structural
2. Dataflow (assign statements)
3. Behavioral (in always block)
   a. Blocking (uses =)
   b. Nonblocking (uses <=)

This chapter has been using two types of variables: *reg* and *wire*. The *reg* variables are assigned values (i.e., left side of equals sign) within an always block while *wire* variables are assigned values outside an always block. The *wire* variables are "nets" (i.e., network connections) representing continuous assignments. The *reg* variables are like registers and flip flops that can lock onto and hold a state between clock ticks irregardless of whether their input signals are varying. Both *reg* and *wire*, along with several other variable types will be described in more detail in the next chapter.

SystremVerilog has another variable type named *logic* that replaces both *reg* and *wire*. It has been recommended that all new Verilog coding use *logic* and avoid any confusion between *reg* and *wire*. I may have used it in this book, but the free Lite edition of Quartus® Prime does not currently support it. In any case, not all code is "new code," and digital designers really should be familiar with all three.

Review Questions

1. Long term maintenance of a digital design will have updates to meet new requirements. What is the advantage of using the * (asterisk) in the sensitivity list instead of a list of variables?
2. Compare posedge(A) to simply A, where A is an input data signal in the sensitivity list of an always block?

3. Check the Internet to find the name of the VHDL construct for behavioral coding that corresponds to the *always* construct in Verilog.
4. *Name a few benefits of the behavior coding style.

Exercises

1. Modify Listing 5.17 for other operations. Try AND (&) and XOR (^). Also try multiplication using the asterisk between SW[4:0] and SW[9:5]. Please compile, download, and test each of these designs.
2. An RS flipflip is coded in Verilog's structural and behavioral styles in Listings 5.18 and 5.19, respectively. Please recode this RS flipflop in the dataflow style.
3. Figure 5.3 is a "first approximation" to being a D flipflip where the data line is sampled and stored when the clock line transitions from low to high (posedge). Please code this circuit in the different Verilog coding styles: structural, dataflow, and behavioral. Download and test each of your three versions of this circuit. Also, why isn't it a true D flipflop like that in Listing 5.20 (i.e., what does it allow that a real D flipflop does not)?

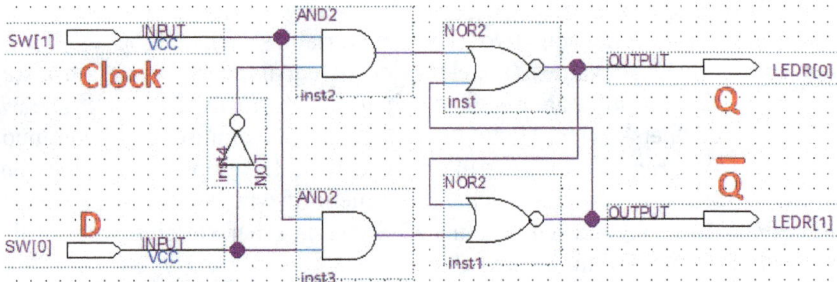

Figure 5.3: First approximation to a D flipflop

— 6 —
Digital Building Blocks

Gates can be combined to form flip flops, and flip flips can be combined to form registers. Register Transfer Level (RTL) is a digital design method commonly used to describe a circuit in terms of data movement between registers. Most computer circuits are synchronous (i.e., clocked), so most of the remainder of this book will involve behavioral style Verilog coding for moving data between registers.

Verilog coding is used in Chapter 6 to build registers, a "bus" structure, multiplexers, demultiplexers, and decoders. Digital building blocks were already introduced graphically in Chapter 2 and will now be rebuilt using Verilog. They will then be used in the following chapters to assemble an ALU (Arithmetic Logic Unit) and a CPU (Central Processing Unit).

Chapter 6 begins by describing the Verilog language more precisely than when it was originally introduced in examples in previous chapters. More depth in behavioral style coding with nonblocking statements will be presented.

Introductions

The following Verilog commands and keywords are introduced in Chapter 6:

- **for:** Command to replicate a group of Verilog statements
- **generate, endgenerate:** Structural and dataflow coding block where "for" and "if" statements are allowed
- **genvar:** Variable within *generate* block for loop count, etc.
- **posedge:** Positive edge (rising) of clock signal
- **negedge:** Negative edge (falling) of clock signal
- **?: :** Conditional operator (similar to a combined if and else)

Compile Time and Run Time "Variables"

Verilog coding involves the use of many variables, and these variables are given names so that we can identify and combine them in the various statements. Some variables represent physical features of the desired digital circuit, such as physical signal lines, gates, and registers. Other variables are used to build the circuit, but do not represent physical entities. For example, there are eight bits in a byte, so the number 8 is used while building the circuit, but it is not part of the circuit itself.

Another type of "variable" is actually a constant, even though it is given a name and looks like a variable. The number of bits in a word is an example. Even though for a given hardware description, it will always be 16, and "16" could actually appear everywhere in the coding, a name such as "wordSize" is used instead. The advantage of representing these constants with names is improved self-documentation of the coding and better long-term maintenance if the "constant" has to be changed in some future application.

Run time, also called execution time, is the operation of the described digital circuit with all its gates and interconnections within the FPGA or CPLD. Compilation is the building of a netlist containing the desired digital circuit from the hardware description in a source code text file. Verilog compilation is often described as having multiple internal steps: translation (convert text to digital constructs), elaboration (generate all component instances), and synthesis (create final netlist for specific FPGA or CPLD). The Netlist is then downloaded (a.k.a., programmed) into the specific FPGA or CPLD to make the running circuit. Some "variables" are for run-time, some for compile-time, and even some are used to substitute constants before compilation.

Identifiers and Keywords

An identifier is a string of characters providing a name for a "key word" within the Verilog language or a user-defined name within a hardware description. It may be composed of upper and lower case letters, digits, dollar sign, and the underline character. It must not begin with a digit and cannot contain blanks. Examples of identifiers:

- **wire:** Verilog keyword (case matters)
- **Time_2_go:** User variable name (case also matters)
- **N:** User variable name. Something longer and more descriptive is better (aides in documentation and maintenance).
- **WIRE:** User variable name. I would not recommend naming a variable WIRE because it can be easily confused with the Verilog keyword *wire* (case sensitive).

Vectors and Arrays

A user-defined variable name can be associated with a single bit, a vector (group of bits, such as a byte or word), or an array (a series of bits or vectors). In the following example, a user-defined name can be defined with both a bit range and element range. Both ranges are optional, and if omitted, default to the value 1.

- **reg**   [bit range] variable_name [array element range];

Examples of a single bit, a vector, and an array:

- **reg openDoor:** Define single-bit variable named openDoor
- **reg [7:0] accum:** Define 8-bit variable named accum
- **reg [7:0] accum [0:255]:** Define an array of 256 8-bit variables named accum[0], accum[1], . . .

The bit range within a vector usually goes from high to low, but that is not required. The reason is that a binary number is really a short notation for a polynomial of powers of 2. For example: 10111_2 is $1 \times 2^4 + 0 \times 2^3 + 1 \times 2^2 + 1 \times 2^1 + 1 \times 2^0$. (i.e., the exponent order is $[4 : 0]$ rather than $[0 : 4]$)

Data Types for Variables

Signal lines and flip flops are physical components in Verilog that have values of 0, 1, or Z (high impedance). The following types of physical components in a digital circuit can be given user-defined variable names.

- **wire:** Connects physical components. It is categorized as a "net" (network connection) used in the Verilog structural and dataflow coding styles. It is the most common variable type used and is the default type assigned by Verilog. It does not hold state (i.e., it is not a flip flop or register).
- **tri:** This is the same as wire, but it implies a tri-state signal line which can have more than one circuit driving it.
- **reg:** This is a Verilog "behavioral" signal line (which can also hold state, as in a register or flip flop). It is given values only within an always block (either combinatorial or sequential logic).
- **logic:** Logic was defined in SystemVerilog as a replacement for both wire and reg. It is currently not supported in the Lite (i.e., free) version of Quartus Prime and will therefore not be demonstrated in this book.

Another type of variable is not associated with a physical construct, but is used to construct physical entities such as registers, flip flops, and signal lines.

- **integer:** Compile-time variables used within always blocks are often needed as loop counters, etc. The integer has no specific size associated with it since it does not represent a physical structure, but its range is large (32 bits, positive and negative two billion).
- **genvar:** Similar to the integer type which is used within an always block for behavioral coding, but is used within *generate* blocks.

Verilog has two other constructs that are similar to variables, but represent constants, and their substitution actually occurs slightly before the compilation:

- **parameter:** Name to substitute for a number or string (such as using PI instead of 3.14156)

- **define:** Macro capability that substitutes a number or string for an identifier just before the compilation.

Scope of Variables

Macros created with the `define command have global scope. A macro's definition is available at any point in the hardware description after it is defined, even inside a different module and even if the define did not occur within any module.

The other "variables," including *parameter*, are local and only retain their assigned values within the module that they are created. The following list describes the scope of three variables in a segment of Verilog code for two modules.

1. module ModA; // create a module
2. parameter P1 = 16;
3. `define D1 = 20;
4. wire W1;
5. endmodule
6. module ModB; // create a second module
7. P1 is not known here
8. D1 is still assigned value 20 here
9. W1 is not known here
10. endmodule

Constants

Constants in Verilog not only have a numeric value, but can also have a size (number of bits). A quote mark separates the size from the value. The value can be expressed in binary, octal, decimal, and hexadecimal indicated by B, O, D, and H, respectively (lower case b, o, d, and h also work). If the base is omitted, decimal is assumed. Examples of several constants:

- **4'B1011:** A 4-bit wide constant containing binary 1011.
- **4'B10:** A 4-bit wide constant containing binary 0010 (assume leading zeroes).
- **6'H10:** A 6-bit wide constant containing hexadecimal 10 (binary 010000).
- **6'D10:** A 6-bit wide constant containing decimal 10 (binary 001010).
- **-6'D10:** A 6-bit wide constant containing negative decimal 10 (binary 110110 which is a two's complement of 001010).
- **8'10:** An 8-bit wide constant containing default decimal 10 (binary 00001010).
- **'B10:** Unsized binary 10 (decimal 2).

- **35:** Unsized decimal 35 (binary 100011).

Constants without a size can be used in Verilog statements where the size is explicitly provided or in statements where size is not needed (for-loop index, for example). Each of the bits can have a value of 0 (low), 1 (high), or Z (high impedance). Verilog simulation also has X (undetermined), but simulation is not used in this book. ASCII characters can also appear as a constant value. Examples using high impedance tri-state or ASCII:

- **4'BZ:** A 4-bit wide constant where the Z is extended into leading bits.
- **"A":** ASCII character within quotation marks (hexadecimal 41).
- **"AB7":** A 24-bit wide ASCII string within quotation marks (hexadecimal 414237).

Data Highway

A bus within a digital circuit is analogous to a limited-access high-speed highway for automobiles:

1. The purpose of the highway is to provide an orderly high performance movement among multiple locations.
2. What controls or limits the speed on the highway?
3. How many "lanes" are present (enables parallel traffic)?
4. Where are the entry points to the highway, and are there signals controlling entry?
5. Where are the exit points from the highway?

In the digital bus, a clock usually controls the speed. A bus can be a single wire, but in computer circuits, data buses of 32 or 64 signal lines are now common. Entry to the bus uses tri-state outputs, multiplexers, decoders, and clock signals. Demultiplexers along with decoders select destinations to receive the data.

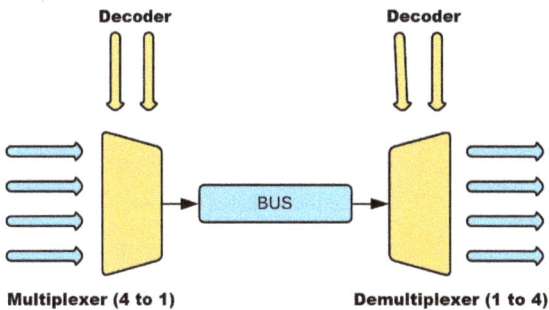

Figure 6.1: Multiplexer controls where data gets on the bus, and demultiplexer controls where it gets off.

Get on the Bus

A bus is a common communications path among multiple digital devices. It is the data path between the CPU and multiple memory and I/O devices. Chapter 2 graphically introduced the bus using tri-state buffers in Figure 2.5. Here in Chapter 6, that same circuit will be rebuilt using the Verilog conditional operator (a.k.a., ternary operator) which is like a shorthand version of combining *if* and *else* statements.

Digital logic has only two states: High and Low, One and Zero, True and False, etc. Right? Actually, some digital output lines can have a "high impedance" (high-Z) third state. In the high-Z state, the gate neither drives nor sinks the bus line. It appears as if the gate is not even connected to the bus line, thereby allowing a different gate to drive the bus without interference. The schematic in Figure 6.2 illustrates a single wire bus having four inputs and one device to display the high/low status of the bus. Take note of the following:

1. The outputs from "data" switches 0, 1, 2, and 3 are wired together.
2. The buffer gates connected to each switch have a third "control" wire coming in from the top.
3. Switches 4, 5, 6, and 7 are connected to the control lines of buffers on switches 0, 1, 2, and 3, respectively.

If all four data switches are High or all four are Low, the output value of the bus is obvious. But what if two switches are High and two Low? Is the value going to be half way? Is there going to be a "fight" or possibly even some type of overload or short circuit? With tri-state, a bus master (i.e., the CPU) can select which <u>one</u> gate will output to the bus line. All of the outputs from the other gates will be inhibited.

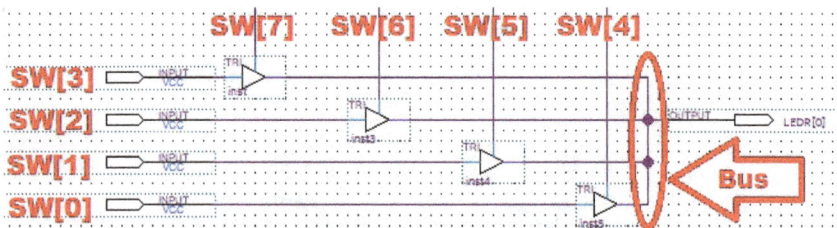

Figure 6.2: Bus where four switches drive one LED

It Looks Like C or Java

Verilog is a hardware description language, but some of its coding looks like a C or Java program. Both the programming languages and Verilog have *for*, *if*, *else*, *case*, and *? :* (the conditional operator). The meanings of these constructs is very similar, but C and Java describe program flow in the "time domain" while Verilog specifies data paths to be constructed in the "space domain."

I will begin with the conditional operator which is a shorthand version of combining *if* with *else* on one command line. It is used in Listing 6.1 to implement the schematic of Figure 6.2. It has the following three parts:

- Condition **?** Do if True **:** Do if False **;**

```
module TopLevel (SW, LEDR);
    input [9:0] SW;
    output [9:0] LEDR;
// condition operator = condition ? value-if-true : value-if-false
    assign LEDR[0] = SW[4] ? SW[0] : 1'bz;
    assign LEDR[0] = SW[5] ? SW[1] : 1'bz;
    assign LEDR[0] = SW[6] ? SW[2] : 1'bz;
    assign LEDR[0] = SW[7] ? SW[3] : 1'bz;
endmodule
```

Listing 6.1: Multiplexer of four input signal lines to one output signal line

Go ahead and compile, download, and run the example in Listing 6.1. Switches 0 through 3 provide the data, while switches 4 through 7 control which <u>one</u> data line gets to drive the bus. The first conditional statement says that if SW[4] is high (value of 1, i.e., true), then the value of SW[0] will be passed onto the bus wire connected to LED[0]. If SW[4] is low (value of 0, i.e., false), then SW[0] is essentially not connected to LED[0] (tri-stated off). Only one control switch (SW[4] through SW[7]) should be high at a time. The data lines (SW[0] through SW[3]) can be in any pattern of highs and lows.

For loops

Another command that Verilog shares with C and Java is the *for* loop. In a programming language, the *for* loop specifies how many times a section of code is to be executed, while in Verilog, it specifies how many copies (a.k.a., instances) of a subcircuit are to be created. A *for* loop has a loop control variable that is initialized for the first pass through the loop, changed on every pass through the loop, and tested to see when the loop is completed.

- for (initialization ; condition to continue ; increment/decrement)

The *for* loop is mostly used in behavioral coding within an *always* block, but it is available for dataflow coding within a *generate* block. The *generate* block, using *genvar* variables, can use the *for* and *if* statements. In Listing 6.1, the *assign* statement appears four times in a similar format that lends itself to be implemented in a loop as shown in Listing 6.2.

```
module TopLevel (SW, LEDR);
    input [9:0] SW;
    output [9:0] LEDR;
    genvar i;
// Demonstrate tri-state and generate
// generate works with if and for. begin block needs a name
// condition operator = condition ? value-if-true : value-if-false
    generate
        for (i=0; i<=3; i=i+1)
            begin:blkname
                assign LEDR[0] = SW[i+4] ? SW[i] : 1'bz;
            end
    endgenerate
endmodule
```

Listing 6.2: Use *generate* to produce same circuit as Listing 6.1.

The *for* loop in Listing 6.2 generates the same four *assign* statements as in Listing 6.1. Go ahead and compile, download, and test the code in Listing 6.2. I'm only showing it to demonstrate that a *for* command can be used in dataflow coding. For only four data lines, I definitely prefer Listing 6.1, but if I had 64 lines, using a *generate* block has advantages not only in writing the initial code, but in long term maintenance.

Multiplexer with Decoder

A multiplexer selects one of several possible input signals to be passed on to its output. There are a couple of concerns regarding the multiplexer circuits in listings 6.1 and 6.2:

1. Exactly one control line must be high (1) at a time. If all are low (0), then the output bus line will be in the high impedance state (which might be OK depending on the digital requirements). If two control lines are high, then two data inputs will be driving the bus line at the same time which will lead to a data error.
2. There are a lot of control lines. Actually, there are as many control lines as data lines. For 4 bits, this is not too much of a problem, but what if there are dozens or hundreds of possible input data lines?

Both of the problems can be avoided by use of a decoder circuit along with the multiplexer as was previously demonstrated in Figure 2.7. Rather than having four control lines selecting "who drives the bus," there are only two.

Switches 4 and 5 provide a binary address to the decoder which then expands the signal out to the four tri-state enable lines on the four buffer gates. The copy of Table 2.1 shows that exactly one input is connected to the bus at a time.

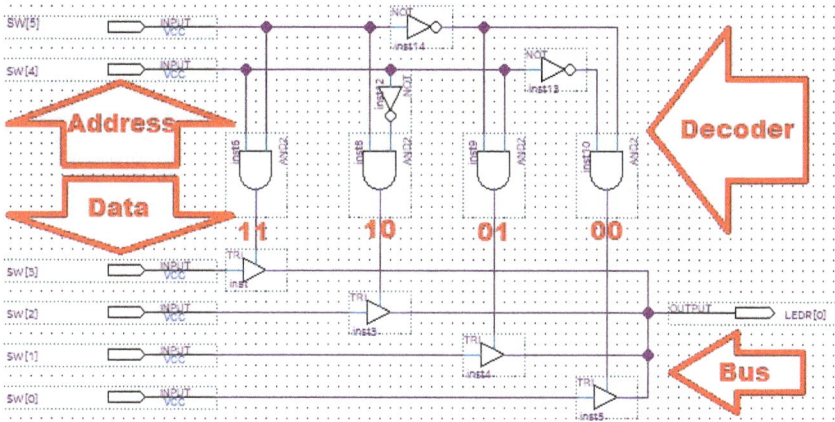

Copy of Figure 2.7: Multiplexer with decoder

SW[5], SW[4]	Switch Signal Selected	Switch Signals Inhibited
0, 0	SW[0]	SW[3], SW[2], SW[1]
0, 1	SW[1]	SW[3], SW[2], SW[0]
1, 0	SW[2]	SW[3], SW[1], SW[0]
1, 1	SW[3]	SW[2], SW[1], SW[0]

Copy of Table 2.1: Multiplexer with decoder selects one of four inputs

The schematic in Figure 2.7 is implemented in Verilog in Listing 6.3. The condition $SW[5:4] == 0$ is true if both switches are low (zeros). Just like in C and Java, the double equals sign is used for the test. A single equals sign would try (and fail) to make an assignment. Note: The two equal signs are adjacent with no blank between them.

```
module TopLevel (SW, LEDR);
   input [9:0] SW;
   output [9:0] LEDR;
// condition operator = condition ? value-if-true : value-if-false
   assign LEDR[0] = SW[5:4] == 0 ? SW[0] : 1'bz;
   assign LEDR[0] = SW[5:4] == 1 ? SW[1] : 1'bz;
   assign LEDR[0] = SW[5:4] == 2 ? SW[2] : 1'bz;
   assign LEDR[0] = SW[5:4] == 3 ? SW[3] : 1'bz;
endmodule
```

Listing 6.3: Verilog version to implement schematic in Figure 2.7

Listing 6.4 also implements Figure 2.7 and is another example of using a *generate* block in the dataflow style of coding. Here's another example where I prefer the simplicity of the code in Listing 6.3 over that of Listing 6.4 for such a short loop.

```
module TopLevel (SW, LEDR);
   input [9:0] SW;
   output [9:0] LEDR;
   genvar i;
   generate
     for (i = 0; i <= 3; i = i + 1)
       begin:blkname
         assign LEDR[0] = SW[5:4] = = i ? SW[i] : 1'bz;
       end
   endgenerate
endmodule
```

Listing 6.4: Demonstrate *for* loop with multiplexer with decoder

Listings 6.3 and 6.4 demonstrate that two input lines to a decoder can select one of four unique outputs. Likewise three input lines can select one of eight outputs, and four input lines can select one out of sixteen unique outputs.

Table 6.1 summarizes decoders having a binary input of N bits which can select one of 2^N outputs.

Number of Input Lines	Number of Output Possibilities
2	4
3	8
4	16
N	2^N

Table 6.1: Number of decoder input and output signal lines

Multiplexer with Decoder (Behavioral Style)

A multiplexer can be created inside an always block using a *for* loop of conditional operators (somewhat like those in Listing 6.4). It can also be implemented using *if* statements as shown in Listing 6.5.

```
module TopLevel (SW, LEDR);
    input [9:0] SW;
    output [9:0] LEDR;
    reg bus_signal;
    integer i;
    assign LEDR[0] = bus_signal;
    always @ (*)
        for (i = 0; i <= 3; i = i + 1)
            if (SW[5:4] = = i) bus_signal = SW[i];
endmodule
```

Listing 6.5: Multiplexer in *always* block

Quite often, multiple *if* statements can be replaced by a *case* statement, which can clearly and cleanly express the desired subcircuits. The *case* statement and its list of possible "cases" have the following characteristics:

1. A *case* statement consists of a condition to be evaluated and a set of possible case values.
2. The condition appears as an expression inside parentheses following the "case" keyword. A list containing possible values and associated subcircuits appears next, followed by an "endcase" keyword. The *case* and *endcase* do not end with semicolons, but each of the subcircuit statements in between do.
3. The condition evaluates to a vector of N bits. There are 2^N possible values and associated subcircuits (see Table 6.1).
4. The Verilog *case* is similar in concept to the "switch," "select," and "case" commands in programming languages. Unlike the switch statement in C, there is no fallthrough in the Verilog *case* statement (i.e., only one case is selected and no "break" is needed).
5. Each of the cases is also a vector (or equivalent binary number) followed by a colon and a subcircuit to be executed.
6. A "default" case can be specified to be used if none of the listed cases matches the condition.
7. If the condition evaluates to a case not provided and no default case is given either, then Verilog generates a latch that will hold the previous state (i.e., nothing changes).

Compile, download, and verify that the multiplexer/decoder implemented with a case statement works the way you expect. The example shown in Listing 6.6 is a trivial example, but many more challenging *case* applications are presented in the following chapters.

6: Digital Building Blocks   109

```
module TopLevel (SW, LEDR);
   input [9:0] SW;
   output [9:0] LEDR;
   reg bus_signal;
   assign LEDR[0] = bus_signal;
   always @ (*)
      case (SW[5:4])
         0: bus_signal = SW[0];
         1: bus_signal = SW[1];
         2: bus_signal = SW[2];
         3: bus_signal = SW[3];
      endcase
endmodule
```

Listing 6.6: Case statement used for decoder.

An even simpler approach for this multiplexer/decoder example is to use an index as shown in listings 6.7 and 6.8. These examples increase the index range to 0 through 7 by using three switches. SW[9:7] becomes a 3-bit decoder to select switches 0 through 7.

Yes, I know I used switch 7 twice, but I ran out of switches on the DE10-Lite. If you have a DE2-115, go with a range of 0 through 15 by using SW[17:14] for the index.

```
module TopLevel (SW, LEDR);
   input [9:0] SW;
   output [9:0] LEDR;
   reg bus_signal;
   assign LEDR[0] = bus_signal;
   always @ (*)
      bus_signal = SW[SW[5:4]];
endmodule
```

Listing 6.7: Use array index for decoder.

```
module TopLevel (SW, LEDR);
   input [9:0] SW;
   output [9:0] LEDR;
   assign LEDR[0] = SW[SW[5:4]];
endmodule
```

Listing 6.8: Array index also works in dataflow style.

Get Off the Bus

Although there is exactly one digital input signal allowed to drive the bus at a time, there can be multiple devices reading the bus simultaneously. However, in most situations, the data on the bus is directed to only one device. Listing 6.9 gets data from the bus, which is basically the reverse of Listing 6.1.

Go ahead and compile, download, and test the hardware description in Listing 6.9. KEY[0] will be driving the bus, and switches SW[0] through SW[7] will control whether LEDR[0] through LEDR[7] receive the signal.

```
module TopLevel (SW, KEY, LEDR);
    input [7:0] SW;
    input [1:0] KEY;
    output [7:0] LEDR;
// condition operator = condition ? value-if-true : value-if-false
    assign LEDR[0] = SW[0] ? KEY[0] : 1'bz;
    assign LEDR[1] = SW[1] ? KEY[0] : 1'bz;
    assign LEDR[2] = SW[2] ? KEY[0] : 1'bz;
    assign LEDR[3] = SW[3] ? KEY[0] : 1'bz;
    assign LEDR[4] = SW[4] ? KEY[0] : 1'bz;
    assign LEDR[5] = SW[5] ? KEY[0] : 1'bz;
    assign LEDR[6] = SW[6] ? KEY[0] : 1'bz;
    assign LEDR[7] = SW[7] ? KEY[0] : 1'bz;
endmodule
```

Listing 6.9: Reading the bus by multiple devices

Listing 6.10 is a rewrite of Listing 6.9 using a *for* loop to generate the assigns. In addition, it changed the "false" result of the conditional operators to a value of 0 (rather than high-Z). Compile, download, and verify Listing 6.10. For a second test modification, change the "false" result to a 1, and see how that works.

```
module TopLevel (SW, KEY, LEDR);
    input [7:0] SW;
    input [1:0] KEY;
    output [7:0] LEDR;
    genvar i;
    generate
        for (i = 0; i <= 7; i = i + 1)
            begin:blkname
                assign LEDR[ i ] = SW[ i ] ? KEY[ 0 ] : 0;
            end
    endgenerate
endmodule
```

Listing 6.10: Alternate coding for circuit in Listing 6.9

6: Digital Building Blocks

Demultiplexer with Decoder (Behavioral Style)

A multiplexer selects one of several possible input signals to be passed on to its output, while a demultiplexer selects one of several possible output locations to receive its input signal. The index decoder approach is demonstrated in Listing 6.11 where the one input "bus" line is provided by KEY[0]. The 3-bit decoder value is in SW[2:0], and the eight possible outputs are LEDs 0 through 7.

```
module TopLevel (SW, KEY, LEDR);
  input [7:0] SW;
  input [1:0] KEY;              // Test line for bus
  output [7:0] LEDR;
  reg [7:0] CS;                 // "Chip Select" for each LEDR
  assign LEDR = CS;
  always @ (*)
    CS[SW[2:0]] = ~KEY[0];
endmodule
```

Listing 6.11: Demultiplexer/decoder using index

Of course, a case statement can build the same decoder (see Listing 6.12). However, case statements normally wouldn't be used in such a simple situation, but instead appear in those where the subcircuits on the right side of the equals sign are more complex.

```
module TopLevel (SW, KEY, LEDR);
  input [7:0] SW;
  input [1:0] KEY;              // Test line for bus
  output [7:0] LEDR;
  reg [7:0] CS;                 // "Chip Select" for each LEDR
  assign LEDR = CS;
  always @ (*)
    case (SW[2:0])
      0: CS[0] = ~KEY[0];
      1: CS[1] = ~KEY[0];
      2: CS[2] = ~KEY[0];
      3: CS[3] = ~KEY[0];
      4: CS[4] = ~KEY[0];
      5: CS[5] = ~KEY[0];
      6: CS[6] = ~KEY[0];
      7: CS[7] = ~KEY[0];
    endcase
endmodule
```

Listing 6.12: Demultiplexer/decoder using case statement

　　　　　　　　　Computer Architecture Tutorial Using an FPGA

Decoder for ARM Machine Code

Decoder circuits are present in several applications in computer architecture besides multiplexers and demultiplexers. Basically, a decoder is used anytime a binary number selects one of a number of possible subcircuits.

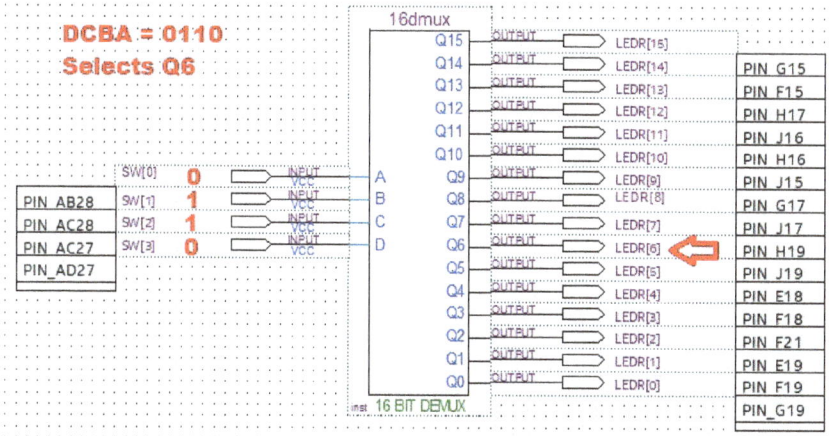

Copy of Figure 2.9: The 4-to-16 decoder circuit demonstrated in Chapter 2

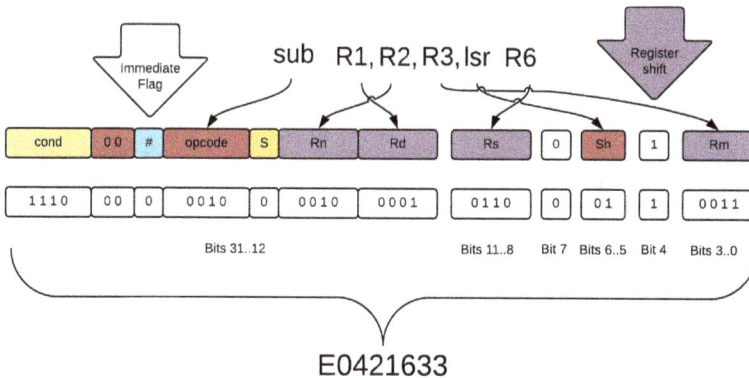

Copy of Figure 12.11: Six 4-bit fields within ARM instruction that can use decoder circuits

The copy of Figure 12.11 shows six fields within the ARM instruction format in which a decoder circuit will select one of sixteen alternatives. Four of the fields point to which one of the sixteen general purpose registers is to be accessed.

Another 4-bit field selects which one of a possible sixteen data processing instructions is to be executed. The sixth 4-bit field indicates whether the instruction should be executed or not depending on the values in status bits. There's even a 2-bit field that indicates which of four possible register shifts is to be performed.

Registers

A register is a group of bits (i.e., flip flops) that holds a number. The use of that number varies by application. Some registers are assigned special purposes, while others are called "general purpose" and can be used either in arithmetic calculations or as pointers into memory.

Figure 6.3 illustrates three 4-bit registers used to calculate a logical AND operation, and Listing 6.13 implements it in Verilog. Compile, download, and verify that the circuit works as expected. Notice how the register contents do not change until the clock (KEY[0]) pulses.

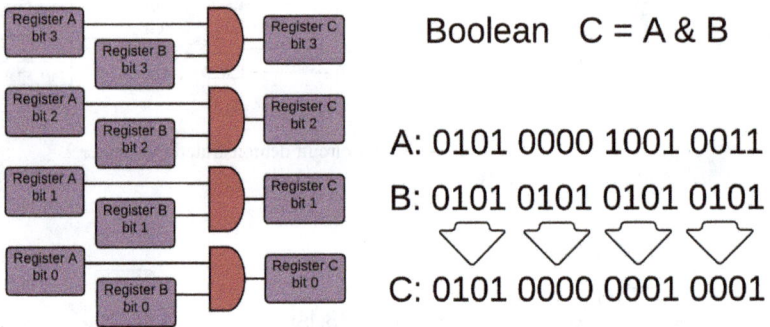

Boolean $C = A$ & B

A: 0101 0000 1001 0011
B: 0101 0101 0101 0101

C: 0101 0000 0001 0001

Figure 6.3: Examples of "bitwise" AND of two values

```
module TopLevel (SW, KEY, LEDR);
   input [7:0] SW;
   input [1:0] KEY;
   output [3:0] LEDR;
   wire [3:0] A, B;
   reg [3:0] C;
   assign A = SW[3:0];
   assign B = SW[7:4];
   assign LEDR[3:0] = C;
   always @ (posedge(~KEY[0]))
      C <= A & B;                    // Logical AND to 4-bit register
endmodule
```

Listing 6.13: Implement logical AND on rising "clock" edge.

The 4-bit parallel logical AND could have been implemented using combinatorial logic in the structural or dataflow styles of coding. An always block with an asterisk in the sensitivity list could have also implemented combinatorial logic. Instead, I chose the synchronous approach which is common among computer circuits.

Since there is only one statement in the always block, it doesn't matter whether it is blocked (=) or nonblocked (<=). However, nonblocking is typically used with synchronous (i.e., clocked) circuits. If this was part of a larger project, there most likely would be additional always blocks, and they would most likely contain nonblocking statements.

There are several reasons synchronous circuits are used in computer architecture: signal "racing" is one, and matching slower devices to faster ones is another.

Highlights and Comparisons

Registers, multiplexers, demultiplexers, decoders, and a bus structure are digital building blocks for constructing a computer. These components will be used in the next several chapters to assemble an ALU (Arithmetic Logic Unit) and a CPU (Central Processing Unit).

Verilog provides many automatic conversions that VHDL does not. A vector of bits can indeed form a binary number, but using that number as an index into an array (i.e., as a decoder) does take a special effort by the Verilog compiler.

Tri-state is a common technique for enabling several devices to write to the same wire or set of wires. It does require a controller (typically the CPU) and an individual control line to each device. For networks, this cabling would become somewhat messy, so a technique known as open-collector is employed.

Review Questions

1. Which of the following can be used in a Verilog *generate* block: *if, else, for, case*?
2. A register can be constructed from what smaller component?
3. *How many input lines are needed for a decoder to select one of a possible 64 different outputs.
4. *What is the difference between *tri* and *wire* variable types?
5. *Convert the following two concatenations to 32-bit hexadecimal values:
   {8'b101, 16'hA47D, 4'd11, 4'o12}, {5'31, 9'hA4, 10'b11, 8'o11}
6. Which variable types can appear on the left side of the equals sign in a Verilog behavioral statement (i.e., inside an *always* block)? Which types can appear on the right side?
7. Which variable types can appear left of the equals sign in a Verilog dataflow *assign* statement? Which types can appear on the right?

Exercises

1. Combine a 4-to-1 multiplexer with 1-to-4 demultiplexer to route any one of four switch values to any one of four LEDs as shown in Figure 6-4. "Build" this circuit using SW[3:0] as the input data, and LEDR[3:0] as the output. The binary number on SW[5:4] selects which of the four input lines, and the binary number on SW[7:6] selects which of the output lines. Use LEDR[9] to show the value on the bus line connecting the switches to the LEDs. Listings 6.7 and 6.8 provide ideas for implementing the multiplexer in Verilog, and Listings 6.11 shows a possible demultiplexer.

2. Build a simple circuit containing a register and ten input data lines from switches. The contents of the switches will be copied into the register when a clock "ticks." Use SW[9:0] as the inputs, LEDR[9:0] to show the current register contents, and KEY[0] as the clock pulse.

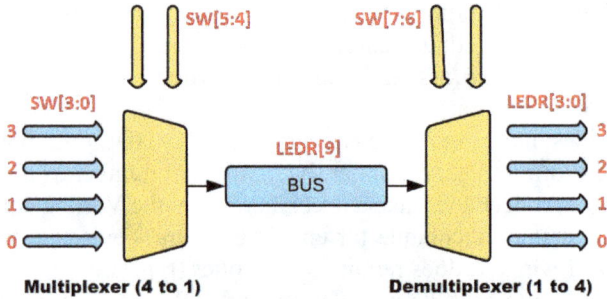

Figure 6.4: Multiplexer and demultiplexer pin assignments for Exercise 1.

Computer Architecture Tutorial Using an FPGA

— 7 —
Arithmetic Logic Unit

Performing calculations and organizing data are the two major activities of computers. From an architectural standpoint, the Arithmetic Logic Unit (ALU) is the center of computation within a computer where the actual calculations such as addition, logical OR, and shifts are performed.

Chapter 7 provides a "hands on" introduction to the set of ARM "data processing" instructions by building a calculator in an FPGA. This Verilog code will then be extended in chapters 8 and 9 with control logic and registers to build a subset of the ARM Central Processing Unit (CPU).

Introductions

No new Verilog commands or keywords are introduced in Chapter 7. However, a limited version of the following ARM "data processing" instructions will be implemented and available for testing:

- **AND:** Logical AND of multiple pairs of bits
- **EOR:** Logical exclusive-OR of multiple pairs of bits
- **SUB:** Subtract an integer value from current register contents
- **RSB:** Reverse Subtract (i.e., negative of SUB)
- **ADD:** Add an integer value to current register contents
- **ORR:** Logical inclusive-OR of multiple pairs of bits
- **MOV:** Load a value into a register
- **BIC:** Clear selected bits in a register
- **MVN:** Load inverted (NOT) value into register

The Machine and Its Language

"Machine Language" generally refers to the numeric codes that instruct a CPU which operation is to be performed and on what values. In order to understand machine language, think of a calculator which has the following features:

- **Display:** Current number being entered or current result of previous operations
- **Operations:** Clear (C), add(+), subtract(-), multiply(*), divide(/), display(=)
- **Data input:** Digits and decimal point, sign(+/-), clear entry (CE), back space (BS)

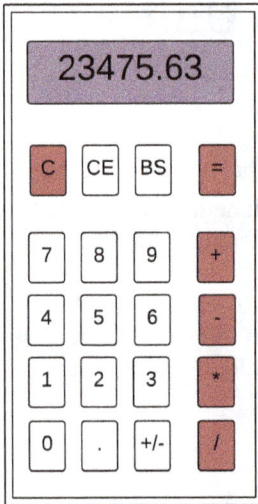

When we use a simple calculator, there is a symbiotic partnership to achieve the final calculation: The calculator does the work, and we provide the directions. For example, we enter the following sequence of instructions to perform the calculation 13×21+6:

1. **Clear**
2. **Add 13**
3. **Multiply 21**
4. **Add 6**
5. **Display**

Figure 7.1: Simple calculator model

Accumulator Register

CPUs contain a small number of fast-access memory units called registers. Depending upon the particular CPU design, the number of registers varies from about five up to nearly one hundred. Some of the registers are accessible to user programs in machine language, and some registers are only accessed by the CPU's electronics to perform its many tasks. The principal data register is generally referred to as the "accumulator." In this calculator example, the accumulator value is that "running total" or accumulated total that is in the display.

Most of the logical and arithmetic operations performed by CPUs are binary operations. Two numbers (called operands) are added, two numbers are multiplied, etc. In the calculator example, the first operand is the value in the display (i.e., the accumulator), and the second operand is the number being entered.

Op-codes

Although a CPU could be constructed to work with the character string names of operations like clear, add, and multiply, it would be somewhat inefficient. Instead, the CPU designers assign an operation code (op-code) to represent each of the available CPU operations. For example, let the following six numbers be

Computer Architecture Tutorial Using an FPGA

assigned to the following six calculator operations:

1. **Clear (C)**: Load zero into the accumulator.
2. **Add (+)**: Add the value of the operand (being entered) to the current accumulator contents.
3. **Subtract (-)**: Subtract the value of the operand (being entered) from the current accumulator contents.
4. **Multiply (*)**: Multiply the value of the operand (being entered) to the current accumulator contents.
5. **Divide (/)**: Divide the current contents of the accumulator by the operand (being entered).
6. **Display (=)**: Copy the current contents of the accumulator to the display line.

If we use the above numeric assignments to translate the previous sequence of instructions to calculate 13×21+6, we will get the following machine code:

Step number	Operation		Op-code with operand
1	Clear	translates to	1 : 0
2	Add 13	translates to	2 : 13
3	Multiply 21	translates to	4 : 21
4	Add 6	translates to	2 : 6
5	Display	translates to	6 : 0

Table 7.1: Translate calculator operations into "machine code."

Note that in this simple model of a calculator being used as a computer, I've represented each instruction as a binomial: an op-code and operand pair. In Table 7.1's translation to machine code, binary operations like Add and Multiply are converted to form "op-code : operand" pairs. Unitary operations, such as Clear and Display, are converted to the form "op-code : 0" because there was no operand. So in this example, we would calculate 13×21+6 by entering "Clear, Add 13, Multiply 21, Add 6, Display" on a calculator, but the corresponding computer program (in machine language) would be the sequence "1 : 0, 2 : 13, 4 : 21, 2 : 6, 6 : 0."

ARM Data Processing Instructions

Rather than continue with an arbitrary set of opcodes, let's use those of the ARM processor. Table 7.2 provides the numeric opcodes associated with nine ARM data processing instructions.

Op Code Value	Op Code Name	Arithmetic or Logical Operation
0	AND	R <= R & op2;
1	EOR Exclusive OR	R <= R ^ op2;
2	SUB	R <= R - op2;
3	RSB Reverse Subtract	R <= op2 - R;
4	ADD	R <= R + op2;
12	ORR inclusive OR	R <= R | op2;
13	MOV Move (i.e., copy)	R <= op2;
14	BIC Bit Clear	R <= R & ~op2;
15	MVN Move NOT	R <= ~op2;

Table 7.2: Nine of the ARM data processing instructions

There are two glaring "holes" in Table 7.2: missing opcodes in the range 5 through 11 and no multiply or divide instructions. The missing opcodes will be included in Chapter 11 when the status register is described. Multiplication will be included in Chapter 13. A division operation is not available in the ARM instruction set, but is in the NEON coprocessor which is presented in Chapter 20.

Calculator Model in FPGA

Listing 7.1 contains Verilog code to build a "calculator" using the ARM data processing opcodes. This code should be compiled and downloaded to your FPGA as was done in the Verilog examples in previous chapters. Don't forget to import the pin assignments file associated with your FPGA board.

```
1.  module TopLevel (SW, KEY, LEDR);
2.     input [9:0] SW;
3.     input [1:0] KEY;
4.     output [9:0] LEDR;
5.     wire w0,w1;
6.     nand(w0,KEY[0],w1);              // Set up RS FF latch
7.     nand(w1,KEY[1],w0);              // as debounced clock.
8.     DataProcIns (SW[9:6], SW[5:0], w0, LEDR[5:0]);
9.  endmodule
10.
11. // Data Processing Instructions
12. module DataProcIns (opcode, op2, clk, display);
13.    input [3:0] opcode;
14.    input [5:0] op2;
15.    input clk;
16.    output [5:0] display;
17.    reg [5:0] R;
18.    assign display = R;
19.    always @ (posedge(clk))
20.       case (opcode)
21.          0: R <= R & op2;            // AND
22.          1: R <= R ^ op2;            // EOR (exclusive OR)
23.          2: R <= R - op2;            // SUB
24.          3: R <= op2 - R;            // RSB (reverse subtract)
25.          4: R <= R + op2;            // ADD
26.          12: R <= R | op2;           // ORR (inclusive OR)
27.          13: R <= op2;               // MOV
28.          14: R <= R & ~op2;          // BIC (bit clear)
29.          15: R <= ~op2;              // MVN (move NOT)
30.          default: R <= 0;            // None of above
31.       endcase
32. endmodule
```

Listing 7.1: ARM "Data Processing" instructions

There are two modules: the calculator itself in lines 11 through 32 and a main module that assigns the actual switches and LEDs to be used.

- Line 1: Module TopLevel is the user interface between the calculator in module DataProcIns and the human operator
- Lines 5 - 7: RS flip flop used to produce "clean" clock pulse
- Line 8: "Instantiate" the DataProcIns module
- Line 17: Accumulator register is 6 bits (I'm naming it "R" to be consistent with ARM nomenclature.)
- Lines 20 - 31: Decoder circuit to select one of 16 possible ARM opcodes
- Line 30: For missing codes 5 - 11, Verilog will set accumulator value to zero.

The calculator will have the following inputs and outputs to perform the ARM operations:

- **Display:** LEDs 0 through 5 will display the accumulator (the result of each operation)
- **Operation:** Switches 6 though 9 indicate which of the operations from Table 7.2 is to be performed
- **Data input:** Switches 0 through 5 give the second operand (first operand is the accumulator value shown in LEDs 0 through 5)
- **Clock Pulse:** Keys 0 and 1 provide the rising and falling edges of clock pulse to initiate each operation

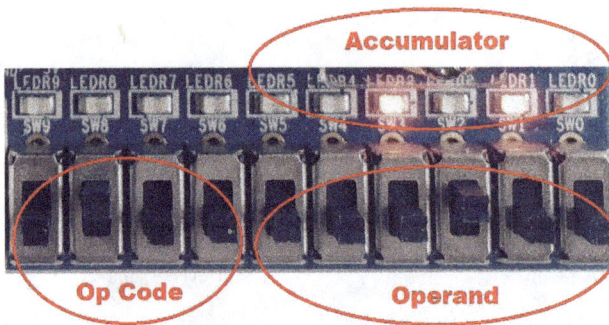

Figure 7.2: "Calculator" inputs and outputs on DE10-Lite development board

Notes:

1. A 4-bit opcode is used because that is exactly the size of an ARM data processing instruction opcode
2. The operand size of 6 bits was chosen because the DE10-Lite only has ten switches, and four are already used for the opcode. The size of the second operand in the ARM instructions is really 12 bits and will be described in later chapters.
3. Two keys are used for the clock pulse to "drive home" the concept of clock edges used in syncronous digital circuits. It also eliminates possible electronic noise from a switch.

The procedure for using the calculator is the following.

1. Put the code for the operation to be performed in switches 6 though 9.
2. Put the value of the operand in switches 0 through 5.
3. Push KEY 1 and let it up, then push KEY 0 to initiate the calculation.
4. The result appears in LEDs 0 through 5.

Go ahead and try a few instructions to see how the calculator works. The opcode

Computer Architecture Tutorial Using an FPGA

decoder table could have been entered in binary as shown in Listing 7.2. These 4-bit values are the ones entered on switches 6 through 9.

```
19.  always @ (posedge(clk))
20.    case (opcode)
21.      4'b0000: R <= R & op2;        // AND
22.      4'b0001: R <= R ^ op2;        // EOR (exclusive OR)
23.      4'b0010: R <= R - op2;        // SUB
24.      4'b0011: R <= op2 - R;        // RSB (reverse subtract)
25.      4'b0100: R <= R + op2;        // ADD
26.      4'b1100: R <= R | op2;        // ORR (inclusive OR)
27.      4'b1101: R <= op2;            // MOV
28.      4'b1110: R <= R & ~op2;       // BIC (bit clear)
29.      4'b1111: R <= ~op2;           // MVN (move NOT)
30.      default: R <= 0;              // None of above
31.    endcase
```

Listing 7.2: Instruction opcodes provided in binary

As an example, calculate 50 - (17 + 12) rounded to nearest multiple of 8. We enter the following sequence of instructions:

1. Move 17 into the accumulator
2. Add 12 to accumulator contents
3. Subtract value in accumulator from 50
4. Add 4 for rounding up
5. Remove modulus 8 to finish rounding

Each of the following switch settings are made followed by a clock pulse (pushing KEY[1] followed by pushing KEY[0]).

Op Code Switches 9 ... 6	Operand Switches 5 ... 0	Comment
1101 (MOV)	010001 (dec. 17)	Move 17 into accumulator
0100 (ADD)	001100 (dec. 12)	Add 12 to accumulator
0011 (RSB)	110010 (dec. 50)	Subtract from 50
0100 (ADD)	000100 (dec. 4)	Round up for "closest 8"
0000 (AND)	111000 (mask)	Finish rounding to +- 8

Table 7.3: Steps to calculate 50 - (17 + 12) rounded to multiple of 8

Hexadecimal Display

Binary is rather clumsy, so let's display what's happening in hexadecimal. Listing 7.3 provides a new display technique, but otherwise, it is identical to the examples in listings 7.1 and 7.2. The DataProcIns module is the same as before.

```
1.  module TopLevel (SW, KEY, LEDR, HEX0, HEX1, HEX2, HEX3, HEX4);
2.      input [9:0] SW;
3.      input [1:0] KEY;
4.      output [9:0] LEDR;
5.      output [7:0] HEX0, HEX1, HEX2, HEX3, HEX4;
6.        function automatic [7:0] digit;
7.          input [3:0] num
8.          case (num)
9.            0: digit = 8'b11000000;          // 0
10.           1: digit = 8'b11111001;          // 1
11.           2: digit = 8'b10100100;          // 2
12.           3: digit = 8'b10110000;          // 3
13.           4: digit = 8'b10011001;          // 4
14.           5: digit = 8'b10010010;          // 5
15.           6: digit = 8'b10000010;          // 6
16.           7: digit = 8'b11111000;          // 7
17.           8: digit = 8'b10000000;          // 8
18.           9: digit = 8'b10010000;          // 9
19.           10: digit = 8'b10001000;         // A
20.           11: digit = 8'b10000011;         // b
21.           12: digit = 8'b11000110;         // C
22.           13: digit = 8'b10100001;         // d
23.           14: digit = 8'b10000110;         // E
24.           15: digit = 8'b10001110;         // F
25.         endcase
26.       endfunction
27.     assign HEX0 = digit(LEDR[3:0]);        // Result
28.     assign HEX1 = digit(LEDR[5:4]);
29.     assign HEX2 = digit(SW[3:0]);          // Operand
30.     assign HEX3 = digit(SW[5:4]);
31.     assign HEX4 = digit(SW[9:6]);          // Op code
32.     wire w0,w1;
33.     nand(w0,KEY[0],w1);                    // Set up RS FF latch
34.     nand(w1,KEY[1],w0);                    // as debounced clock.
35.     DataProcIns (SW[9:6], SW[5:0], w0, LEDR[5:0]);
36. endmodule
```

Listing 7.3: Display opcode, operand, and result in hexadecimal

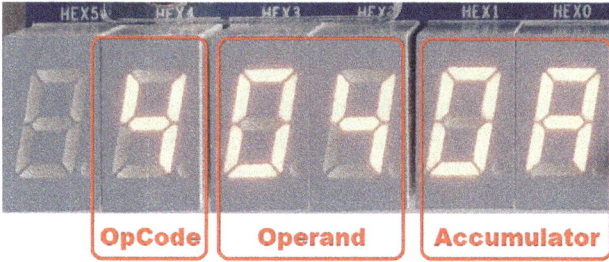

Figure 7.3: Sample output from running Listing 7.3

Figure 7.3 illustrates a typical output from running the code in Listing 7.3. Listing 7.4 contains all of the code from Listing 7.3, but provides the opcode cases in hexadecimal as shown below. The numeric cases can be provided in binary, octal, decimal, or hexadecimal.

```
46.  always @ (posedge(clk))
47.    case (opcode)
48.      4'h0: R <= R & op2;        // AND
49.      4'h1: R <= R ^ op2;        // EOR (exclusive OR)
50.      4'h2: R <= R - op2;        // SUB
51.      4'h3: R <= op2 - R;        // RSB (reverse subtract)
52.      4'h4: R <= R + op2;        // ADD
53.      4'hC: R <= R | op2;        // ORR (inclusive OR)
54.      4'hD: R <= op2;            // MOV
55.      4'hE: R <= R & ~op2;       // BIC (bit clear)
56.      4'hF: R <= ~op2;           // MVN (move NOT)
57.      default: R <= 0;           // None of above
58.    endcase
```

Listing 7.4: Instruction opcodes provided in hexadecimal

Highlights and Comparisons

Chapter 7 models the operation of a CPU using a calculator and opcodes from the ARM processor. The instruction format used in Chapter 7 consists of a 4-bit opcode and 6-bit operand. As new features are include in the CPU model in the next few chapters, this initial 10-bit format will be expanded to 21 bits in Chapter 9 and the full 32-bit ARM format in Chapter 11.

Review Questions

1. *What is the largest unsigned number that can be placed in the operand field of the current 10-bit instruction format?
2. *If the opcode field was five bits instead of four, how many opcodes would be possible?
3. Check the Internet for a short description of a "debounced switch."
4. Almost every *always* block has a *begin/end* sequence. Why does Listing 7.1 have no *begin/end*?

Exercises

1. Use the calculator from Listing 7.3 to add the sequence: 5, 13, 17, and 22. Of course, you have to convert those numbers from decimal to binary before entering them.
2. Again using the calculator, add the numbers 30 and 35. What is wrong with the answer? Why are no errors detected?

— 8 —
Central Processing Unit

When I ask my students when do they think the first computers were developed, I usually get answers like the 1980s or even the 1940s. Very few know that the basic architecture of almost all computers used in the past 70 years dates back to the 1830s with Charles Babbage's Analytical Engine.

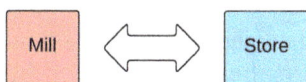

Figure 8.1: Babbage's Analytical Engine

Babbage's Analytical Engine consisted of two principal components:

- Mill: The hardware that did the work (arithmetic and logic operations)
- Store: Data storage for intermediate values present in working variables

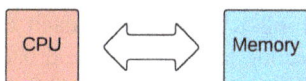

Figure 8.2: "Modern" computer hardware nomenclature

Babbage's mill and store correspond to today's computers:

- CPU: The Central Processing Unit performs the arithmetic and logic operations.
- Memory: Data storage for intermediate values present in working variables

By no means am I implying there existed a positive progression of concepts and devices from Babbage's day to today. The computer pioneers in the 1940s recreated much of what was lost for nearly one hundred years. I personally observed the microcomputer software industry in the 1980s recreate much of the same material as the mainframe developers did in the 1960s.

Introductions

- **parameter** Verilog command that defines a name to be used as a substitute for a constant, such as PI for 3.1416.

State Machine

In the previous chapter, the ALU was manually controlled, and each instruction was provided by switch settings. Now, we will put the instructions in memory and have an automated process to execute their sequence.

Listing 8.1 provides the foundation for a CPU that will be built over the next several chapters. It will eventually process most ARM instructions, but at this point, it could be the basis of almost any CPU. Basically, this Verilog code provides a "state machine" that switches among three states:

1. **Fetch:** Get next the instruction from the computer's memory
2. **Decode:** Separate parts of the instruction to determine what is to be done
3. **Execute:** Actually do the work

The first thing to notice is that the top level module name is "CPU_UI" instead of "TopLevel" which was used in previous chapters. I chose the name TopLevel originally because we were focusing on the structure of Verilog coding, but it is now time to switch to a more application-related name. You should restart the Quartus Prime software "New Project Wizard" to make a new project as was done in the first few chapters. Of course you could name your top level design any valid character string, but the name must match (including upper/lower case) that which is entered on the first page of the project wizard (see figures 1.7 and 3.2).

Compile and download the code in Listing 8.1. It has two push button KEYs as input and two LEDs as output. As was done in Chapter 7, a clock pulse is generated by pushing KEY[1] followed by pushing KEY[0]. With each clock pulse, the state will change from 1) fetch to 2) decode to 3) execute and back to 1) fetch, etc. The current state will be displayed in binary in LEDs 8 and 9. Later in this chapter, we will switch to a built-in 50 MHz clock to automatically and quickly switch among the states.

- Line 1: Module name is "CPU_UI" (CPU User Interface)
- Lines 4 - 6: Same RS flip flop used in Chapter 7 to provide clean clock pulses
- Line 7: If both KEYs are pushed at the same time, a "reset" signal will be generated that will initialize the state to "fetch" (i.e., a value of 1)
- Line 8: The current state will be held in a two-bit register.
- Line 9 - 11: Rather than just calling the states 1, 2, and 3, better documentation of the code is achieved by using the names fetch decode and execute.
- Line 12: The current state will be displayed in LEDs 8 and 9.
- Line 13: This state machine will only change states on a clock edge: either the "clk" signal switches from low to high or the "reset" signal switches from low to high.
- Lines 15, 16: If a reset signal occurs, the state machine will be re-

initialized to "fetch."

- Lines 18 - 25: A decoder is constructed to specify what is to be done when the "clk" ticks.

```
1.  module CPU_UI (KEY, LEDR);
2.    input [1:0] KEY;
3.    output [9:0] LEDR;
4.    wire clk,w1,reset;
5.    nand(clk,KEY[0],w1);              // Set up RS FF latch
6.    nand(w1,KEY[1],clk);             // as debounced clock.
7.    and(reset,~KEY[0],~KEY[1]);      // Reset if both keys pushed
8.    reg [1:0] CPU_state;
9.    parameter fetch = 2'b01;
10.   parameter decode = 2'b10;
11.   parameter execute = 2'b11;
12.   assign LEDR[9:8] = CPU_state;
13.   always @ (posedge(clk), posedge(reset))
14.     begin
15.       if (reset)
16.         CPU_state <= fetch;
17.       else
18.         case (CPU_state)
19.           fetch:                    // Get next instruction in program
20.             CPU_state <= decode;
21.           decode:                   // Disassemble the instruction
22.             CPU_state <= execute;
23.           execute:                  // Perform desired operation
24.             CPU_state <= fetch;
25.         endcase
26.     end
27. endmodule
```

Listing 8.1: Fetch, decode, and execute states

Note that the CPU_state variable is two bits which results in actually four states: 00, 01, 10, and 11. Only three are provided in the code (fetch, decode, and execute), so Verilog creates a latch that will hold the state value in the event a 00 occurs. Remember that we are really building circuits, not programs, and all possible options must be accommodated.

We will be using synchronous (i.e., "clocked") circuits, so clock edges are in the sensitivity list of the always statement, and signals are assigned using non-blocking (<=) statements.

The Verilog *parameter* statement as shown on lines 9, 10 and 11 allows names to be used instead of numbers. This leads to 1) better self documentation and 2) ease of updating the source code. The scope of the *parameter* value does not extend beyond the module in which it is defined.

- **parameter** name = constantValue ;

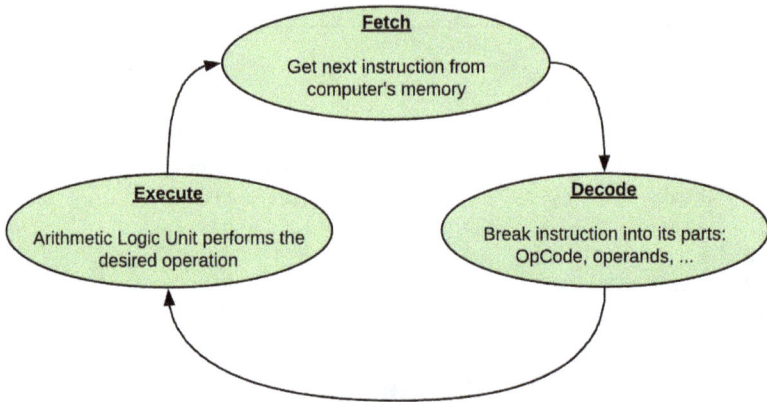

Figure 8.3: Fetch, Decode, Execute cycle

Fetch: Get Next Instruction from Memory

CPUs have a register that points to the next instruction in memory that is to be executed. Historically, about half the CPU architectures refer to their instruction pointers as IP registers, while half refer to them as PC (Program Counter) registers. Some architectures even allow "load/move" instructions to alter the PC or IP, but even if they do, I generally recommend not doing so.

Listing 8.2 has enhancements over Listing 8.1 to implement the "fetch circuit."

- Line 13: The instruction register will contain a copy of the next instruction in memory.
- Line 14: Verilog will allocate 64 "units" of memory, each containing 12 bits. This memory will hold the "machine code" program. Notice that I will expand the 6-bit operand field from Chapter 7 to eight bits (two hex digits), thereby making the instruction size twelve bits.
- Lines 15, 16: A 6-bit program counter addresses the 64 memory locations, and its value is displayed in LED[5:0].
- Line 21: A reset initializes the program counter to the first instruction.
- Lines 28, 29: The next instruction is loaded, and the program counter is updated during the Fetch state.

Compile and download the code in Listing 8.2. As in Listing 8.1, each clock pulse (push Key[1] followed by Key[0]) will change the state from 1) fetch to 2) decode to 3) execute and back to 1) fetch, etc. In addition to the current state being displayed in LEDs 8 and 9, the PC will be displayed in LEDs 0 through 5.

```
1.   module CPU_UI (KEY, LEDR);
2.     input [1:0] KEY;
3.     output [9:0] LEDR;
4.     wire clk,w1,reset;
5.     nand(clk,KEY[0],w1);                // Set up RS FF latch
6.     nand(w1,KEY[1],clk);                // as debounced clock.
7.     and(reset,~KEY[0],~KEY[1]);         // Reset if both keys pushed
8.     reg [1:0] CPU_state;
9.     parameter fetch = 2'b01;
10.    parameter decode = 2'b10;
11.    parameter execute = 2'b11;
12.    assign LEDR[9:8] = CPU_state;
13.    reg [11:0] IR;                       // Instruction register
14.    reg [11:0] progMem [0:63];           // Program memory
15.    reg [5:0] PC;                        // Program Counter
16.    assign LEDR[5:0] = PC;               // (a.k.a., Instruction Pointer)
17.    always @ (posedge(clk), posedge(reset))
18.      begin
19.        if (reset)
20.          begin
21.            PC <= 0;
22.            CPU_state <= fetch;
23.          end
24.        else
25.          case (CPU_state)
26.            fetch:                       // Get next instruction in program
27.              begin
28.                IR <= progMem[PC];       // Load next instruction
29.                PC <= PC + 1;            // Point to following instruction
30.                CPU_state <= decode;
31.              end
32.            decode:                      // Disassemble the instruction
33.              CPU_state <= execute;
34.            execute:                     // Perform desired operation
35.              CPU_state <= fetch;
36.          endcase
37.      end
38.  endmodule
```

Listing 8.2: Fetch next instruction from memory

Decode: Divide the Instruction Into Components

Our "machine code" is currently a very simple one: a 4-bit operation code and an 8-bit operand value. Table 8.1 is basically a copy of Table 7.3 which was a sequence of instructions to perform a simple calculation: 50 - (17 + 12) rounded to nearest multiple of 8. Instead of entering this program manually on switches, it will now be entered in memory.

Instruction Sequence	Op Code High order 4 bits of instruction	Operand Low order 8 bits of instruction	Comment
0	D (MOV is 13)	11 (decimal 17)	Move 17 into accumulator
1	4 (ADD is 4)	0C (decimal 12)	Add 12 to accumulator
2	3 (RSB is 3)	32 (decimal 50)	Subtract accumulator from 50
3	4 (ADD is 4)	04 (decimal 4)	Round up for "closest 8"
4	0 (AND is 0)	11111000 (mask)	Finish rounding for modulus 8

Table 8.1: ARM data processing instructions

Listing 8.3 includes a program in memory and the instruction decode circuit.

- Lines 17, 18: Internal registers to hold the operation code and operand
- Lines 23 - 28: Machine code program in memory
- Lines 42, 43: Decode circuit (put opcode and operand into internal registers)

Compile and download the code in Listing 8.3. It will appear to run the same as Listing 8.2. Internally, however, it will also be performing the decode state, but nothing will be done in the execute state at this time.

```
1.  module CPU_UI (KEY, LEDR);
2.    input [1:0] KEY;
3.    output [9:0] LEDR;
4.    wire clk,w1,reset;
5.    nand(clk,KEY[0],w1);          // Set up RS FF latch
6.    nand(w1,KEY[1],clk);          // as debounced clock.
7.    and(reset,~KEY[0],~KEY[1]);   // Reset if both keys pushed
```

```verilog
8.    reg [1:0] CPU_state;
9.    parameter fetch = 2'b01;
10.   parameter decode = 2'b10;
11.   parameter execute = 2'b11;
12.   assign LEDR[9:8] = CPU_state;
13.   reg [11:0] IR;                        // Instruction register
14.   reg [11:0] progMem [0:63];            // Program memory
15.   reg [5:0] PC;                         // Program Counter
16.   assign LEDR[5:0] = PC;                // (a.k.a., Instruction Pointer)
17.   reg [3:0] opCode;                     // Operation code
18.   reg [7:0] op2;                        // Operand data
19.   always @ (posedge(clk), posedge(reset))
20.     begin
21.       if (reset)
22.         begin
23.           progMem [0] <= 12'hD11;       // MOV 17
24.           progMem [1] <= 12'h40C;       // ADD 12
25.           progMem [2] <= 12'h332;       // RSB from 50
26.           progMem [3] <= 12'h404;       // ADD 4
27.           progMem [4] <= 12'h0F8;       // AND 8'b11111000
28.           progMem [5] <= 12'h400;       // ADD 0
29.           PC <= 0;
30.           CPU_state <= fetch;
31.         end
32.       else
33.         case (CPU_state)
34.           fetch:                        // Get next instruction in program
35.             begin
36.               IR <= progMem[PC];        // Load next instruction
37.               PC <= PC + 1;             // Point to following instruction
38.               CPU_state <= decode;
39.             end
40.           decode:                       // Disassemble the instruction
41.             begin
42.               opCode <= IR[11:8];       // Opcode is in upper 4 bits
43.               op2 <= IR[7:0];           // Operand is in lower 8 bits
44.               CPU_state <= execute;
45.             end
46.           execute:                      // Perform desired operation
47.               CPU_state <= fetch;
48.         endcase
49.     end
50. endmodule
```

Listing 8.3: Instructions in program memory

Execute: Actually Do the Work

It's finally time to do the work, and display the results. Listing 8.4 contains the same hexadecimal display function previously used in Listing 7.3, and it also incorporates the circuit for performing the arithmetic and logic operations associated with each opcode.

- Line 1: Include five 7-segment hex digits for display
- Lines 4..25: Function to display 4-bit value as a hex digit
- Line 41: 8-bit accumulator ("user accessible" working register)
- Lines 42..46: Assign hex digits to display the current accumulator value, op-code, and operand

```
1.   module CPU_UI (KEY, LEDR, HEX0, HEX1, HEX2, HEX3, HEX4);
2.     input [1:0] KEY;
3.     output [9:0] LEDR;
4.     output [7:0] HEX0, HEX1, HEX2, HEX3, HEX4;
5.       function automatic [7:0] digit;
6.         input [3:0] num
7.         case (num)
8.           0: digit = 8'b11000000;              // 0
9.           1: digit = 8'b11111001;              // 1
10.          2: digit = 8'b10100100;              // 2
11.          3: digit = 8'b10110000;              // 3
12.          4: digit = 8'b10011001;              // 4
13.          5: digit = 8'b10010010;              // 5
14.          6: digit = 8'b10000010;              // 6
15.          7: digit = 8'b11111000;              // 7
16.          8: digit = 8'b10000000;              // 8
17.          9: digit = 8'b10010000;              // 9
18.          10: digit = 8'b10001000;             // A
19.          11: digit = 8'b10000011;             // b
20.          12: digit = 8'b11000110;             // C
21.          13: digit = 8'b10100001;             // d
22.          14: digit = 8'b10000110;             // E
23.          15: digit = 8'b10001110;             // F
24.        endcase
25.      endfunction
26.    wire clk,w1,reset;
27.    nand(clk,KEY[0],w1);                       // Set up RS FF latch
28.    nand(w1,KEY[1],clk);                       // as debounced clock.
29.    and(reset,~KEY[0],~KEY[1]);                // Reset if both keys pushed
```

Listing 8.4 A: Fetch-Decode-Execute with hexadecimal display

```
30.    reg [1:0] CPU_state;
31.    parameter fetch = 2'b01;
32.    parameter decode = 2'b10;
33.    parameter execute = 2'b11;
34.    assign LEDR[9:8] = CPU_state;
35.    reg [11:0] IR;                        // Instruction register
36.    reg [11:0] progMem [0:63];            // Program memory
37.    reg [5:0] PC;                         // Instruction Pointer
38.    assign LEDR[5:0] = PC;                // (a.k.a., Instruction Pointer)
39.    reg [3:0] opCode;                     // Operation code
40.    reg [7:0] op2;                        // Operand data
41.    reg [7:0] R;                          // Accumulator register
42.    assign HEX0 = digit(R[3:0]);          // Display accumulator
43.    assign HEX1 = digit(R[7:4]);
44.    assign HEX2 = digit(op2[3:0]);        // Operand
45.    assign HEX3 = digit(op2[7:4]);
46.    assign HEX4 = digit(opCode);          // Op code
47.    always @ (posedge(clk), posedge(reset))
48.      begin
49.        if (reset)
50.          begin
51.            progMem [0] <= 12'hD11;        // MOV 17
52.            progMem [1] <= 12'h40C;        // ADD 12
53.            progMem [2] <= 12'h332;        // RSB from 50
54.            progMem [3] <= 12'h404;        // ADD 4
55.            progMem [4] <= 12'h0F8;        // AND 8'b11111000
56.            progMem [5] <= 12'h400;        // ADD 0
57.            IP <= 0;
58.            CPU_state <= fetch;
59.          end
60.        else
61.          case (CPU_state)
62.            fetch:                         // Get next instruction in program
63.              begin
64.                IR <= progMem[PC];         // Load next instruction
65.                PC <= PC + 1;              // Point to following instruction
66.                CPU_state <= decode;
67.              end
68.            decode:                        // Disassemble the instruction
69.              begin
70.                opCode <= IR[11:8];        // Opcode is in upper 4 bits
71.                op2 <= IR[7:0];            // Operand is in lower 8 bits
72.                CPU_state <= execute;
73.              end
```

Listing 8.4 B: Instructions in program memory

Listing 8.4 contains the entire fetch-decode-execute process including a hexadecimal display.

- Lines 76..87: Same arithmetic and logic operations from Listing 7.1

```
74.          execute:                // Perform desired operation
75.             begin
76.                case (opCode)
77.                   0: R <= R & op2;        // AND
78.                   1: R <= R ^ op2;        // EOR (exclusive OR)
79.                   2: R <= R - op2;        // SUB
80.                   3: R <= op2 - R;        // RSB (reverse subtract)
81.                   4: R <= R + op2;        // ADD
82.                   12: R <= R | op2;       // ORR (inclusive OR)
83.                   13: R <= op2;           // MOV
84.                   14: R <= R & ~op2;      // BIC (bit clear)
85.                   15: R <= ~op2;          // MVN (move NOT)
86.                   default: R <= 0;        // None of above
87.                endcase
88.                CPU_state <= fetch;
89.             end
90.          endcase
91.       end
92.    endmodule
```

Listing 8.4 C: Execute with hex display output

Compile and download the code in Listing 8.4. Run through the same program sequence as before. Push both keys at the same time to initialize the PC register to zero, then push KEY[1] followed by KEY[0] to move to the next state.

Figure 8.4: Display register contents

The following "program trace" will display on HEX[4] through HEX[0] for each fetch-decode-execute state of each instruction step of the program. The PC will also be displayed in binary on LEDR[5:0].

- **d1111:** Instruction MOV 11 is D11, and accumulator = 11
- **40C1d:** Instruction ADD 0C is 40C, and accumulator = 1D
- **33215:** Instruction RSB 32 is 332, and accumulator = 15
- **40419:** Instruction ADD 04 is 404, and accumulator = 19
- **0F818:** Instruction AND F8 is 0F8, and accumulator = 18
- **40018:** Instruction ADD 00 is 400, and accumulator = 18

Clock and Breakpoint

Are you tired of pushing buttons six times to execute each instruction? Let's use a real clock signal that is available on the Terasic DE2 and DE10-Lite boards. However, it will still be useful to stop once at each instruction as the CPU progresses through a machine language program.

- Lines 1, 3: clk = CLOCK_50 is included and replaces manual clock that was using KEY[0] and KEY[1].

```
1.  module CPU_UI (KEY, CLOCK_50, LEDR, HEX0, HEX1, HEX2, HEX3,
    HEX4);
2.    input [1:0] KEY;
3.    input CLOCK_50;
4.    output [9:0] LEDR;
5.    output [7:0] HEX0, HEX1, HEX2, HEX3, HEX4;
6.      function automatic [7:0] digit;
7.        input [3:0] num
8.        case (num)
9.          0: digit = 8'b11000000;                  // 0
10.         1: digit = 8'b11111001;                  // 1
11.         2: digit = 8'b10100100;                  // 2
12.         3: digit = 8'b10110000;                  // 3
13.         4: digit = 8'b10011001;                  // 4
14.         5: digit = 8'b10010010;                  // 5
15.         6: digit = 8'b10000010;                  // 6
16.         7: digit = 8'b11111000;                  // 7
17.         8: digit = 8'b10000000;                  // 8
18.         9: digit = 8'b10010000;                  // 9
19.         10: digit = 8'b10001000;                 // A
20.         11: digit = 8'b10000011;                 // b
21.         12: digit = 8'b11000110;                 // C
22.         13: digit = 8'b10100001;                 // d
23.         14: digit = 8'b10000110;                 // E
24.         15: digit = 8'b10001110;                 // F
25.       endcase
26.     endfunction
```

Listing 8.5 A: Hexadecimal display with 50 MHz clock included

```
27.    wire clk,w1,reset;
28.    assign clk = CLOCK_50;
29.    and(reset,~KEY[0],~KEY[1]);           // Reset if both keys pushed
30.    reg [1:0] CPU_state;
31.    parameter fetch = 2'b01;
32.    parameter decode = 2'b10;
33.    parameter execute = 2'b11;
34.    assign LEDR[9:8] = CPU_state;
35.    reg [11:0] IR;                         // Instruction register
36.    reg [11:0] progMem [0:63];            // Program memory
37.    reg [5:0] PC;                          // Program Counter
38.    assign LEDR[5:0] = PC;                // (a.k.a., Instruction Pointer)
39.    reg [3:0] opCode;                      // Operation code
40.    reg [7:0] op2;                         // Operand data
41.    reg [7:0] R;                           // Accumulator register
42.    assign HEX0 = digit(R[3:0]);          // Display accumulator
43.    assign HEX1 = digit(R[7:4]);
44.    assign HEX2 = digit(op2[3:0]);        // Operand
45.    assign HEX3 = digit(op2[7:4]);
46.    assign HEX4 = digit(opCode);          // Op code
47.    reg run;                               // Flag indicating CPU is running
48.    always @ (posedge(clk), posedge(reset))
49.      begin
50.        if (reset)
51.          begin
52.            progMem [0] <= 12'hD11;       // MOV 17
53.            progMem [1] <= 12'h40C;       // ADD 12
54.            progMem [2] <= 12'h332;       // RSB from 50
55.            progMem [3] <= 12'h404;       // ADD 4
56.            progMem [4] <= 12'h0F8;       // AND 8'b11111000
57.            progMem [5] <= 12'h400;       // ADD 0
58.            PC <= 0;
59.            CPU_state <= fetch;
60.          end
61.        else
62.          case (CPU_state)
63.            fetch:                         // Get next instruction in program
64.              begin
65.                IR <= progMem[PC];         // Load next instruction
66.                PC <= PC + 1;              // Point to following instruction
67.                CPU_state <= decode;
68.              end
```

Listing 8.5 B: Same code as before, but old "clock" is removed.

An interlock is added to the decode state (lines 73 through 79) that will halt program execution until KEY[0] is pushed momentarily followed by KEY[1] being pushed momentarily.

```
69.                 decode:                      // Disassemble the instruction
70.                     begin
71.                         opCode <= IR[11:8];     // Opcode is in upper 4 bits
72.                         op2 <= IR[7:0];         // Operand is in lower 8 bits
73.                         if (~KEY[0])
74.                             run <= 0;
75.                         if (~KEY[1] && ~run)    // KEY[1] and run must both be low
76.                             begin
77.                                 run <= 1;
78.                                 CPU_state <= execute;
79.                             end
80.                     end
81.                 execute:                     // Perform desired operation
82.                     begin
83.                         case (opCode)
84.                             0: R <= R & op2;     // AND
85.                             1: R <= R ^ op2;     // EOR (exclusive OR)
86.                             2: R <= R - op2;     // SUB
87.                             3: R <= op2 - R;     // RSB (reverse subtract)
88.                             4: R <= R + op2;     // ADD
89.                             12: R <= R | op2;    // ORR (inclusive OR)
90.                             13: R <= op2;        // MOV
91.                             14: R <= R & ~op2;   // BIC (bit clear)
92.                             15: R <= ~op2;       // MVN (move NOT)
93.                             default: R <= 0;     // None of above
94.                         endcase
95.                         CPU_state <= fetch;
96.                     end
97.                 endcase
98.             end
99.     endmodule
```

Listing 8.5 C: Same code as Listing 8.4 except for interlock in decode state.

The procedure for "single stepping" through each line of the program:

1. Start the program with a system reset: Holding down KEY[0] while momentarily pushing KEY[1] will reset the PC to address 0. The CPU will run until it is stopped in the decode state. Note: The accumulator, R, does not get changed by the reset, but neither do data registers in real CPUs get changed by a reset.

2. The display shows the current accumulator value (HEX1, HEX0) and the next instruction to be executed (opcode in HEX4 and operand in HEX3 and HEX2). The PC address shown in LEDR[5:0] has already been incremented to point to the next instruction to be executed.

3. To execute each of the remaining instructions in sequence, momentarily push KEY[0] followed by momentarily pushing KEY[1].

Highlights and Comparisons

Almost every CPU designed in the past 70 years runs through the three basic states: fetch, decode, execute. Many CPUs have a few additional states, and I will add one as well in the next chapter.

Review Questions

1. Name two advantages of using the *parameter* construct in Verilog coding?
2. *Each state of the instruction cycle (fetch, decode, and execute) is composed of its own circuit which is parallel to and mostly independent of the other states' circuits. What do the electronic circuits that perform the fetch and decode states do when the execute state is active?
3. *As described in this chapter, convert the instruction "ORR 13" to hexadecimal machine code.
4. *As described in this chapter, convert the hexadecimal machine code 'h123 back into a text version of the instruction.

Exercises

1. Write a program in machine langage to sum five numbers. Convert it to the hexadecimal code and enter it as a replacement for lines 52 through 57 in Listing 8.5. Compile this new Listing 8.5 code, download it, and run the program.
2. Write another program as was done in the previous exercise, but this time calculate the vertical parity of five numbers. This can be done by doing an exclusive (EOR) on each of the numbers.

— 9 —
Register Set

What about the second half of Babbage's Analytical Engine? What about the store, which is the component we now call memory?

Over the decades, computer memory has been available in many formats, but from a programmer's perspective, memory is essentially only an ocean of bits. These bits have been stored magnetically in cores (rings or toroids), tapes, disks, relays, and even "bubbles" within a chip. They have been stored electrically in capacitors and in circuits made from transistors or vacuum tubes.

Memory types differ by speed, cost, and location. Registers are the fastest, but few in number, and they are located within the CPU, itself.

Figure 9.1: "Core" memory was popular during the 1950s, 1960s, and 1970s (image magnified 10 times).

Chapter 9 extends the simple CPU from Chapter 8 to include a set of sixteen general purpose registers. The instruction format must now be expanded to contain three register fields: two operands and a destination. This 21-bit format will be executed in programs in Chapter 10 and increased again to a final 32-bit format in Chapter 11.

Introductions

A "macro" is a text substitution that is similar to "copy and paste." It is introduced here in Chapter 9 and will be used often in the following chapters to substitute names for text strings and numeric codes.

- **define**: Verilog macro command specifying text substitution

Memories

The memory media type determines the size, speed, price, access mode, and electrical interface requirements. It's no wonder that computer architecture has been complicated by the variety of memories that has been used over the decades. Chapters 14, 16, 17, and 21 will describe accessing main memory from ARM and NEON instructions. Here in Chapter 9, the focus is on memory within the CPU referred to as registers.

From a functional standpoint, memory can be divided into the following categories:

- RAM: This Random Access Memory is the main working memory of a computer system where variables are stored. Since the mid 1970s, this has been "solid state," (i.e., transistors), but was previously magnetic "core" as shown in Figure 9.1. Relays, vacuum tubes, and electrical delay lines were used before core was common in the 1950s.
- Disk: Relatively fast large capacity memory storage
- Tape: Inexpensive long term, large capacity, usually off line for backup
- Registers: A small number of fast memory cells within the CPU that are directly accessible by machine code instructions
- Cache: Blocks of memory within the CPU that contain copies of data blocks in "RAM" memory.

Instruction Format with Registers

Most arithmetic and logical operations involve three variables, which I'll refer to as D, N, and M (such as D = N - M). So where are these variables located? Today, our computer systems have billions of bytes of storage in main memory, but only about a couple dozen registers. Obviously, our variables are stored in main memory and only temporarily copied into the registers when we are using them in calculations.

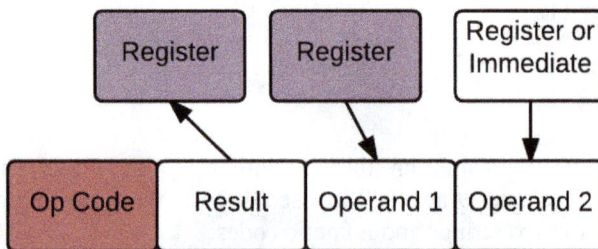

Figure 9.2: Instruction format in a RISC architecture

Computer Architecture Tutorial Using an FPGA

RISC computers, such as the ARM, basically have the instruction format as illustrated in Figure 9.2. It is an expanded format from that of Chapter 8 because we must now identify register ID numbers for the first operand (N), second operand (M), and result (D). In the previous format, the second operand had to be a constant, but now that we have multiple registers, it can be either a constant or a register.

The set of ARM data processing instructions actually has the general format of $D = N - M \times 2^S$ where the minus sign indicating subtraction is just one of the sixteen possible "data processing" operation codes. In Chapter 12, the second operand of $M \times 2^S$ which contains a bit-shift operation will be fully explained, but for now we will continue to look at it as either a register or an immediate constant.

- **D**: Destination register to hold the result
- **N**: First operand register: In subtraction, it is the minuend (quantity from which another quantity is to be subtracted)
- **M**: Second operand: In subtraction, it is the subtrahend (quantity to subtract). There are two possible formats for M:
    1. **M** is in a register
    2. **M** is an immediate constant: Note: It's more complicated than a single binary integer, but we'll hold the full explanation until Chapter 12.

Expanded Instruction Format

The data processing instruction format developed in Chapter 8 consisted of 12 bits: a 4-bit opcode and an 8-bit immediate operand value. That format will now be expanded to 21 bits as shown in Figure 9.3. Eight of the new bits are needed to indicate the sixteen possible registers for Rd and Rn. The ninth new bit is the "immediate" flag bit which indicates whether the second operand is in a register (Rm) or is an immediate constant.

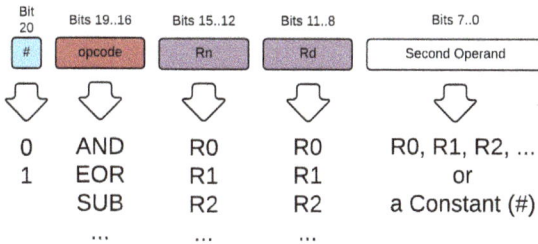

Bit 20	Bits 19..16	Bits 15..12	Bits 11..8	Bits 7..0
#	opcode	Rn	Rd	Second Operand
0	AND	R0	R0	R0, R1, R2, ...
1	EOR	R1	R1	or
	SUB	R2	R2	a Constant (#)
	...	...	...	

Figure 9.3: Instruction format used in the examples in chapters 9 and 10

Listing 9.1 demonstrates the new 21-bit instruction format and contains the following two modules:

- CPU_UI: Control module that transfers switch contents to module ProgMod and displays output in hexadecimal
- ProgMod: Module containing "program" (sequence of data processing instructions)

```
1.   module CPU_UI (SW, KEY, LEDR, HEX0, HEX1, HEX2, HEX3, HEX4, HEX5);
2.      input [9:0] SW;
3.      input [1:0] KEY;
4.      output [9:0] LEDR;
5.      output [7:0] HEX0, HEX1, HEX2, HEX3, HEX4, HEX5;
6.      function automatic [7:0] digit;
7.         input [3:0] num;
8.         case (num)
9.            0: digit = 8'b11000000;              // 0
10.           1: digit = 8'b11111001;              // 1
11.           2: digit = 8'b10100100;              // 2
12.           3: digit = 8'b10110000;              // 3
13.           4: digit = 8'b10011001;              // 4
14.           5: digit = 8'b10010010;              // 5
15.           6: digit = 8'b10000010;              // 6
16.           7: digit = 8'b11111000;              // 7
17.           8: digit = 8'b10000000;              // 8
18.           9: digit = 8'b10010000;              // 9
19.           10: digit = 8'b10001000;             // A
20.           11: digit = 8'b10000011;             // b
21.           12: digit = 8'b11000110;             // C
22.           13: digit = 8'b10100001;             // d
23.           14: digit = 8'b10000110;             // E
24.           15: digit = 8'b10001110;             // F
25.        endcase
26.     endfunction
27.     wire [20:0] IR;
28.     assign HEX0 = digit(IR[3:0]);              // Rm or constant
29.     assign HEX1 = digit(IR[7:4]);
30.     assign HEX2 = digit(IR[11:8]);             // Rd
31.     assign HEX3 = digit(IR[15:12]);            // Rn
32.     assign HEX4 = digit(IR[19:16]);            // OpCode
33.     assign HEX5 = digit(IR[20]);               // Immediate flag
34.     assign LEDR = SW;
35.     ProgMod (SW[7:0], ~KEY[0], ~KEY[1], IR);
36.  endmodule
37.
```

Listing 9.1 A: Module to demonstrate 21-bit instruction format

- Line 35: Create instance of module ProgMod where the switches provide a memory address and KEY[1] provides the "clock."
- Line 38: The ProgMod module has two input "clocks": "reset" to initialize the memory and "clk" to fetch the contents of a particular memory address.
- Line 42: Reservation for 256 "memory" slots of 21-bit instructions
- Line 46: Fetch 21-bit instruction when "clk" has rising pulse.
- Lines 49, 50: Machine code for two examples of 21-bit instructions is initialized at addresses 0 and 1.

```
38.  module ProgMod (address, reset, clk, instr);
39.    input [7:0] address;
40.    input reset, clk;
41.    output [20:0] instr;
42.    reg [20:0] progMem[0:255];          // 21 bits per instruction
43.    reg [20:0] IR;                       // Instruction Register
44.    assign instr = IR;
45.    always @ (posedge(clk))
46.      IR <= progMem[address];
47.    always @ (posedge(reset))
48.      begin
49.        progMem[0] <= 21'h126513;        // SUB R5,R6,19 @ op2 is a constant
50.        progMem[1] <= 21'h022103;        // SUB R1,R2,R3 @ op2 is in a register
51.      end
52.  endmodule
```

Listing 9.1 B: Module containing sequence of 21-bit data processing instructions

Please compile and download Listing 9.1. Verify that the two 21-bit instructions have been loaded into program memory addresses 0 and 1. For each address, first set the address on the switches, then push KEY[1] to clock in the data.

Between lines 50 and 51, add a few more 21-bit instructions of your own for different memory addresses, and then recompile, download, and verify.

Figure 9.4: Hexadecimal display of instruction format

Having a separate module for the ARM machine code "program" will be a common practice in the remainder of this book. Here, we are just testing the module. Note: This memory is part of the FPGA and not in a separate memory chip on the FPGA circuit board. In Chapter 10, we will use the ProgMod module to "fetch" each instruction to be decoded and executed.

Assembly Language

Adding three register fields and an immediate flag increased the complexity of the machine code format. Using hexadecimal in the ProgMod module to enter machine code instructions is a considerable improvement over binary. However, in the next few chapters, more instructions, a fourth register field, shift operations, and status bits will complicate the format of the "data processing" instructions even more. Instructions in the ARM CPU, such as multiplication and branching, do not fit within the ARM data processing opcode field and will therefore add even more complexity to the "simple" machine language format.

This problem of entering machine code instructions has existed for decades and has been addressed by assembly language as well as higher level language compilers. Substituting mnemonic names for opcodes and registers is the first advantage of assembly language over the entry of machine code as a series of binary or even hexadecimal digits.

My objective is to create an assembler within Verilog that is very close to that which programmers are familiar. See Appendix E for details and examples of using the very popular GNU assembler. Two features of the Verilog compiler will provide ARM instruction input in a symbolic format:

1. The preprocessor *macro* capability will provide conversion of mnemonics such as ADD and SUB to opcodes 4 and 2, respectively.
2. The *parameter* statement will provide conversion of mnemonics such as R3 and R4 to register ID numbers 3 and 4, respectively.

Macro processing is performed during a first pass through the source code file before the Verilog compiler synthesizes the hardware description. It is similar to the "cut and paste" function in an editor. It differs from the *parameter* statement in that the macro can contain almost anything, such as half of a Verilog command. I will be using it to convert a string such as "ADD" to "asdp (4'd4," which will be the beginning of a Verilog function call.

Basically, an assembler performs the type of conversion shown in Figure 9.5. The subtraction operator "sub" is replaced by its opcode value 2. The sixteen general purpose register names, R0 through R15, are replaced by their respective ID numbers, 0 through 15. If the second operand is a constant, such as 19, then its value is placed in the instruction and the immediate flag must be set.

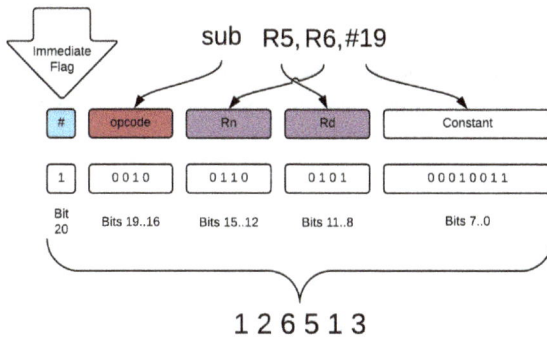

Figure 9.5: Second operand is a constant (immediate flag = 1)

If the second operand is contained in a register, then the format illustrated in Figure 9.6 is needed. The register ID number is placed in the instruction, and the immediate flag bit must be zero.

The macro preprocessor in Verilog is rather simple and not currently capable of distinguishing the different formats of the second operand. In upcoming chapters, the ARM format is going to get much more complicated, so the approach I am taking is to have the macro preprocessor build a Verilog function call, and the function will do most of the conversion to ARM machine code.

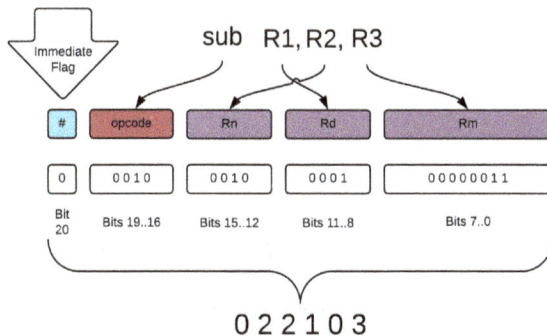

Figure 9.6: Second operand is in a register (immediate flag = 0)

Using the Concatenation Operator

An assembler would convert the text line "sub R5,R6,#19" to the machine code 21'h126513. We can do that using the concatenation operator as shown in Figure 9.7. The instruction fields are joined by the { } concatenation operator. The macro preprocessor, parameter definitions, a function I have named "asdp" (assemble data processing), and the concatenation operator will work together in Listing 9.2 to approximate the conversion performed by an assembler program. The following steps are performed:

1. "sub R5,R6,#19" will be written as "`SUB R5,R6,19 `_" in the Verilog code
2. "`SUB R5,R6,19 `_" will be converted to "asdp(4'd2, R5, R6, 19);" by the macro preprocessor
3. "asdp(4'd2, R5, R6, 19);" will be converted to 21'h126513 by Verilog function "asdp."

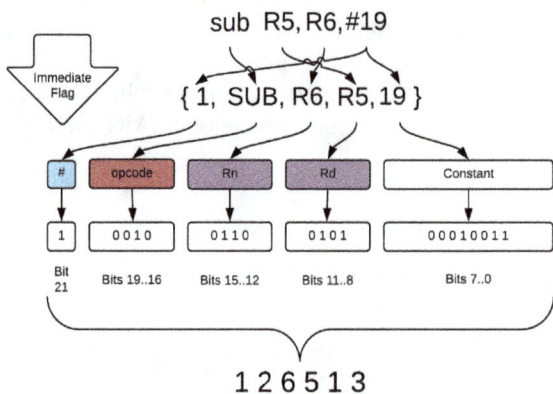

Figure 9.7: Concatenation operator combines fields of a data processing instruction

- `SUB and `_ are macros that are converted by the preprocessor to "asdp(4'd2," and ");" respectively. The R5, R6, and 19 fields are not changed by the preprocessor.
- Parameters R5 and R6 are equated to 16'h1005 and 16'h1006, respectively. They were assigned these values rather than the obvious 5 and 6 so that the "asdp" function can distinguish between register ID numbers and immediate value constants.
- The "asdp" function concatenates the fields to complete the 21 bit hexadecimal machined code 126513.

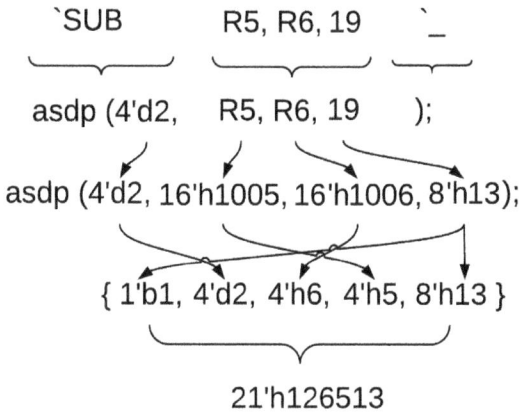

Figure 9.8: Macro, function, and concatenation used to build a data processing instruction

```
38.   `define AND asdp (4'd0,        // [Rd] = [Rn] AND Op2
39.   `define EOR asdp (4'd1,        // [Rd] = [Rn] Exclusive OR Op2
40.   `define SUB asdp (4'd2,        // [Rd] = [Rn] - Op2
41.   `define RSB asdp (4'd3,        // [Rd] = Op2 - [Rn]
42.   `define ADD asdp (4'd4,        // [Rd] = [Rn] + Op2
43.   `define ORR asdp (4'd12,       // [Rd] = [Rn] Inclusive OR Op2
44.   `define MOV asdp (4'd13,       // [Rd] = [Rn]
45.   `define BIC asdp (4'd14,       // [Rd] = [Rn] AND NOT Op2
46.   `define MVN asdp (4'd15,       // [Rd] = NOT [Rn]
47.   `define _ );                   // End of instruction
48.
```

Listing 9.2 A: Ten macros used to provide "assembly language" coding format

Listing 9.2 produces the same result as Listing 9.1, except the ARM data processing instructions are entered in assembly language text format instead of hexadecimal. The first 37 lines contain the CPU_UI control module and are identical to that appearing in Listing 9.1 and will not be shown here. Lines 38 through 47 define the macros for each of the operation codes:

- The macro definition begins with the single back quote character (not an apostrophe) followed by the word "define."
- The name of the macro then follows. Note: It's case sensitive.
- The characters to be substituted for the macro name then follow (up to the end of line or a double-slash comment)
- A macro call consists of the back quote followed by the macro name. Once defined, a macro can be called within any module. It's not like a function that has to be defined within the module it is used.

```
49.  module ProgMod (address, clk, reset, instr);
50.    input [7:0] address;
51.    input clk, reset;
52.    output [20:0] instr;
53.    parameter R0 = 16'h1000;              // General purpose registers
54.    parameter R1 = 16'h1001;
55.    parameter R2 = 16'h1002;
56.    parameter R3 = 16'h1003;
57.    parameter R4 = 16'h1004;
58.    parameter R5 = 16'h1005;
59.    parameter R6 = 16'h1006;
60.    parameter R7 = 16'h1007;
61.    parameter R8 = 16'h1008;
62.    parameter R9 = 16'h1009;
63.    parameter R10 = 16'h100A;
64.    parameter R11 = 16'h100B;
65.    parameter R12 = 16'h100C;
66.    parameter R13 = 16'h100D;             // a.k.a. "SP"
67.    parameter R14 = 16'h100E;             // a.k.a. "LR"
68.    parameter R15 = 16'h100F;             // a.k.a. "PC"
69.
70.    function [20:0] asdp ();
71.      input [15:0] opcode,Rd,Rn,Rm;
72.      if (Rm < 'h1000)
73.        asdp = {1'b1,opcode[3:0],Rn[3:0],Rd[3:0],Rm[7:0]};
74.      else
75.        asdp = {1'b0,opcode[3:0],Rn[3:0],Rd[3:0],Rm[7:0]};
76.    endfunction
77.
78.    reg [20:0] progMem[0:255];            // 21 bits per instruction
79.    reg [20:0] IR;                        // Instruction Register
80.    assign instr = IR;
81.    always @ (posedge(clk))
82.      IR <= progMem[address];
83.    always @ (posedge(reset))
84.    begin
85.      progMem[0] <= `SUB   R5, R6, 19  `_
86.      progMem[1] <= `SUB   R1, R2, R3  `_
87.    end
88.  endmodule
```

Listing 9.2 B: "Assembly language" style for entering machine code instruction

- Line 72: Function asdp has two possible outputs depending on whether the second operand is a constant or in a register (ID numbers biased by 13'h1000).

XOR from AND, OR, and NOT

Logical gates can be built from other gates. Starting in Chapter 2 (Figure 2.4), an exclusive OR function is constructed from two OR gates, two NOT gates, and an AND gate. This circuit was then recoded using the structural, dataflow, and behavioral Verilog approaches in listings 5.9, 5.10, and 5.11, respectively. In Listing 9.3, the same logic is being "programmed" in ARM assembly language.

```
82.  always @ (posedge(clk))
83.    IR <= progMem[address];
84.  always @ (posedge(reset))
85.    begin
86.      progMem[0] <= `MOV   R1, 0, 'b0101  `_      // Move 0b0101 into R1
87.      progMem[1] <= `MOV   R2, 0, 'b0011  `_      // Move 0b0011 into R2
88.      progMem[2] <= `BIC   R11, R1, R2  `_        // Not(R2) & (R1) => R11
89.      progMem[3] <= `BIC   R12, R2, R1  `_        // Not(R1) & (R2) => R12
90.      progMem[4] <= `ORR   R3, R11, R12  `_       // (R11) | (R12) => R3
91.      progMem[5] <= `EOR   R4,R1,R2  `_           // Exclusive OR instruction
92.    end
```

Listing 9.3: Assembly language program to do exclusive OR

Go ahead and compile, download, and run the code in Listing 9.3. Go through each of the six ARM instructions and verify that the machine code generated by the macro and function match the values in Table 9.1. Note: The program address is now eight bits, so make sure all of SW[7:0] is down to get address 0.

Program Address	Assembly Language	Machine Code (hexadecimal)
0	`MOV   R1, 0, 'b0101  `_	1d0105
1	`MOV   R2, 0, 'b0011  `_	1d0203
2	`BIC   R11, R1, R2  `_	0E1b02
3	`BIC   R12, R2, R1  `_	0E2C01
4	`ORR   R3, R11, R12  `_	0Cb30C
5	`EOR   R4,R1,R2  `_	011402

Table 9.1: Machine code from assembly language of XOR program

Here in Chapter 9, we are only looking at the machine code that is generated, but in Chapter 10 we will be running the program on the ARM imitation.

Intel or AT&T Syntax

I have represented a subtraction such as $D = N - M$ in assembly language as "SUB   Rd, Rn, Rm." This order of parameters and coding format is now commonly referred to as "Intel Syntax." If the order of the result and operand registers is reversed (something like "SUB   Rn, Rm, Rd"), then we would have "AT&T Syntax." There are also a few other differences between the two assembly language formats.

If you like, you can easily change to the "AT&T Syntax" by modifying the "asdp" task. I find the "Intel Syntax" to be more common. Actually, the "Intel Syntax" format was common for assembly language programming on mainframes such as the Univac 1108 and CDC 6600 years before the development of UNIX by AT&T and even the existence of the Intel Corporation.

Any Other Instructions?

Computers are great for doing the same thing over and over again, but on different sets of input data. Sometimes, we write a program to perform the calculations differently, depending on the type and values of the data being processed. Making these decisions, as well as knowing when to exit these repetitive loops, is done by jump instructions (a.k.a., branch instructions) and conditional test instructions. We'll examine these techniques beginning in Chapter 11.

What's Missing?

In this chapter, the ARM instruction format has been "built" consisting of 21 bits that include an operation code, three registers, constant data, and an immediate flag. This capability has allowed the running of some simple machine language programs. What's still missing as we build up to the full 32-bit ARM instruction?

1. More instructions: There is the obvious gap in the opcode numbering sequence of the data processing instructions. Common instructions, like multiplication, have not been described, either. Chapters 11 through 19 add more instructions.
2. The second operand, which is either register Rm or a constant, needs to be more fully explained. Chapter 12 shows that the second operand can have two registers, two constants, and a shift operation, all within the 12-bit second operand.
3. There is no way to change program flow, or change which instructions are to be executed based on status information. Chapters

11 and 15 address these issues.

4. Moving data between registers and main memory has not been described. Chapters 14, 16, and 21 will describe instructions that access memory.

5. Moving data to and from external devices is described in Chapter 17.

6. The PC (Program Counter) is incremented by "1" for each instruction in the current imitation of the ARM instruction set. This will eventually be changed to 4, since it takes four 8-bit bytes to make one 32-bit ARM instruction.

Highlights and Comparisons

Complex Instruction Set Computers (CISC) have an instruction format that was becoming popular around 1980. Their instruction formats allowed the first operand (N in the D = N - M example) to come from either a register or memory. The value for the second operand (M) could come from a register, memory, or be an immediate constant. The result of the calculation (D) could then be placed into either a register or written into main memory.

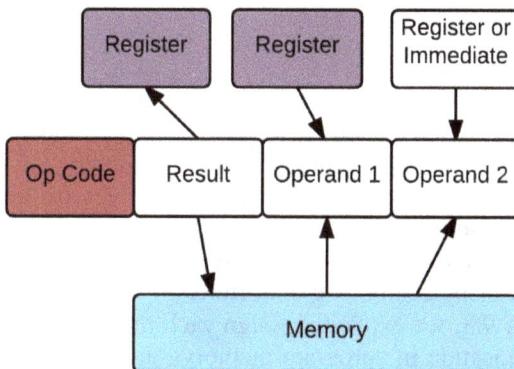

Figure 9.9: Instruction format in a CISC architecture

This instruction format actually enables D = N - M to be performed in a single CPU instruction and not even use any user accessible registers. The problem was one of execution speed, and it was being addressed in the early 1980s at the University of California, Berkeley, and Stanford University. The conclusion was to develop Reduced Instruction Set Computer (RISC) architectures.

The primary difference between a RISC architecture, such as that in the ARM, and a CISC architecture is that most instructions would only access data in registers, not in main memory. This difference accelerates program execution in two ways: The instructions are simpler in format thereby being easier to decode and of course getting data from registers is always faster than getting data from main memory.

This RISC idea was not new in the 1980s, but was present in the CDC 6600 mainframe supercomputer of the 1960s. Its arithmetic, logic, and shifting instructions worked only with data in its 60-bit registers. There were individual instructions that only moved data between main memory and the registers, but performed no calculations. The ARM also has instructions that solely move data between main memory and the general purpose registers (See Chapter 14).

Review Questions

1. *How does the Verilog macro capability (*define* statement) compare to that of a Verilog *function*?
2. Why are there 16 general purpose registers rather than a different number like 12 or 20?
3. *What is the scope (range of definition) of a *define* macro compared to that of the *parameter* statement?
4. Check the Internet to find at least two other differences between RISC and CISC architectures.

Exercises

1. In Chapter 8, instructions were entered in hexadecimal, so it was very convenient to have fields line up on 4-bit boundaries. That's one reason I put the immediate bit at the beginning of the instruction rather than the end. Assembly language frees us from those hex-aligned boundaries. Temporarily, modify the code in Listing 9.3 so that the immediate bit is adjacent to the second operand (where it is located in many computer architectures).
2. In Listing 9.3, we explicitly assign each machine code instruction to the next location in "program memory." Instead, modify the code to use an integer variable named IP, so each instruction will look like progMod[IP] <= `MOV R1,0,R2. Three changes will be necessary:
   a. Variable IP must be defined before it is used (*integer* IP;).
   b. The number in brackets (such as [0], [1], ...) for each assembly language type statement must all be changed to [IP].
   c. The IP must then be incremented after each statement by IP = IP + 1;

— 10 —
Instruction Cycle

An Arithmetic Logic Unit (ALU) was introduced in Chapter 7 to perform most ARM data processing instructions. A program in memory automated the execution of a series of these instructions in Chapter 8. Multiple data registers were added to the instruction format in Chapter 9, and programs using those instructions will be executed here in Chapter 10.

A fourth CPU state, Write Back, will be implemented to update the multiple data registers. This approach will be extended even further in the chapters covering memory access and I/O operations.

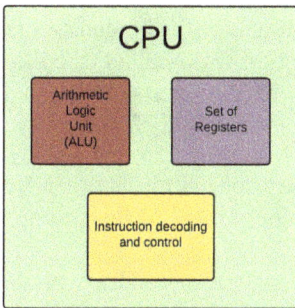

A Central Processing Unit basically consists of three major components:

- ALU: The hardware that does the work (arithmetic and logic operations)
- Control: Program instruction fetching, decoding, execution, and write back
- Set of Registers: Data being calculated and pointers into memory

Figure 10.1: Basic components of a CPU

Introductions

A Verilog *task* is similar to a *function* in that it can generate a subcircuit within a module, but it can also create other outputs.

- **task, endtask:** Generate multiple outputs, but no "return" value

Listing File Format

The ARM imitation is a work in progress within this book and with each iteration it is getting closer to being a full ARM emulation. As we go deeper, the ARM instruction format is getting more complex, so I will be relying more on "assembly language" to generate the machine code, and this also adds to the complexity of the Verilog coding.

The increased complexity means we will be needing more modules. I will keep all the modules needed for a particular demonstration in the same listing file so that there will be no problem of getting wrong versions as we progress from simple concepts to operational models. At the beginning of each file, I will add a small directory header saying which computer architecture feature is being demonstrated and which modules are present.

Chapter 10 has two listing files that build upon the work done in Chapter 9:

1. Listing file 9.3 will be modified to change the asdp function to a Verilog task.
2. The XOR example "program" from Listing 9.3 will be executed, and several register contents will be monitored with each step of the program.

Verilog Tasks

Verilog functions were introduced in Chapter 4 to simplify the construction of hexadecimal displays. In Chapter 9, the asdp function was created to build CPU machine code from text lines, somewhat like what is done in assembly language. A Verilog task is similar to a function, but has the following differences:

1. A function returns a value that can be used within another Verilog statement. A task does not return a value, and it "stands alone" as a statement.
2. A function creates exactly one output (its return value), while a task can generate multiple outputs (signals and variables).

The asdp Task

The asdp function started in Chapter 9 is a "work in progress." It will now be changed to a task for the following reasons:

1. I want the assembly language statements to look more like "assembly language statements." In other words, I don't want to continue to put "progMem[ ] = " before every line of code.
2. I have not yet provided the full ARM format. Specifically, not only will ARM instructions be shown to be 32 bits in Chapter 11, but they will each be stored as four 8-bit bytes in Chapter 14.

The asdp task will be modified again in the next two chapters, and more tasks will be created to convert other ARM instructions like multiplication and shifting to machine code. Of course, I could have started with the final version of asdp, but this book's approach is to gradually build a working imitation of the ARM architecture while introducing more features of Verilog.

Listing 10.1 contains 115 lines of Verilog code and is a slight update to Listing 9.3 in order to change asdp from a function to a task. The beginning of the listing contains the following directory header:

```
1. // Listing 10.1 uses 9 ARM Data Processing instructions
2. // 1) Assembles 21-bit "ARM-like" Data Processing instructions
3. // 2) Dumps 21-bit words from memory in hexadecimal
4. //
5. // Modules and macros contained in this file:
6. // 1) CPU_UI: User Interface that dumps 21-bit words from program memory
7. // 2) Macros for assembling 9 ARM data processing instructions
8. // 3) ProgMod: Memory containing "ARM" program
```

Listing 10.1 A: Directory in Listing_10-1.txt file

Listing 10.1 contains two modules and the set of macros. The CPU_UI user interface module and macros are the same as in Listing 9.3.

The ProgMod module contains the assembly language program and has been upgraded to the new task format. Its list of parameters is the same as before and will remain the same in the upcoming chapters even though both the ARM machine code and test programs will become more complex.

- **address**: Points to memory location of instruction to be fetched
- **clk**: When the argument to the clk parameter transitions from 0 to 1, module ProgMod will copy the contents of the selected memory address to the output vector instr.
- **reset**: Used to initialize the memory contents
- **instr**: Output vector (type wire) to receive contents from memory

```
64. //
65. //---------------- Memory containing "ARM" program ----------------
66. //
67. module ProgMod (address, clk, reset, instr);
68.     input [7:0] address;
69.     input clk, reset;
70.     output [20:0] instr;
71.     parameter R0 = 16'h1000;          // General purpose register set names
72.     parameter R1 = 16'h1001;
```

Listing 10.1 B: Parameters of ProgMod module

Every machine code instruction, and every assembly language line of code for that matter, is assigned to a specific address in memory. To drive home this point, "progMem[ ] = " precedes every line of assembler code in Listing 9.3. Real assembly language programming is not like that. It is the assembler's responsibility to assign each line of code to sequential addresses in memory. A Verilog compile-time integer variable, IP, will now be used to assign machine code to sequential addresses in program memory.

- Line 88: Memory address for next assembly language instruction.
- Line 89: Beginning of new asdp task
- Lines 92, 94: Building the next instruction and putting it into memory (right side of equals sign is the same as in Listing 9.3)
- Line 95: Update memory address for next assembly language instruction.
- Line 106: Initialize memory for first of several machine code instructions
- Lines 107 - 112: Assembly language XOR demonstration program

```
86.   parameter R15 = 16'h100F;          // a.k.a. "PC"
87.
88.   integer IP;                        // Instruction Pointer for "assembler"
89.   task asdp ();
90.      input [15:0] opcode,Rd,Rn,Rm;
91.      if (Rm < 'h1000)
92.         progMem[IP] = {1'b1,opcode[3:0],Rn[3:0],Rd[3:0],Rm[7:0]};
93.      else
94.         progMem[IP] = {1'b0,opcode[3:0],Rn[3:0],Rd[3:0],Rm[7:0]};
95.      IP = IP + 1;
96.   endtask
97.
98.   reg [20:0] progMem[0:25];           // 21 bits per instruction
99.   reg [20:0] IR;                      // Instruction Register
100.  assign instr = IR;
101.
102.  always @ (posedge(clk))
103.     IR <= progMem[address];
104.  always @ (posedge(reset))
105.  begin
106.     IP = 0;
107.     `MOV  R1, 0, 'b0101  `_          // Move 0b0101 into R1
108.     `MOV  R2, 0, 'b0011  `_          // Move 0b0011 into R2
109.     `BIC  R11 ,R1, R2  `_            // Not(R2) & (R1) => R11
110.     `BIC  R12, R2, R1  `_            // Not(R1) & (R2) => R12
111.     `ORR  R3, R11, R12  `_           // (R11) | (R12) => R3
112.     `EOR  R4, R1, R2  `_             // Exclusive OR instruction
113.     progMem[25] = 0;
114.  end
115. endmodule
```

Listing 10.1 C: List XOR program in memory

Compile and download the code in Listing 10.1. Please verify that the asdp task is building the same series of machine code instructions that were generated by Listing 9.3. In particular, check that addresses 0 though 5 contain the same machine code instructions as shown in Table 9.1. This will validate that the IP variable is working to automatically assign an address for each instruction.

XOR Demonstration Program

Listing file 10.2 contains a program to demonstrate that the exclusive OR logical operation can be calculated by a sequence of ANDs, NOTs, and inclusive OR operations. This was first performed graphically in Figure 2.4 and then again using the Verilog structural coding style in Listing 5.9. Basically, M ^ N = ~M & N | ~N & M.

The directory header of Listing 10.2 states there are five modules present. The ProgMod module containing the "ARM like" program and its associated macros are the same as in Listing 10.1. There is a new CPU module and DataProcIns (ALU) module to support multiple general purpose registers and a new user interface to dump the contents of multiple registers at the same time.

```
1.  // Listing 10.2 demonstrates execution of 21-bit instructions
2.  // 1) Assembles 21-bit "ARM like" Data Processing instructions
3.  // 2) Executes 21-bit "XOR" example program
4.  // 3) Dumps register contents in hexadecimal
5.  //
6.  // Modules and macros contained in this file:
7.  // 1) CPU_UI: User Interface that dumps string of hex digits
8.  // 2) CPU: 21-bit "ARM like" CPU with 16 8-bit general purpose registers
9.  // 3) DataProcIns: 9 "ARM like" DP instructions using 8-bit registers
10. // 4) Macros for assembling 9 ARM data processing instructions
11. // 5) ProgMod: Memory containing "ARM" "XOR" application program
```

Listing 10.2 A: Header file for 21-bit XOR demonstration execution

Four States

Let's begin with the CPU module. In previous chapters, there was only one accumulator register, so there was really only one place to put the result of a calculation. However, now with sixteen general purpose registers, I need another decoder and will make a new "Write Back" state for storing calculation results.

1. **Fetch:** Get next instruction from computer's memory.
2. **Decode:** Separate parts of instruction to determine what is to be done.
3. **Execute:** Actually do the work.
4. **Write Back:** Update the destination register.

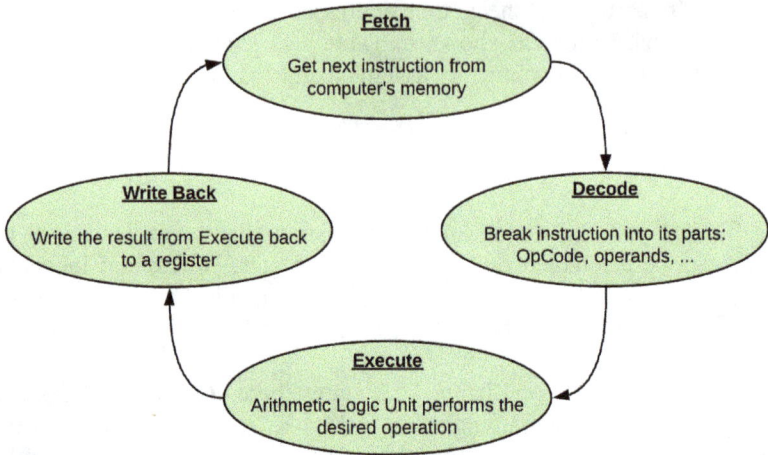

Figure 10.2: Write Back state added to Fetch, Decode, Execute

Write Back: Update Register With the Result

Listing 10.2 B shows the instruction cycle now including Write Back inserted between Execute and Fetch.

- Lines 118,119: A reset initializes the program counter (PC) register to zero and the CPU state to "fetch." Note: Register R15 will now be used as the PC. This is what is done in the real ARM 32-bit architecture.
- Lines 125,126: The fetch state updates the PC to point to the next instruction and changes the state to "decode," but notice that it did not directly fetch anything from memory. How did the instruction register get loaded?
- Lines 132-138: The interlock code is still present that allows stepping through the program one instruction at a time.
- Line 141: It looks like the "execute" state isn't doing anything but switching control to the "write back" state.
- Line 144: The new "write back" state puts the results from the CPU operation into the destination register. Question: What caused the results to be generated? Note: It looks like there is no code within the "execute" state.

```
112.    // Instruction cycle and reset
113.
114.    always @ (posedge(clk), posedge(reset))
115.      begin
116.        if (reset)
117.          begin
118.            R[PC] <= 0;                    // Boot address is 0000
119.            CPU_state <= fetch;
120.          end
121.        else
122.          case (CPU_state)
123.            fetch:                         // Get next instruction in program
124.              begin
125.                R[PC] <= R[PC] + 1;
126.                CPU_state <= decode;
127.              end
128.            decode:                        // Disassemble the instruction
129.              begin
130.                RnVal <= R[RnID];          // Source data register
131.                RmVal <= R[RmID];          // 2nd op data register
132.                if (~KEY[0])               // Instruction cycle interlock
133.                  run <= 0;
134.                if (~KEY[1] && ~run)
135.                  begin
136.                    run <= 1;
137.                    CPU_state <= execute;
138.                  end
139.              end
140.            execute:                       // Perform desired operation
141.              CPU_state <= writeBack;
142.            writeBack:                     // Update specific register
143.              begin
144.                R[RdID] <= RdValDP;        // Destination register
145.                CPU_state <= fetch;
146.              end
147.          endcase
148.      end
149. endmodule
150.
```

Listing 10.2 B: Instruction cycle for 21-bit ARM imitation

The above Verilog coding shows how the CPU imitation moves among the four states. We will next examine how variables, such as RnVal and RnID, are defined in Listing 10.2 C.

```
55.  //--------------- "ARM" CPU imitation ---------------
56.  //
57.  module CPU (KEY, reset, clk, LEDR, hexDisp);
58.     input [1:0] KEY;
59.     input reset, clk;
60.     output [9:0] LEDR;
61.     output [23:0] hexDisp;
62.     assign hexDisp[3:0] = R[1];
63.     assign hexDisp[7:4] = R[2];
64.     assign hexDisp[11:8] = R[11];
65.     assign hexDisp[15:12] = R[12];
66.     assign hexDisp[19:16] = R[3];
67.     assign hexDisp[23:20] = R[4];
68.     assign LEDR = R[15];
69.
72.     reg run;                          // Flag indicating CPU is running
73.
74.     // CPU instruction cycle
75.
76.     reg [1:0] CPU_state;
77.     parameter fetch = 2'b01;
78.     parameter decode = 2'b10;
79.     parameter execute = 2'b11;
80.     parameter writeBack = 2'b00;
81.
82.     // Instruction format
83.
84.     wire [3:0] opCode;                 // Operation code value
85.     wire iFlag;                        // Immedite data flag
86.     wire [7:0] op2;                    // Second operand raw code
87.     wire [7:0] RdValDP;                // Destination register value
88.     reg [7:0] RnVal;                   // Source data register value
89.     reg [7:0] RmVal;                   // 2nd op data register value
90.     wire [3:0] RdID;                   // Destination register number
91.     wire [3:0] RnID;                   // Source register number
92.     wire [3:0] RmID;                   // Second operand register number
93.     wire [20:0] IR;                    // Instruction register
94.     reg [7:0] op2Val;                  // Numeric value of second operand
95.     assign iFlag = IR[20];            // Immediate data flag
96.     assign opCode = IR[19:16];        // Opcode is in upper 4 bits
97.     assign RnID = IR[15:12];          // Source data register
98.     assign RdID = IR[11:8];           // Destination register
99.     assign RmID = IR[3:0];            // 2nd operand register
100.    assign op2 = IR[7:0];            // Operand is in lower 8 bits
```

Listing 10.2 C: Define variables used in Execute state

Listing 10.2 C defines the variables used in the CPU module as it fetches, decodes, executes, and writes back. Listing 10.2 D shows where the CPU module instantiates module ProgMod which does the instruction fetch and module DataProcIns which is the ALU that performs the calculations.

- Line 57: The CPU module has three inputs: a system clock, a reset "button," and KEYs for single-stepping through each instruction. It has two outputs for displaying what is happening inside the CPU.
- Line 61: This is a 24-bit return value that will be displayed as six hexadecimal digits.
- Lines 62 - 67: For this example, the 24-bit display will show the lower 4 bits of six registers: R1, R2, R11, R12, R3, and R4.
- Line 68: The PC, which is register R15 in 32-bit ARM processors, will be displayed on the LEDs in binary.
- Lines 76 - 80: The fourth CPU state, write back, is now included along with the previous states of fetch, decode, and execute.
- Lines 84 - 94: Variables for holding the components of the data processing instruction format
- Lines 95 - 100: The decode state will break the instruction into multiple components
- Line 104: There are sixteen 8-bit registers. These will be expanded to 32 bits in Chapter 11.
- Line 105: The PC is located in general purpose register 15.
- Line 109: The ProgMod module is instantiated with four arguments: 1) address of instruction to fetch, 2) pulse when a instruction fetch is needed, 3) system reset pulse, and 4) a return location to receive data.
- Line 110: The DataProcIns module is instantiated with seven arguments: 1) operation code, 2) first operand value, 3) second operand if in register, 4) second operand if immediate, 5) immediate flag bit, 6) pulse when operation is to be performed, and 7) a "return" location to receive data.

```
101.
102.    // General purpose registers R0 - R15
103.
104.    reg [7:0] R[0:15];                    // 16 registers
105.    parameter PC = 15;                    // Program Counter
106.
107.    // Instantiate program memory and ALU processing
108.
109.    ProgMod (R[PC], CPU_state==fetch, reset, IR);
110.    DataProcIns (opCode, RnVal, RmVal, op2, iFlag, CPU_state==execute, RdValDP);
```

Listing 10.2 D: Instantiate the modules needed by the CPU module

- Line 155: Six inputs and one output. Note: These "wires" contain the actual data and are not pointers to memory or registers containing the data.
- Line 156: The desired ARM operation to be performed
- Line 157: First operand. This is an 8-bit value, but will be expanded to 32 in the next chapter.
- Lines 158 - 160: Two possible inputs for the second operand value.
- Line 161: The "clk" indicates when this module will perform the calculation. It is not the system "clock" because it will only "tick" during the execution state of a data processing instruction.
- Line 162: Value calculated by ALU module

```
151.  //
152.  //---------------- ALU for ARM data processing instructions ----------------
153.  //
154.  // Note: All arguments are values (i.e, not register ID numbers)
155.  module DataProcIns (opCode, Rn, Rm, op2raw, iFlag, clk, Rd);
156.     input [3:0] opCode;              // Data processing instruction opcode
157.     input [7:0] Rn;                  // Source register contents
158.     input [7:0] Rm;                  // Possible 2nd operand register contents
159.     input [7:0] op2raw;              // Second operand "as is"
160.     input iFlag;                     // Immediate value flag
161.     input clk;                       // Pulse to produce calculation
162.     output reg [7:0] Rd;             // Value to return
163.     wire [7:0] op2;                  // Calculated value of second operand
164.     assign op2 = (iFlag) ? op2raw : Rm;
165.     always @ (posedge(clk))
166.        case (opCode)
167.           0: Rd <= Rn & op2;         // AND
168.           1: Rd <= Rn ^ op2;         // EOR (exclusive OR)
169.           2: Rd <= Rn - op2;         // SUB
170.           3: Rd <= op2 - Rn;         // RSB (reverse subtract)
171.           4: Rd <= Rn + op2;         // ADD
172.           12: Rd <= Rn | op2;        // ORR (inclusive OR)
173.           13: Rd <= op2;             // MOV
174.           14: Rd <= Rn & ~op2;       // BIC (bit clear)
175.           15: Rd <= ~op2;            // MVN (move NOT)
176.           default: Rd <= 0;          // None of above
177.        endcase
178.  endmodule
179.
```

Listing 10.2 E: Execute with hex display output

The above ALU is basically the same as that introduced in Chapter 7, except it has been modified to support multiple registers. Up until this point, there was only one accumulator, and it provided both the contents of the first operand as well as the destination for the result. Now we have two operands coming from possibly two separate locations, and there is a third register to receive the result. Listing 10.2 F gives the user interface.

```
13.  //
14.  //---------------- User Interface ----------------
15.  //
16.  module CPU_UI (KEY, CLOCK_50, LEDR, HEX0, HEX1, HEX2, HEX3, HEX4, HEX5);
17.     input [1:0] KEY;
18.     input CLOCK_50;
19.     output [9:0] LEDR;
20.     output [7:0] HEX0, HEX1, HEX2, HEX3, HEX4, HEX5;
21.     function automatic [7:0] digit;
22.        input [3:0] num;
23.        case (num)
24.           0: digit = 8'b11000000;          // 0
25.           1: digit = 8'b11111001;          // 1
26.           2: digit = 8'b10100100;          // 2
27.           3: digit = 8'b10110000;          // 3
28.           4: digit = 8'b10011001;          // 4
29.           5: digit = 8'b10010010;          // 5
30.           6: digit = 8'b10000010;          // 6
31.           7: digit = 8'b11111000;          // 7
32.           8: digit = 8'b10000000;          // 8
33.           9: digit = 8'b10010000;          // 9
34.           10: digit = 8'b10001000;         // A
35.           11: digit = 8'b10000011;         // b
36.           12: digit = 8'b11000110;         // C
37.           13: digit = 8'b10100001;         // d
38.           14: digit = 8'b10000110;         // E
39.           15: digit = 8'b10001110;         // F
40.        endcase
41.     endfunction
42.     wire [23:0] hexDisp;
43.     wire reset;
44.     and(reset,~KEY[0],~KEY[1]);            // Reset if both keys pushed
45.     assign HEX0 = digit(hexDisp[3:0]);
46.     assign HEX1 = digit(hexDisp[7:4]);
47.     assign HEX2 = digit(hexDisp[11:8]);
48.     assign HEX3 = digit(hexDisp[15:12]);
49.     assign HEX4 = digit(hexDisp[19:16]);
50.     assign HEX5 = digit(hexDisp[23:20]);
51.     CPU (KEY, reset, CLOCK_50, LEDR, hexDisp);
52.  endmodule
```

Listing 10.2 F: User interface to control "ARM like" processor

Compare Computation of ~A&B | ~B&A to XOR

The logical gates can be built from other gates. As pointed out in Chapter 9 and previous chapters, an exclusive OR function can be constructed from two AND gates, two NOT gates, and an OR gate. This identity comes from looking at the four rows of the truth table for the "exclusive or" (EOR in ARM mnemonics). From Table 1.27, the logical output of the EOR is "1" only if row 2 is "1" OR if row 3 is "1."

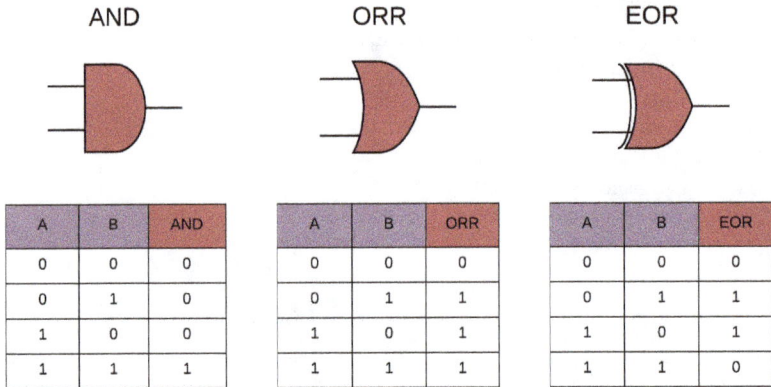

AND				ORR				EOR		
A	B	AND		A	B	ORR		A	B	EOR
0	0	0		0	0	0		0	0	0
0	1	0		0	1	1		0	1	1
1	0	0		1	0	1		1	0	1
1	1	1		1	1	1		1	1	0

Copy of Figure 1.27: Truth tables for AND, inclusive OR, and exclusive OR

The ARM program in Listing 10.2 G computes all four rows of the truth table together. Bit column 3 does the top row (0^0=0). The other three rows, (0^1=1), (1^0=1), and (1^1=0) are computed in bit columns 2, 1, and 0, respectively. In other words, the "exclusive or" of binary 0101 with 0011 will be 0110 according to the truth table.

237.	`MOV  R1,0,'b0101  `_	// Move 0b0101 into R1
238.	`MOV  R2,0,'b0011  `_	// Move 0b0011 into R2
239.	`BIC  R11,R1,R2  `_	// Not(R2) & (R1) => R11
240.	`BIC  R12,R2,R1  `_	// Not(R1) & (R2) => R12
241.	`ORR  R3,R11,R12  `_	// (R11)	(R12) => R3
242.	`EOR  R4,R1,R2  `_	// Exclusive OR instruction

Listing 10.2 G: Program to compare EOR instruction to calculated XOR

Go ahead and compile, download, and run the code in Listing 10.2. Step through each of the six ARM instructions and verify that the values in the registers match that of Table 10.1. The procedure for "single stepping" through each line of the program is the same as in Chapter 8:

1. Start the program with a system reset: Holding down KEY[0] while momentarily pushing KEY[1] will reset the PC to address 0. The CPU will run until it is stops in the decode state. The register contents do not get changed by the reset, but neither do data registers in real CPUs get changed by a reset. That's why I have exes (xxxx) listed in Table 10.1 for values of registers that could have any "left over" values from previous use.
2. The display shows the low order four bits of six registers as shown by the hex display in Figure 10.3. Although all eight bits are being used in each calculation, we are only interested in the values of the lower four. Look at lines 45 through 50 and 62 through 68 of Listing 10.2 to see how the register contents have been connected to the seven-segment displays.
3. To execute each of the remaining instructions in sequence, momentarily push KEY[0] followed by momentarily pushing KEY[1].

Figure 10.3: Display lower four bits from registers R4, R3, R12, R11, R2, and R1

Next Address on LEDs	Instruction to be executed	R4, R3, R12, R11, R2, R1 on HEX5, . . . HEX0
001	`MOV R1, 0, 'b0101 `_	x x x x x x
010	`MOV R2, 0, 'b0011 `_	x x x x x 5
011	`BIC R11 ,R1, R2 `_	x x x x 3 5
100	`BIC R12, R2, R1 `_	x x x 4 3 5
101	`ORR R3, R11, R12 `_	x x 2 4 3 5
110	`EOR R4, R1, R2 `_	x 6 2 4 3 5
111	0	6 6 2 4 3 5

Table 10.1 Register contents while single stepping through XOR program

Highlights and Comparisons

Chapter 10 added a fourth processor state and changed the asdp function to a task. In the next chapter, the asdp task will have an upgrade to accommodate the handling of status information.

Review Questions

1. *Why is the ProgMem module "called" where it is instead of where it "logically" makes sense down in the fetch state?
2. In lines 84 through 94 of Listing 10.2 C, most of the variables are given type "wire" except for RnVal and RnVal. Why are these defined as type "reg"?
3. Compare features of tasks, functions, and modules.
4. *What logical identity can be used to generate the logical AND operation from the other logical operations?
5. What logical identity can be used to generate the OR operation from the other logical operations?

Exercises

1. Modify the ARM program on lines 237 through 242 of Listing 10.2 to calculate the AND operation and compare it to a calculated value (see Question 4 above).
2. Modify the same lines from Listing 10.2 to write a short ARM program to sum a series of numbers such as 1+2+3+...
    a. Initialize R0 to 0, and R1 to 1 using two MOV instructions.
    b. The third instruction will ADD the contents of R1 to R0.
    c. The fourth instruction will ADD the immediate value of 1 to register R1.
    d. The fifth instruction will load an immediate value of 2 into register R15 (the PC). This will force the next instruction to be executed to be the one that adds the contents of R1 to R0. This is an infinite loop because we have no way of stopping it, until we have status bits described in Chapter 11.

Computer Architecture Tutorial Using an FPGA

— 11 —
Status Register

Nearly every CPU design has had a program status register which provides status regarding instructions that were previously executed:

1. Did the previous arithmetic or logic operation result equal zero?
2. Was the previous result positive or negative?
3. Did the previous result fit within the register size.
4. Was there a possible error, such as the sum of two positive numbers resulting in a negative number?

Chapter 11 will complete the set of sixteen ARM data processing instructions and include support for status bits in the program status register. Chapter 7 began the Verilog-based imitation of the ARM processor with a 7-bit instruction format and nine data processing instructions. In order to include multiple registers in Chapter 9, the instruction format was expanded to 21 bits.

Here in Chapter 11, the data processing instruction format will be expanded to its full 32 bits in order to include status bit processing. Note: Most ARM processors also support a special restricted 16-bit instruction format known as "Thumb mode," and it will be described in Chapter 19.

Introductions

No new Verilog commands or keywords are introduced in Chapter 11. However, the following seven instructions that explicitly use the status bits will complete the set of ARM "data processing" instructions:

- **ADC:** Add an integer plus Carry
- **SBC:** Subtract an integer including Carry
- **RSC:** Reverse Subtract including Carry
- **TST:** Test for specific bits present (like AND)
- **TEQ:** Test for all corresponding bits matching (like XOR)
- **CMP:** Compare two values (like SUB)
- **CMN:** Compare Register to Negative value (like ADD)

The ADC, SBC, and RSC are the same operations as the ADD, SUB, and RSB instructions, except they include the "carry" status bit. The TST, TEQ, CMP, and CMN instructions are very similar to the AND, XOR, SUB, and ADD instructions in setting the status bits, but they differ in that no data registers are modified.

Current Program Status Register (CPSR)

The status bits generally result from the completion of an instruction involving a 32-bit register. In the ARM, the CPSR (Current Program Status Register) includes N, Z, C, and V status flags:

N	**Negative**: Previous operation result was negative (i.e., bit 31 = 1)
Z	**Zero**: Previous operation result was zero (i.e., bits 31..0 = 0)
C	**Carry**: Previous operation resulted in a value that exceeded 32 bits.
V	**Overflow**: Previous operation resulted in a possible "sign" error.

Table 11.1: Status bits in the CPSR

If an instruction has a result where all 32 bits of a 32-bit register are zero, then the Z-bit is set. The negative (N-bit) matches the high-order sign bit: "one" indicates a negative number, and "zero" implies a positive number.

Since the early days of digital computing, "carry" has been a common technique to string together arithmetic circuits. It was demonstrated in Chapter 5, Listing 5.16, to perform multibit addition.

"Overflow" sounds like an instruction resulted in error, but under some conditions, there is no problem at all. It depends on whether the software is doing signed arithmetic or not.

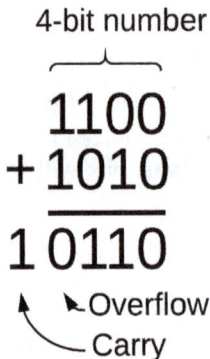

Figure 11.1 shows the addition of two 4-bit numbers resulting in an overflow and carry. The focus is on the high order bits. Regardless of whether the register size is 4 bits or 32 bits, the concepts are the same.

The 4-bit addition of binary 1100 and 1010 sets the following status bits:

- Sign bit N: Clear
- Zero bit Z: Clear
- Carry bit C: Set
- Overflow bit V: Set

4-bit number

$$
\begin{array}{r}
1100 \\
+\ 1010 \\
\hline
1\ 0110
\end{array}
$$

Overflow
Carry

Figure 11.1: Setting the C (carry) and V (overflow) status bits

Integers and whole numbers use exactly the same arithmetic circuits. In other words, 16-bit whole numbers having a range of 0 to 65,535 and 16-bit integers having a range of -32,768 to +32,767 use exactly the same circuit to add any two numbers in their ranges. The hardware circuit doesn't even "know" if the software is considering the values to be signed integers or only positive whole numbers. That is the beauty of using either one's complement or two's complement to represent negative numbers. See Appendix C for details.

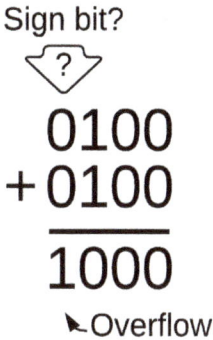

The 4-bit example in Figure 11.2 has a whole number range of 0 to 15 and an integer range of -8 to +7 using two's complement. If a decimal 4 is added to itself ($0100_2 + 0100_2$), the result is binary 1000 and both the sign and overflow status bits are set.

If the software application doing the addition considers the values to be signed integers, we definitely have an error. However, if the software considers this to be a whole number in the range of 0 to 15, the sign and overflow flags can be ignored.

Sign bit?

$$0100$$
$$+0100$$
$$\overline{1000}$$

Overflow

Figure 11.2: Overflow doesn't always mean an error

- If 4-bit whole number: 4 + 4 = 8
- If 4-bit integer: 4 + 4 = -8

32-bit Data Processing Instruction Format

Figure 11.3 illustrates the four new or modified fields that will expand the 21-bit format used in Chapter 10 to the full 32-bit ARM instruction format:

- Bits 28 - 31: Status required before executing any instruction
- Bits 26, 27: ID code indicating a "data processing" instruction
- Bit 20: Execution of this instruction will change status bits
- Bits 8..11: Second operand increased to 12 bits

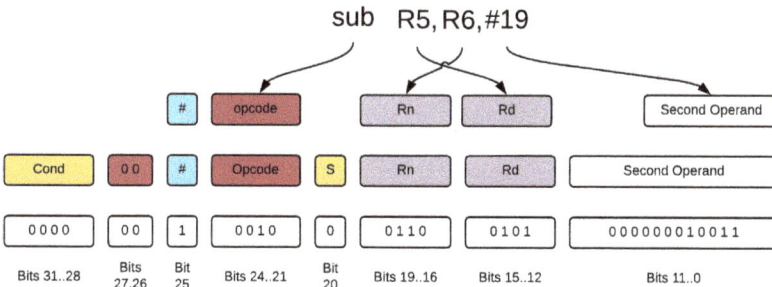

sub R5, R6, #19

| # | opcode | | Rn | Rd | Second Operand |

| Cond | 0 0 | # | Opcode | S | Rn | Rd | Second Operand |

| 0000 | 0 0 | 1 | 0010 | 0 | 0110 | 0101 | 000000010011 |
| Bits 31..28 | Bits 27,26 | Bit 25 | Bits 24..21 | Bit 20 | Bits 19..16 | Bits 15..12 | Bits 11..0 |

Figure 11.3: Data processing instruction format expanded to full 32 bits

Because we will be doing a lot of work with 32-bit values, we will need code to examine memory and register contents. The main "CPU_UI" module in Listing 11.1 displays a 32-bit value on the 7-segment displays as eight 4-bit hexadecimal digits. It works with both the DE10-Lite and DE2 models because it only requires four 7-segment displays.

```
12.  module CPU_UI (SW, KEY, LEDR, HEX0, HEX1, HEX2, HEX3, HEX4, HEX5);
13.    input [4:0] SW;
14.    input [1:0] KEY;
15.    output [4:0] LEDR;
16.    output [7:0] HEX0, HEX1, HEX2, HEX3, HEX4, HEX5;
17.    function automatic [7:0] digit;
18.      input [3:0] num;
19.      case (num)
20.        0: digit = 8'b11000000;                    // 0
21.        1: digit = 8'b11111001;                    // 1
22.        2: digit = 8'b10100100;                    // 2
23.        3: digit = 8'b10110000;                    // 3
24.        4: digit = 8'b10011001;                    // 4
25.        5: digit = 8'b10010010;                    // 5
26.        6: digit = 8'b10000010;                    // 6
27.        7: digit = 8'b11111000;                    // 7
28.        8: digit = 8'b10000000;                    // 8
29.        9: digit = 8'b10010000;                    // 9
30.        10: digit = 8'b10001000;                   // A
31.        11: digit = 8'b10000011;                   // b
32.        12: digit = 8'b11000110;                   // C
33.        13: digit = 8'b10100001;                   // d
34.        14: digit = 8'b10000110;                   // E
35.        15: digit = 8'b10001110;                   // F
36.      endcase
37.    endfunction
38.    wire [31:0] IR;
39.    assign HEX0 = KEY[1] ? digit(IR[19:16]) : digit(IR[3:0]);
40.    assign HEX1 = KEY[1] ? digit(IR[23:20]) : digit(IR[7:4]);
41.    assign HEX2 = KEY[1] ? digit(IR[27:24]) : digit(IR[11:8]);
42.    assign HEX3 = KEY[1] ? digit(IR[31:28]) : digit(IR[15:12]);
43.    assign HEX4 = 8'hFF;
44.    assign HEX5 = 8'hFF;
45.    assign LEDR = SW;
46.    ProgMod (SW[4:0], ~KEY[1], ~KEY[0], IR);
47.  endmodule
48.
```

Listing 11.1 A: This CPU_UI module displays 8 hex digits in two parts

- Line 12: Module name is "CPU_UI" as before even though it is now only used as a test display for 32-bit values.

- Lines 39 - 42: Conditional operators display either the high-order four digits (KEY[1] is up) or the low-order four digits.
- Lines 43, 44: Turn off the LEDs in unused 7-segment displays.
- Line 52: Module ProgMod will contain the sequence of ARM instructions.
- Line 57: Thirty two 32-bit memory locations will be allocated.
- Line 61: The clock must "tick" to update the display after the address switches are changed.
- Line 63: The "memory" contents are initialized with "reset"
- Line 65: Memory address 0 contains the hex pattern: 01234567.
- Line 66: Memory address 1 contains the hex pattern: 89ABCDEF.

```
52.  module ProgMod (address, clk, reset, instr);
53.      input [4:0] address;
54.      input clk, reset;
55.      output [31:0] instr;
56.
57.      reg [31:0] progMem[0:31];           // 32 bits per instruction
58.      reg [31:0] IR;                       // Instruction Register
59.      assign instr = IR;
60.
61.      always @ (posedge(clk))
62.        IR <= progMem[address];
63.      always @ (posedge(reset))
64.      begin
65.        progMem[0] <= 32'h01234567;
66.        progMem[1] <= 32'h89ABCDEF;
67.      end
68.  endmodule
```

Listing 11.1 B: Module ProgMod provides series of 32-bit values

Figure 11.4: Controls for running memory dump in Listing 11.1

Figure 11.4 shows the test setup for viewing 32-bit numbers on the 7-segment displays.

1. Push KEY[0] to initialize.
2. Set the switches to desired memory address (0 or 1).
3. Push KEY[1] to update and show the 4 low-order (rightmost) hex digits. Let it up to show the 4 high-order digits.
4. Repeat steps 2 and 3 to see the contents of other addresses.

The DE2-115 has eight 7-segment displays. Listing 11.2 shows a version of the CPU_UI module that can display the whole 32 bits at once. Of course, the ProgMod module is the same as in Listing 11.1. I will be including the two-step display method from Listing 11.1 in upcoming listings because it works for both four and eight digit display boards. However, if you have an eight digit display, I recommend copying the code from Listing 11.2 where needed.

```
12.  module CPU_UI (SW, KEY, LEDR, HEX0, HEX1, HEX2, HEX3, HEX4, HEX5,
     HEX6, HEX7);
13.    input [4:0] SW;
14.    input [1:0] KEY;
15.    output [4:0] LEDR;
16.    output [7:0] HEX0, HEX1, HEX2, HEX3, HEX4, HEX5, HEX6, HEX7;
17.    function automatic [7:0] digit;
18.      input [3:0] num;
19.      case (num)
20.        0: digit = 8'b11000000;                              // 0
21.        1: digit = 8'b11111001;                              // 1
22.        2: digit = 8'b10100100;                              // 2
23.        3: digit = 8'b10110000;                              // 3
24.        4: digit = 8'b10011001;                              // 4
25.        5: digit = 8'b10010010;                              // 5
26.        6: digit = 8'b10000010;                              // 6
27.        7: digit = 8'b11111000;                              // 7
28.        8: digit = 8'b10000000;                              // 8
29.        9: digit = 8'b10010000;                              // 9
30.        10: digit = 8'b10001000;                             // A
31.        11: digit = 8'b10000011;                             // b
32.        12: digit = 8'b11000110;                             // C
33.        13: digit = 8'b10100001;                             // d
34.        14: digit = 8'b10000110;                             // E
35.        15: digit = 8'b10001110;                             // F
36.      endcase
37.    endfunction
38.    wire [31:0] IR;
39.    assign HEX0 = digit(IR[3:0]);
40.    assign HEX1 = digit(IR[7:4]);
41.    assign HEX2 = digit(IR[11:8]);
42.    assign HEX3 = digit(IR[15:12]);
43.    assign HEX4 = digit(IR[19:16]);
44.    assign HEX5 = digit(IR[23:20]);
45.    assign HEX6 = digit(IR[27:24]);
46.    assign HEX7 = digit(IR[31:28]);
47.    assign LEDR = SW;
48.    ProgMod (SW[4:0], ~KEY[1], ~KEY[0], IR);
49.  endmodule
```

Listing 11.2: Display all eight hex digits at one time on the DE2 or DE2-115.

The instruction format is starting to get rather complicated, and we will be relying more on "assembly language" style input to generate the 32-bit instructions. The use of Verilog macros, tasks, functions, and parameters to imitate assembly language began in Chapter 9. Listing 11.3 extends the assembly language input from Listing 10.1 to the 32-bit format, and includes the hexadecimal dump capability of Listing 11.1.

All 117 lines of Listing 11.3 are included in the GitHub download file, but the first 52 are not listed below because they contain the directory header and the CPU_UI hexadecimal dump module as in Listing 11.1. The ProgMod module generates 32-bit ARM instructions from assembly language style text lines. Lines 53 through 89 (shown below) are the same as those from Listing 9.2.

```
53.    `define AND asdp (4'd0,      // [Rd] = [Rn] AND (2nd operand)
54.    `define EOR asdp (4'd1,      // [Rd] = [Rn] Exclusive Or (2nd operand)
55.    `define SUB asdp (4'd2,      // [Rd] = [Rn] - (2nd operand)
56.    `define RSB asdp (4'd3,      // [Rd] = (2nd operand) - [Rn]
57.    `define ADD asdp (4'd4,      // [Rd] = [Rn] + (2nd operand)
58.    `define ORR asdp (4'd12,     // [Rd] = [Rn] Inclusive OR (2nd operand)
59.    `define MOV asdp (4'd13,     // [Rd] = [Rn]
60.    `define BIC asdp (4'd14,     // [Rd] = [Rn] AND NOT (2nd operand)
61.    `define MVN asdp (4'd15,     // [Rd] = NOT [Rn]
62.    `define _ );                 // End of instruction

68.    module ProgMod (address, clk, reset, instr);
69.        input [4:0] address;
70.        input clk, reset;
71.        output [31:0] instr;
72.
73.        parameter R0 = 16'h1000;     // General purpose register set names
74.        parameter R1 = 16'h1001;
75.        parameter R2 = 16'h1002;
76.        parameter R3 = 16'h1003;
77.        parameter R4 = 16'h1004;
78.        parameter R5 = 16'h1005;
79.        parameter R6 = 16'h1006;
80.        parameter R7 = 16'h1007;
81.        parameter R8 = 16'h1008;
82.        parameter R9 = 16'h1009;
83.        parameter R10 = 16'h100A;
84.        parameter R11 = 16'h100B;
85.        parameter R12 = 16'h100C;
86.        parameter R13 = 16'h100D;     // a.k.a. "SP"
87.        parameter R14 = 16'h100E;     // a.k.a. "LR"
88.        parameter R15 = 16'h100F;     // a.k.a. "PC"
89.
```

Listing 11.3 A: Macros and parameters from Listing 9.2

```
90.    integer IP;
91.    task asdp ();
92.      input [15:0] opcode,Rd,Rn,Rm;
93.      if (Rm < 'h1000)
94.        progMem[IP] <=
             {4'hE,2'h0,1'b1,opcode[3:0],1'b0,Rn[3:0],Rd[3:0],Rm[11:0]};
95.      else
96.        progMem[IP] <=
             {4'hE,2'h0,1'b0,opcode[3:0],1'b0,Rn[3:0],Rd[3:0],Rm[11:0]};
97.      IP = IP + 1;
98.    endtask
99.
100.   reg [31:0] progMem[0:25];          // 32 bits per instruction
101.   reg [31:0] IR;                     // Instruction Register
102.   assign instr = IR;
103.
104.   always @ (posedge(clk))
105.     IR <= progMem[address];
106.   always @ (posedge(reset))
107.   begin
108.     IP = 0;
109.     `MOV   R1, 0, 'b0101   `_        // Move 0b0101 into R1
110.     `MOV   R2, 0, 'b0011   `_        // Move 0b0011 into R2
111.     `BIC   R11, R1, R2    `_         // Not(R2) & (R1) => R11
112.     `BIC   R12, R2, R1    `_         // Not(R1) & (R2) => R12
113.     `ORR   R3, R11, R12   `_         // (R11) | (R12) => R3
114.     `EOR   R4, R1, R2     `_         // Exclusive OR instruction
115.     progMem[25] <= 0;
116.   end
117.  endmodule
```

Listing 11.3 B: Assembly language program to do exclusive OR

Go ahead and compile and download Listing 11.3. Run the switches through the six addresses and verify the hexadecimal values provided in Table 11.2.

Program Address	Assembly Language	Machine Code (hexadecimal)
0	`MOV   R1, 0, 'b0101   `_	E3A01005
1	`MOV   R2, 0, 'b0011   `_	E3A02003
2	`BIC   R11, R1, R2   `_	E1C1B002
3	`BIC   R12, R2, R1   `_	E1C2C001
4	`ORR   R3, R11, R12   `_	E18B300C
5	`EOR   R4,R1,R2   `_	E0214002

Table 11.2: ARM machine codes for six instructions.

Listing 11.3 has the same XOR assembly language program on lines 109 through 114 as in the listings in Chapter 10, but it generates different machine code because of the expansion to the 32-bit format. See the changes on lines 94, 96, 100, and 101.

The ARM Machine Code Instruction

Listing 11.3 only expands the size of the ARM instruction to 32 bits. It does nothing special for the "cond" (bits 28..31) and "s" (bit 20) fields that are now included in the data processing instruction format. Basically, the "cond" field dictates whether an instruction is to be executed or ignored, while the "s" field indicates whether the status bits should be updated when and if the instruction is executed.

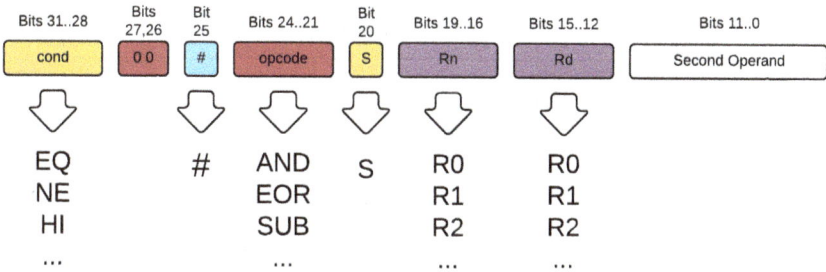

Figure 11.5: Positions in machine code

Bits	Name	Contents
31..28	Cond	Only execute this instruction on condition of the value in the NZCF flags
27..26	00	These two bits being zero distinguish the set of 16 data processing instructions from other ARM instructions
25	#	Immediate operand flag
24..21	Opcode	Which operation (add, sub, and, orr, eor, ...)
20	S	Indicates that this instruction will modify the condition codes
19..16	Rn	ID number of register containing the first operand
15..12	Rd	ID number of register to receive the result
11..0	op2	Three formats possible for second operand (see Chapter 12)

Table 11.3: Enhancements to the instruction format to include status bit processing

The "second operand" has thus far been described as either a constant or the contents of register Rm, depending on the value of the immediate bit 20. However, it is actually quite a bit more complicated, and it will be described in detail in Chapter 12.

The condition code field relies on the current value of either one or two status bits. Table 11.4 provides the meaning of the sixteen possible condition patterns.

Cond code	Assembly mnemonic	Necessary status bits	Meaning
0	EQ	Z	EQual (equals zero)
1	NE	!Z	Not Equal
2	CS or HS	C	Carry Set (unsigned Higher or Same)
3	CC or LO	!C	Carry Clear / unsigned LOwer
4	MI	N and !C	MInus
5	PL	!N	PLus (positive or zero)
6	VS	V	oVerflow Set
7	VC	!V	oVerflow Clear
8	HI	!C and !Z	HIgher (unsigned)
9	LS	!C or Z	Lower or Same (unsigned)
10	GE	N = V	Greater than or Equal (signed)
11	LT	N != V	Less Than (signed)
12	GT	!Z and (N = V)	Greater Than (signed)
13	LE	Z or (N != V)	Less than or Equal (signed)
14	AL	Always	Default (same as omitted)
15	Never	Reserved	Code 15 for future ARMs

Table 11.4: List of ARM assembly language condition codes

The general format of an ARM operation code in an assembly language instruction is `OpCode`Cond`S

1. The "OpCode" is simply the mnemonic for the operation to be performed by the current instruction (i.e., add, sub, mul, mov, ...).
2. The "Cond" field is optional and indicates which combination of condition status bits has to be set (or clear) for the current instruction to be executed or ignored. Note: In Table 11.3, "Z" means the Z-flag

Computer Architecture Tutorial Using an FPGA

must be set (value=1), "!Z" means the Z-flag must be clear (value=0), "C" means the C-flag must be set, "!C" means the C-flag must be clear, etc.

3. The "S" field is optional and indicates whether the NZCV condition status bits are to be modified by the execution of the current instruction.

Most assemblers concatenate the OpCode, Cond, and S fields. They use SUBEQS instead of `SUB`EQ`S. Note: Some ARM assemblers reverse the order of the "Cond" and "S" fields (SUBSEQ instead of SUBEQS). See Exercise 2 at the end of this chapter.

Let's use the subtraction instruction as an example for showing the multiple formats available in the ARM. In each of the examples in Table 11.5, the constant value of 7 is subtracted from the contents of register R6 with the result delivered to register R5.

SUB	R5,R6,#7	@ Always do the subtraction, but do not change the NZCV flags.
SUBS	R5,R6,#7	@ Always do the subtraction, and change flags depending on the result of the subtraction.
SUBEQ	R5,R6,#7	@ Only do the subtraction if the Z-flag is set (i.e., a previous result was zero), but do not change NZCV flags due to this subtraction.
SUBEQS	R5,R6,#7	@ Only do the subtraction if the Z-flag is set, and then also change NZCV based on subtraction results.

Table 11.5: Combinations of SUB operator with condition bits

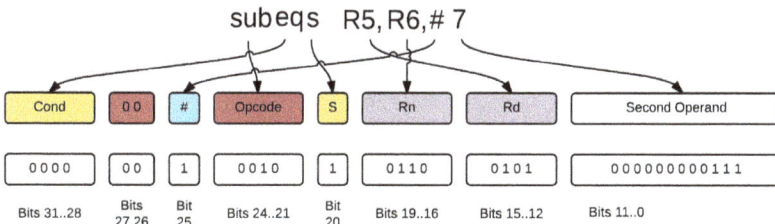

Figure 11.6: ARM 32-bit instruction using status bits = hex 02565007

The assembler constructs the machine code instruction 'h02565007 by "filling in" the fields of the instruction word with "sub" = 0010, "eq" = 0000, "s" = 1, and "#" = 1. The operands and destination register ID are in the lower 20 bits of the instruction and consist of R5, R6, and the immediate constant of 7 as shown in Figure 11.6.

Instructions Using Status Bits

The implementation and testing of the new 32-bit format will be done in three steps here in Chapter 11.

1. Listing 11.4 contains the new modifications to the "assembly language" asdp task to provide the setting and testing of the status bits along with support for the seven new instructions. The 32-bit memory dump will be used to observe a variety of instructions converted from assembly language to machine code.
2. Listing 11.5 provides the Verilog code to execute the 32-bit machine code for all sixteen data processing instructions, including support for the setting and testing of the status bits. XOR program is executed.
3. Listing 11.6 demonstrates a running ARM application program that generates factorials.

```
50.  //
51.  //---------------- Macro definitions for assembly language ----------------
52.  //
53.  `define AND asdp (4'b0000,      // [Rd] = [Rn] AND (2nd operand)
54.  `define EOR asdp (4'b0001,      // [Rd] = [Rn] Exclusive Or (2nd operand)
55.  `define SUB asdp (4'b0010,      // [Rd] = [Rn] - (2nd operand)
56.  `define RSB asdp (4'b0011,      // [Rd] = (2nd operand) - [Rn]
57.  `define ADD asdp (4'b0100,      // [Rd] = [Rn] + (2nd operand)
58.  `define ADC asdp (4'b0101,      // [Rd] = [Rn] + (2nd operand) + C
59.  `define SBC asdp (4'b0110,      // [Rd] = [Rn] (2nd operand) + C - 1
60.  `define RSC asdp (4'b0111,      // [Rd] = (2nd operand) - [Rn] + C - 1
61.  `define TST asdp (4'b1000,      // [Rn] AND (2nd operand) => status bits
62.  `define TEQ asdp (4'b1001,      // [Rn] Exclusive Or (2nd operand) => stats bits
63.  `define CMP asdp (4'b1010,      // [Rn] + (2nd operand) => status bits
64.  `define CMN asdp (4'b1011,      // [Rn] - (2nd operand) => s tatus bits
65.  `define ORR asdp (4'b1100,      // [Rd] = [Rn] Inclusive OR (2nd operand)
66.  `define MOV asdp (4'b1101,      // [Rd] = [Rn]
67.  `define BIC asdp (4'b1110,      // [Rd] = [Rn] AND NOT (2nd operand)
68.  `define MVN asdp (4'b1111,      // [Rd] = NOT [Rn]
69.  `define _1 ,0,0,0,0,0,0,0);     // End instruction of 1 field
70.  `define _2 ,0,0,0,0,0,0);       // End instruction of 2 fields
71.  `define _3 ,0,0,0,0,0);         // End instruction of 3 fields
72.  `define _4 ,0,0,0,0);           // End instruction of 4 fields
73.  `define _5 ,0,0,0);             // End instruction of 5 fields
74.  `define _6 ,0,0);               // End instruction of 6 fields
75.  `define _7 ,0);                 // End instruction of 7 fields
76.  `define _8 );                   // End instruction of 8 fields
```

Listing 11.4 A: Beginning and ending for macros that call the asdp task

Listing 11.4 supports the "assembly language" of the seven new instructions and status bits in the 32-bit ARM format. Note the following changes (also, the first 49 lines contain the same hexadecimal memory dump code as in Listing 11.1):

- Lines 58 - 64: The remaining seven data processing opcodes, 5 through 11, are now included.
- Lines 69 - 76: Because the opcode field can now have either one, two, or three parts (such as, `SUB`EQ`S), the number of arguments sent to the asdp task can change from line to line. Verilog tasks and functions must have a fixed number of arguments. In order to keep the number of arguments fixed at 8 and allow for omitted fields, various endings are available. Examine the assembly code in Listing 11.4 F to see examples.
- Lines 78 - 94: The condition code and associated mnemonics for testing the status bits are included. These are the same as in the real ARM processors.
- Line 94: The AL condition is the most common, but it is normally not explicitly included since it is the default.
- Line 95: Condition code 15 ('b1111) is an ARM reserved code that should never be used. Therefore, I "repurposed" the number 15, and assigned the status update flag "S" to this code. Task asdp will convert it to bit 20 in the instruction format.

78.	`define EQ 4'b0000,	// EQual (zero); Z set
79.	`define NE 4'b0001,	// Not Equal (non-zero); Z clear
80.	`define HS 4'b0010,	// Unsigned Higher or Same; C set -- also "CS"
81.	`define CS 4'b0010,	// Carry set
82.	`define LO 4'b0011,	// Unsigned LOwer; C clear --also "CC"
83.	`define CC 4'b0011,	// Carry clear
84.	`define MI 4'b0100,	// MInus or negative; N set
85.	`define PL 4'b0101,	// PLus or positive; N clear
86.	`define VS 4'b0110,	// Overflow; V Set
87.	`define VC 4'b0111,	// No overflow; V Clear
88.	`define HI 4'b1000,	// Unsigned HIgher; C set and Z clear
89.	`define LS 4'b1001,	// Unsigned Lower or Same; C clear or Z set
90.	`define GE 4'b1010,	// Signed Greater than or Equal to; N equals V
91.	`define LT 4'b1011,	// Signed Less Than; N not same as V
92.	`define GT 4'b1100,	// Signed Greater Than; Z clear and N equals V
93.	`define LE 4'b1101,	// Signed Less than or Equal; Z set or N not same as V
94.	`define AL 4'b1110,	// ALways; any status bits OK -- usually omitted
95.	`define S 4'b1111,	// Status update (code for NeVer, i.e., reserved)

Listing 11.4 B: ARM conditions for instruction execution

Assembly Language Omitted Fields

A simple assembly language that supported a fixed format was introduced in Chapter 9. Chapter 10 extended that support to 32-bit machine code, but assemblers should also be more flexible by allowing omitted fields.

```
97.   //
98.   //---------------- Memory containing "ARM" program ----------------
99.   //
100.  module ProgMod (address, clk, reset, instr);
101.    input [4:0] address;
102.    input clk, reset;
103.    output [31:0] instr;
104.
105.    parameter R0 = 16'h1000;          // General purpose register set names
106.    parameter R1 = 16'h1001;
107.    parameter R2 = 16'h1002;
108.    parameter R3 = 16'h1003;
109.    parameter R4 = 16'h1004;
110.    parameter R5 = 16'h1005;
111.    parameter R6 = 16'h1006;
112.    parameter R7 = 16'h1007;
113.    parameter R8 = 16'h1008;
114.    parameter R9 = 16'h1009;
115.    parameter R10 = 16'h100A;
116.    parameter R11 = 16'h100B;
117.    parameter R12 = 16'h100C;
118.    parameter R13 = 16'h100D;          // a.k.a. "SP"
119.    parameter R14 = 16'h100E;          // a.k.a. "SP"
120.    parameter R15 = 16'h100F;          // a.k.a. "PC"
121.    integer IP;
122.
123.  // ----- Tasks and Functions that implement the "assembler" -----
124.
125.  // Task asdp is called by the data processing opcode macros `SUB, `AND, ...
126.  // The number of parameters will vary between three and eight.
127.  // `SUB`EQ`S R1,R2,R3,LSR,R4 `_8 // General format with 8 parameters
128.  // `SUB R1,R2 `_3 // Many parameters are optional
129.
130.    task asdp ();
131.      input [15:0] P0,P1,P2,P3,P4,P5,P6,P7;
132.      progMem[IP] <= asdp1(P0,P1,P2,P3,P4,P5,P6,P7);
133.      IP = IP + 1;
134.    endtask
```

Listing 11.4 C: Task asdp calls function asdp1, then increments IP.

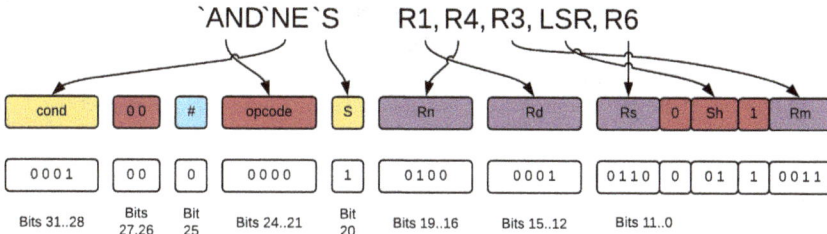

Figure 11.7: ARM assembler using macros

I will continue the theme of making this "ARM assembly language within Verilog" to look more like the assembly language format commonly used with real assemblers. Figure 11.7 shows how complicated the data processing instruction format will become in Chapter 12. Assembly language programmers do not want to fill in every one of these fields, but rely on defaults whenever practical.

For example, most lines of machine code within a program are always executed, but programmers do not want to put an AL on almost every line. Likewise, programmers do not need the status codes generated for every instruction, so they prefer to omit the "S" unless it is needed.

Chapter 12 will show that every data processing instruction can also shift a register, but here again, programmers do not shift on every instruction. Finally, in many cases, the first operand register, Rn, is the same as the destination register, Rd, and programmers prefer to just provide one register ID , and let the assembler duplicate it for the other.

In order to incorporate the new fields for status and also provide the flexibility of using default values to omit arguments, the asdp task will call upon the following three functions:

1. **asdp1**: Function **asdp1** receives all eight arguments that were built for the **asdp** task by the macros ADD, SUB, AND, etc. including the ending macros such as _8, _7, _6, etc. Function **asdp1** places the condition codes (EQ, NE, etc.) into the machine code instruction, including checking for omitted arguments for default cases of AL and ~S. Function **asdp2** is then called to fill the remainder of the machine code with the opcode, registers, and immediate constant.

2. **asdp2**: Function **asdp2** receives six arguments: the operation code (now 0 through 15), the destination register Rd, and a flexible format that could be either a constant, a register (Rn), or a register and a constant. There will also be two arguments of value zero that will not be used until the next chapter covering shift instructions Function **asdp2** places the opcode into the machine code. Function **asdp3** is then called to fill in the registers and immediate constant.

3. **asdp3**: Function **asdp3** receives three arguments: destination register Rd, first operand register Rn, and second operand (either a constant or register Rm). It finishes filling the 32-bit ARM machine code format.

```
136.  // Function asdp1 constructs the opcode, condition, and status fields.
137.     // `SUB Format 1: Always do subtraction, but don't set status
138.     // `SUB`S Format 2: Always do subtraction, and also update status
139.     // `SUB`EQ Format 3: Do subtraction only if Z-flag, but don't set status
140.     // `SUB`EQ`S Format 4: Do subtraction and update status only if Z-flag set
141.
142.     function [31:0] asdp1 ();
143.        input [15:0] P0,P1,P2,P3,P4,P5,P6,P7;
144.        if (P1 >= R0)                      // Format 1
145.           asdp1 = 'hE<<28 | asdp2(P0,P1,P2,P3,P4,P5);
146.        else
147.           if (P1 == 'hF)                  // Update? (S flag), Format 2
148.              asdp1 = 'hE<<28 | 1<<20 | asdp2(P0,P2,P3,P4,P5,P6);
149.           else
150.              if (P2 != 'hF)               // Format 3
151.                 asdp1 = P1<<28 | asdp2(P0,P2,P3,P4,P5,P6);
152.              else                         // Format 4
153.                 asdp1 = P1<<28 | 1<<20 | asdp2(P0,P3,P4,P5,P6,P7);
154.     endfunction
```

Listing 11.4 D: Function asdp1 fills in Cond and S, then calls asdp2 to finish up.

Figure 11.8 shows the progression of arguments from the asdp task down through the asdp3 function. There will be one more layer, asdp4, in the next chapter in which register shifting is included.

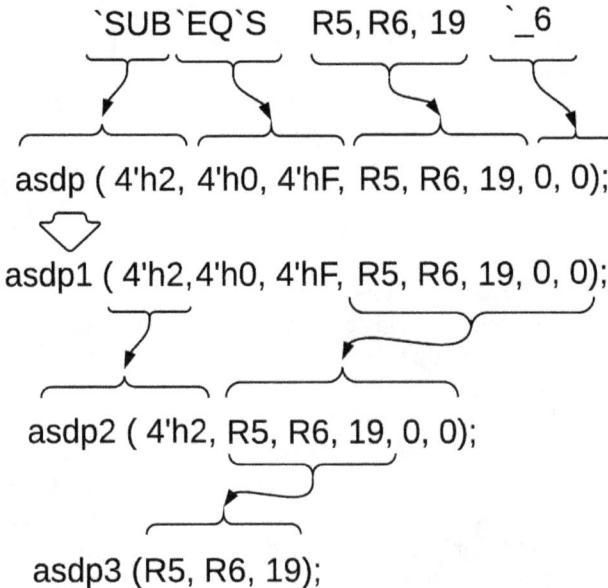

Figure 11.8: Arguments passed from asdp task down through asdp3 function

```
156.  // Function asdp2 constructs the instruction's operand fields for function asdp1.
157.      // `SUB R1,7 Format 1: Rd = Rd - constant
158.      // `SUB R1,R2 Format 2: Rd = Rd - Rm
159.      // `SUB R1,R2,7 Format 3: Rd = Rn - constant
160.      // `SUB R1,R2,R3 Format 4: Rd = Rn - Rm
161.
162.      function [31:0] asdp2 ();
163.          input [15:0] opCode, Q1, Q2, Q3, Q4, Q5;
164.          if (opCode[3:2]=='b10)                     // TST, TEQ, CMP, CMN
165.              asdp2 = opCode<<21 | 1<<20 | asdp3(0, Q1, Q2);
166.          else
167.              if (Q3==0)
168.                  if (Q2<R0 | opCode==13)            // R1,7 or mov R1,7
169.                      asdp2 = opCode<<21 | asdp3(Q1, 0, Q2);
170.                  else                               // R1,R2
171.                      asdp2 = opCode<<21 | asdp3(Q1, Q1, Q2);
172.              else                                   // R1, R2, R3 or 7
173.                  asdp2 = opCode<<21 | asdp3(Q1, Q2, Q3);
174.      endfunction
175.
176.  // Function asdp3 fills in the Rd, Rn, and 12-bit second operand field Rm/constant.
177.      // Note: This function will be more complicated in Chapter 12 with "shifts"
178.
179.      function [31:0] asdp3 ();
180.          input [15:0] Rd,Rn,Rm;
181.          if (Rm<R0)                                 // Rm is a constant?
182.              asdp3 = 1<<25 | Rd[3:0]<<12 | Rn[3:0]<<16 | Rm[11:0];
183.          else
184.              asdp3 = Rd[3:0]<<12 | Rn[3:0]<<16 | Rm[11:0];
185.      endfunction
```

Listing 11.4 E: Functions asdp2 and asdp3 fill in most parts of the 32-bit ARM machine code.

Sixteen ARM Data Processing Instructions

The data processing opcode field, being four bits, can support sixteen different opcodes. I started the ARM imitation in Chapter 7 with only nine of these sixteen possible codes, bypassing those operations specifically accessing the status bits.

Now that the remaining seven codes are being included, along with the fields for testing and setting of the status bits, the macros and asdp task are producing machine code instructions that exactly match those of the ARM processor. The purpose of the code in Listing 11.4 is to study that 32-bit format, including instructions using the status bits.

```
187.    reg [31:0] progMem[0:25];        // 32 bits per instruction
188.    reg [31:0] IR;                   // Instruction Register
189.    assign instr = IR;
190.
191.    always @ (posedge(clk))
192.        IR <= progMem[address];
193.    always @ (posedge(reset))
194.    begin
195.        IP = 0;
196.        `MOV    R1,'b0101   `_3       // 0: E3A01005 Move 0b0101 into R1
197.        `MOV    R2,'b0011   `_3       // 1: E3A02003 Move 0b0011 into R2
198.        `BIC    R11,R1,R2   `_4       // 2: E1C1B002 Not(R2) & (R1) => R11
199.        `BIC    R12,R2,R1   `_4       // 3: E1C2C001 Not(R1) & (R2) => R12
200.        `ORR R3,R11,R12     `_4       // 4: E18B300C (R11) | (R12) => R3
201.        `EOR    R4,R1,R2    `_4       // 5: E0214002 Exclusive OR instruction
202.        `SUB    R5,R6,7     `_4       // 6: E2465007 Always subtract, no status
203.        `SUB`S  R5,R6,7     `_5       // 7: E2565007 Always subtract, with status
204.        `SUB`EQ  R5,R6,7 `_5          // 8: 02465007 If Z-flag, subtract, no status
205.        `SUB`EQ`S  R5,R6,7 `_6        // 9: 02565007 If Z-flag, subtract, with status
206.        `MOV    R5,25    `_3          // A: E3A05019 Rd = constant
207.        `MOV    R5,R6    `_3          // B: E1A05006 Rd = Rm
208.        `ADD    R5,R6,25    `_4       // C: E2865019 Rd = Rn + constant
209.        `ADD    R5,R6,R9    `_4       // D: E0865009 Rd = Rn + Rm
210.        `TST    R1,'b0101   `_3       // E: E3110005 Test 0b0101 into R1
211.        `CMP    R2,25    `_3          // F: E3520019 Compare
212.        `ADC    R11,R1    `_3         // 10: E0ABB001 Add with carry
213.        `SBC`EQ   R12,R2    `_4       // 11: 00CCC002 Subtract with carry
214.        progMem[25] <= 0;
215.    end
216. endmodule
```

Listing 11.4 F: Create instructions containing status bits for testing

Lines 196 through 213 of Listing 11.4 F contain a list of 32-bit ARM instructions. They do not form a program, but are provided as a demonstration of the machine code format and the assembly language features. I recommend compiling Listing 11.4, downloading it, and examining each of the 32-bit instructions that are generated.

1. Use the same sequence of setting the address on switches SW[4:0], followed by pushing KEY[1] as was done before (see Figure 11.4).
2. The comment section on each line of the above assembly code contains the 32-bit machine code in hexadecimal. It should match what is seen at each address while testing on the DE10-Lite or DE2-115.
3. Notice how the `_3 macro completes an assembly line that has three

elements (opcode plus two more). Likewise, ` _4 completes assembly text having four parts. This "ending" for assembly language coding will be used in the remainder of this book to overcome the restriction that Verilog tasks cannot have omitted arguments.

Table 11.6 summarizes all 16 data processing instructions. Three of the new opcodes, ADC, SBC, and RSC, use the status bits in their calculations. All sixteen instructions can set the status bits if the S flag bit is set in the instruction.

Op Code Value	Op Code Name	Arithmetic or Logical Operation
0	AND	R <= R & op2;
1	EOR Exclusive OR	R <= R ^ op2;
2	SUB	R <= R - op2;
3	RSB Reverse Subtract	R <= op2 - R;
4	ADD	R <= R + op2;
5	ADC	R <= R + op2 + C;
6	SBC	R <= R - op2 + C - 1;
7	RSC	R <= op2 - R + C - 1;
8	TST	Set status for R & op2
9	TEQ	Set status for R ^ op2
10	CMP	Set status for R - op2
11	CMN	Set status for R + op2
12	ORR inclusive OR	R <= R | op2;
13	MOV Move (i.e., copy)	R <= op2;
14	BIC Bit Clear	R <= R & ~op2;
15	MVN Move NOT	R <= ~op2;

Table 11.6: All 16 ARM data processing instructions

ADC, SBC, and RSC

The ADC, SBC, and RSC are the same operations as the ADD, SUB, and RSB instructions, except they include the "carry" status bit generated in a previous instruction. These instructions have traditionally been used for performing high precision arithmetic on very large numbers. For example, 64-bit double precision can be performed by adding the low-order 32 bits of each number using the ADDS (carry bit must be generated). The high-order 32 bits are then added using the ADC instruction. Using this procedure, addition of numbers having any multiple of 32-bits can be performed.

- **ADC:** Add an integer plus Carry
- **SBC:** Subtract an integer including Carry
- **RSC:** Reverse Subtract including Carry

How does this double and quad high precision compare to floating point format? As described in Chapter 22, floating point is similar to scientific notation which contains a base value and an exponent. It is used to represent very large as well as very small numbers. In a way, floating point numbers have a more analog nature to them than digital nature.

Although floating point would be used in the vast majority of scientific and engineering applications, it has the following three weaknesses that prohibit its usefulness in some special scientific applications.

1. The floating point format does not have enough precision. Even double precision floating point has a fixed number of bits in its significant (a.k.a, mantissa) needed to represent a very precise measurement.
2. Every floating point operation that is performed in a chain of calculations may decrease the precision of the final result.
3. Floating point arithmetic is much slower that integer arithmetic.

TST, TEQ, CMP, CMN

The TST, TEQ, CMP, and CMN instructions are very similar to the AND, XOR, SUB, and ADD instructions in setting the status bits, but they differ in that no data registers are modified:

- **TST:** Test for specific bits present (like AND)
- **TEQ:** Test for all bits same value (like XOR)
- **CMP:** Compare two values (like SUB)
- **CMN:** Compare Register to Negative value (like ADD)

The TST instruction is handy for testing whether a single bit is set or clear. It can also be used to test for any of a combination of bits being set or for all of the selected bits being clear. What about the other three instructions that appear to test if two 32-bit values as a whole are equal? How are they different, and which should be used? This question appears at the end of this chapter, but a clue is obtained by comparing XOR to SUB to ADD along with information on two's complement described in Appendix C.

The purpose of these four instructions is to set the status bits. An interesting note is that some assemblers will flag CMPS or TEQS as an error because the S status bit is implied by the opcode itself. The assembler developed in this book is not that sophisticated, so both `CMP and `CMP`S produce the same instruction (the S bit is set, but no error is noted). For consistency sake, I recommend not putting the `S on the TST, TEQ, CMP, and CMN instructions.

Modules in This ARM Imitation With Status

The text files containing the Verilog source code are starting to get rather large. Typically, an application would keep these modules in separate files within the project that can be added or included. In this book, the architecture and testing procedure is evolving from a very simple format to one that is much more complex. It would be very easy to get the wrong version of modules, so I keep everything for a particular demonstration or test "locked together" in a single file. For production work, please don't do this, except maybe in the very early prototyping phase of a project.

Listing 11.5 provides the Verilog code to execute the 32-bit machine code for all sixteen data processing instructions, including support for the setting and testing of the status bits. It does not fully test the new features, but simply reaffirms that the previous capability is still present. The same XOR assembly language example as in Listing 10.2 is used, and it will operate the same way "from a user's perspective." However, internally, it will be using the 32-bit ARM machine code format. The following five Verilog modules are included in Listing 11.5:

1. **CPU_UI:** The User Interface is the same as in Listing 10.2. Due to an upgrade in the CPU module, the LEDs will now display both the PC and the N, Z, C, and V status bits.
2. **CPU:** The registers in the CPU module are increased to 32 bits. The Current Program Status Register (CPSR) and N, Z, C, and V status bits are created. The instruction cycle checks the status bits before completing each instruction.
3. **DataProcIns:** The ADC, SBC, RSC, TST, TEQ, CMP, and CMN operations are included to finish the set of 16 data processing instructions. Status bits can be generated for every instruction.
4. **Macros:** This set of macros is the same as in Listing 11.4 and supports the assembly language for all 16 data processing instructions, including status bit setting and testing.

5. **ProgMod:** The XOR assembly language program is the same as in Listings 11.3 and 10.2.

```
1.   // Listing 11.5 demonstrates XOR program using 32-bit registers
2.   // 1) Assembles 32-bit ARM Data Processing instructions
3.   // 2) Executes 32-bit "XOR" example program
4.   // 3) Dumps contents of 6 registers in hexadecimal
5.   //    (lower 4 bits of R1, R2, R11, R12, R3, R4)
6.   //
7.   // Modules and macros contained in this file:
8.   // 1) CPU_UI: User Interface that dumps 32-bit words from program memory
9.   // 2) CPU: 32-bit "ARM like" CPU with all 16 data processing instructions
10.  // 3) DataProcIns: 16 "ARM like" DP instructions using 32-bit registers
11.  // 4) Macros for assembling 16 ARM data processing instructions
12.  // 5) ProgMod: Memory containing parameters and functions from
13.  //    Listing 11.4 and also has the XOR demonstration from Listing 10.2
```

Listing 11.5 A: Directory header for 32-bit XOR test program (File: Listing_11-5.txt).

```
385.   always @ (posedge(reset))
386.   begin
387.     IP = 0;
388.     `MOV   R1,'b0101   `_3      // 0: E3A01005 Move 0b0101 into R1
389.     `MOV   R2,'b0011   `_3      // 1: E3A02003 Move 0b0011 into R2
390.     `BIC   R11,R1,R2   `_4      // 2: E1C1B002 Not(R2) & (R1) => R11
391.     `BIC   R12,R2,R1   `_4      // 3: E1C2C001 Not(R1) & (R2) => R12
392.     `ORR   R3,R11,R12  `_4      // 4: E18B300C (R11) | (R12) => R3
393.     `EOR   R4,R1,R2    `_4      // 5: E0214002 Exclusive OR instruction
394.     progMem[25] = 0;
395.   end
```

Listing 11.5 B: XOR assembly language program in module ProgMod

The CPU module shown in Listing 11.5 C has been upgraded from the version in Listing 10.2. The inclusion of the CPSR register and increasing the size of the general purpose registers R0 through R15 to 32 bits account for most of the changes.

- Line 70: The lower 6 bits on the LED display will be the memory address of the ARM instruction that will be executed next.
- Line 71: The upper 4 bits of the LED display will contain the Current Program Status Register (CPSR) bits N, Z, C, and V.
- Line 75: This "run" variable is the same as in Listing 10.2. It allows the program to be executed one instruction at a time.
- Line 89: The SU flag bit indicates if the status bits will be updated by the current instruction.

- Line 90: Every instruction now has a 4-bit field indicating which status bits must be set or clear for the instruction to execute.
- Lines 92 - 94: Register size increased to 32 bits.

```
56.  //
57.  //---------------- "ARM" CPU imitation ----------------
58.  //
59.  module CPU (KEY, reset, clk, LEDR, hexDisp);
60.    input [1:0] KEY;
61.    input reset, clk;
62.    output [9:0] LEDR;
63.    output [23:0] hexDisp;
64.    assign hexDisp[3:0] = R[1];
65.    assign hexDisp[7:4] = R[2];
66.    assign hexDisp[11:8] = R[11];
67.    assign hexDisp[15:12] = R[12];
68.    assign hexDisp[19:16] = R[3];
69.    assign hexDisp[23:20] = R[4];
70.    assign LEDR[5:0] = PCir[5:0];
71.    assign LEDR[9:6] = CPSR[31:28];
72.
73.    // Clock, reset, and breakpoint
74.
75.    reg run;                      // Flag indicating CPU is running
76.
77.    // CPU instruction cycle
78.
79.    reg [1:0] CPU_state;
80.    parameter fetch = 2'b01;
81.    parameter decode = 2'b10;
82.    parameter execute = 2'b11;
83.    parameter writeBack = 2'b00;
84.
85.    // Instruction format
86.
87.    wire [3:0] opCode;            // Operation code value
88.    wire iFlag;                   // Immediate data flag
89.    wire SU;                      // Update status flag
90.    wire [3:0] cond;              // Condition code in instruction.
91.    wire [11:0] op2;              // Second operand raw code
92.    wire [31:0] RdVal_DP;         // Destination register value
93.    reg [31:0] RnVal;            // Source data register value
94.    reg [31:0] RmVal;            // 2nd op data register value
95.    wire [3:0] RdID;              // Destination register number
96.    wire [3:0] RnID;              // Source register number
97.    wire [3:0] RmID;              // Second operand register number
```

Listing 11.5 C: Variables and assigns for 32-bit CPU imitation

```
98.   wire [31:0] IR;                          // Instruction register
99.   assign cond = IR[31:28];                 // Conditions needed to execute
100.  assign iFlag = IR[25];                   // Immediate data flag
101.  assign opCode = IR[24:21];               // Opcode is in upper 4 bits
102.  assign SU = IR[20];                      // Update status flag
103.  assign RnID = IR[19:16];                 // Source data register
104.  assign RdID = IR[15:12];                 // Destination register
105.  assign RmID = IR[3:0];                   // 2nd operand register
106.  assign op2 = IR[11:0];                   // Operand is in lower 12 bits
107.
108.  // General purpose registers R0 - R15
109.
110.  reg [31:0] R[0:15];                      // 16 registers
111.  parameter PC = 15;                       // Program Counter
112.
113.  // Current Program Status Register (CPSR)
114.
115.  reg [31:0] CPSR;
116.  wire [31:0] CPSR_DP;
117.  wire N_flag, Z_flag, C_flag, V_flag;
118.  assign N_flag = CPSR[31];                // Negative condition
119.  assign Z_flag = CPSR[30];                // Zero condition
120.  assign C_flag = CPSR[29];                // Carry condition
121.  assign V_flag = CPSR[28];                // Overflow condition
122.  reg ok2exe;
123.  reg [31:0] PCir;
124.  wire RdVal_DPU;
125.
126.  // Instantiate program memory and ALU processing
127.
128.  ProgMod (R[PC], CPU_state==fetch, reset, IR);
129.  DataProcIns (opCode, RnVal, RmVal, op2, iFlag, CPU_state==execute, CPSR,
      RdVal_DP, RdVal_DPU, CPSR_DP);
```

Listing 11.5 D: Variables and assigns for 32-bit CPU imitation

Listing 11.5 D shows new variables that were added for the processing of status bit information.

- Line 98: Instruction register is now 32-bits (same as in the ARM).
- Line 99: Condition of status bits needed for execution
- Line 102: Status should be updated if the S bit is set in the instruction.
- Line 110: There are now sixteen 32-bit general purpose registers
- Line 111: Register R15 is the PC in the 32-bit ARM architecture.
- Line 115: "Current" Program Status Register (value before instruction executes)
- Line 116: New CPSR value obtained after execution of instruction in the DataProcIns module.
- Lines 117 - 121: Flags within the CPSR are given the same position

as in a real 32-bit ARM CPSR

- Line 122: When ok2exe is true, the status bits in the CPSR match the requirements in the cond field of the current instruction.
- Line 123: The PCir variable contains the address of the current instruction that is ready to be executed. This variable is not needed for the CPU to run, but only included for the LED display.
- Line 124: The RdVal_DPU signals when the DataProcIns module has a new value for the Rd register. The TST, TEQ, CMP, and CMN instructions only set the status bits and do not update any of the general purpose registers R0 through R15.

```
130.
131.    // Instruction cycle and reset
132.
133.    always @ (posedge(clk), posedge(reset))
134.      begin
135.        if (reset)
136.          begin
137.            R[PC] <= 0;                    // Boot address is 0000
138.            CPU_state <= fetch;
139.          end
140.        else
141.          case (CPU_state)
142.            fetch:                         // Get next instruction in program
143.              begin
144.                PCir <= R[PC];
145.                R[PC] <= R[PC] + 1;
146.                CPU_state <= decode;
147.              end
148.            decode:                        // Disassemble the instruction
149.              begin
150.                RnVal <= R[RnID];          // Source data register
151.                RmVal <= R[RmID];          // 2nd op data register
152.                if (~KEY[0])               // Instruction cycle interlock
153.                  run <= 0;
154.                if (~KEY[1] && ~run)
155.                  begin
156.                    run <= 1;
```

Listing 11.5 E: Instruction cycle "fetch" state in 32-bit version

- Line 144: PCir is saved for LED display.
- Lines 152 - 156: This is the same interlock that enables stepping through a program one instruction at a time. In Listing 11.6, this will be changed to stop the program only at specific addresses or even allow it to run at full speed.

```
157.  case (cond)
158.     0: ok2exe <= Z_flag;               // EQual (zero); Z set
159.     1: ok2exe <= ~Z_flag;              // Not Equal (non-zero); Z clear
160.     2: ok2exe <= C_flag;               // Carry Set (Unsigned Higher or Same)
161.     3: ok2exe <= ~C_flag;              // Carry Clear (Unsigned LOwer)
162.     4: ok2exe <= N_flag;               // MInus or negative; N set
163.     5: ok2exe <= ~N_flag;              // PLus or positive; N clear
164.     6: ok2exe <= V_flag;               // Overflow; V Set
165.     7: ok2exe <= ~V_flag;              // No overflow; V Clear
166.     8: ok2exe <= C_flag & ~Z_flag;     // Unsigned HIgher; C set and Z clear
167.     9: ok2exe <= ~C_flag | Z_flag;     // Unsigned Lower or Same
168.     10: ok2exe <= N_flag == V_flag;    // Signed Greater than or Equal to
169.     11: ok2exe <= N_flag != V_flag;    // Signed Less Than
170.     12: ok2exe <=                      // Signed Greater Than
                ~Z_flag & (N_flag==V_flag);
171.     13: ok2exe <=                      // Signed Less than or Equal
                Z_flag | (N_flag!=V_flag);
172.     14: ok2exe <= 1;                   // ALways; any status bits ok2exe
173.  endcase
```

Listing 11.5 F: Status bit conditions implemented from Table 11.4

```
174.                    CPU_state <= execute;
175.              end
176.          end
177.      execute:                   // Perform desired operation
178.          CPU_state <= ok2exe ? writeBack : fetch;
179.      writeBack:                 // Update specific register
180.          begin
181.              if (RdVal_DPU)<= R[RdID] RdVal_DP;   // Destination register
182.              if (SU) CPSR <= CPSR_DP;
183.              CPU_state <= fetch;
184.          end
185.      endcase
186.  end
187.  endmodule
```

Listing 11.5 G: Execute and Write Back states of the 32-bit instruction cycle

- Lines 157 - 173: Variable "ok2exe" will be set to true (i.e., a value of 1), if the current status bits match what is required in the condition field of the current instruction.
- Line 178: By skipping over the write back state, the Verilog conditional operator "ignores" instructions based on the condition code and status bits.

- Line 181: Some instructions (ADD, ORR, ...) update a data register, while others (CMP, TST, ...) do not.
- Line 182: The CPSR only gets updated if the S bit is set in the instruction.

```
189.  //
190.  //--------------- ALU for ARM data processing instructions ---------------
191.  //
192.  module DataProcIns (opCode, Rn, Rm, op2raw, iFlag, clk, CPSR, Rd,
      RdVal_DPU, CPSR_DP);
193.    input [3:0] opCode;            // Data processing instruction opcode
194.    input [31:0] Rn;              // Source register contents
195.    input [31:0] Rm;             // Possible 2nd operand register contents
196.    input [11:0] op2raw;         // Second operand "as is"
197.    input iFlag;                // Immediate value flag
198.    input clk;                 // Pulse to produce calculation
199.    output reg [31:0] Rd;       // Value to return
200.
201.    // Current Program Status Register (CPSR)
202.
203.    input [31:0] CPSR;
204.    output [31:0] CPSR_DP;
205.    output RdVal_DPU;
206.    reg C;
207.    wire N, Z, V, CI;
208.    assign CI = CPSR[29];          // Carry (In) condition
209.    assign CPSR_DP[31] = N;        // Negative condition
210.    assign CPSR_DP[30] = Z;        // Zero condition
211.    assign CPSR_DP[29] = C;        // Carry (Out) condition
212.    assign CPSR_DP[28] = V;        // Overflow condition
213.    reg RdVal_DP;
214.
215.    wire [31:0] op2;               // Calculated value of second operand
216.    assign op2 = (iFlag) ? op2raw : Rm;
217.    assign RdVal_DPU = opCode < 8 | opCode > 11;
218.    assign N = Rd[31];
219.    assign Z = (Rd) ? 0 : 1;
```

Listing 11.5 H: 32-bit Arithmetic Logic Unit for data processing instructions

The data processing instructions both use and set the status bits.

- Line 208: The "carry in" (generated in a previous instruction)is used by the ADC, SBC, and RSC instructions.
- Lines 209 - 212: The N, Z, C and V status bits will be set in this module, but whether these new status bits update the CPSR will be determined in the CPU module which decides based on the presence of the S bit in the current instruction.

- Lines 223 - 238: Adding or subtracting two 32-bit numbers generates a 33 bit result. Using the { } concatenation operator, this 33-bit result is divided between the output carry status bit in C and the 32-bit output value in register Rd.

```
220.    assign V = Rd[31] ^ Rd[30];
221.    always @ (posedge(clk))
222.      case (opCode)
223.        0: {C,Rd} <= Rn & op2;           // AND
224.        1: {C,Rd} <= Rn ^ op2;           // EOR (exclusive OR)
225.        2: {C,Rd} <= Rn - op2;           // SUB
226.        3: {C,Rd} <= op2 - Rn;           // RSB (reverse subtract)
227.        4: {C,Rd} <= Rn + op2;           // ADD
228.        5: {C,Rd} <= Rn + op2 + CI;      // ADC Add with carry
229.        6: {C,Rd} <= Rn - op2 - CI;      // SBC Subtract with carry
230.        7: {C,Rd} <= op2 - Rn - CI;      // RSC (reverse SUB with carry)
231.        8: {C,Rd} <= Rn & op2;           // TST Test (like AND)
232.        9: {C,Rd} <= Rn ^ op2;           // TEQ Test Equal (like EOR)
233.        10: {C,Rd} <= Rn - op2;          // CMP Compare (like SUB)
234.        11: {C,Rd} <= Rn + op2;          // CMN Compare Negative (like ADD)
235.        12: {C,Rd} <= Rn | op2;          // ORR (inclusive OR)
236.        13: {C,Rd} <= op2;               // MOV
237.        14: {C,Rd} <= Rn & ~op2;         // BIC (bit clear)
238.        15: {C,Rd} <= ~op2;              // MVN (move NOT)
239.      endcase
240.    endmodule
```

Listing 11.5 I: Data processing instructions setting the C carry status bit.

Compile, download, and test the CPU imitation in Listing 11.5. It contains the same XOR program from Chapter 10. It uses none of the special features that are added for handing status information, but I used it as a first test of 32-bit processing simply to demonstrate "upward compatibility." Computer architecture designers have always felt strong pressure from the marketplace to maintain as much compatibility with previous designs as possible.

The User Interface is the same as in Listing 10.2: It will display the lower four bits of registers R1, R2, R11, R12, R3, and R4 on 7-segment displays. The LEDs will now display both the PC and the N, Z, C, and V status bits. KEYs 0 and 1 are used to initialize the CPU and step through the program just like it was while testing Listing 10.2.

Practical Example: Factorial Calculation

The factorial of a whole number is calculated as the product of that number and all the non-zero whole numbers below it. For example, six factorial is 6 ! = 6 × 5 × 4 × 3 × 2 × 1 (product equals 720). Twelve factorial is the largest that can be calculated using 32-bit registers (12 ! = 479,001,600 or 32h'1C8CFC00).

I will be using the factorial example here and also in the next few chapters because it will demonstrate many of the features of a running CPU program. The factorial will be calculated by a loop that starts with the "number" and then decrements that number on each pass through the loop until 1 is reached. A variety of instructions will be used, thereby demonstrating many features of an operating CPU.

Program Loops

Computers are great for doing repetitive operations. A loop is a "process" that can be performed multiple times until a "decision" is made to move onto something else. Examples of processes and decisions:

- Process: Eating one mouthful of food at lunch
- Decision: Is there any more food on my plate?

- Process: Grading one student's exam
- Decision: Are there any more exams to grade?

A program loop consists of three parts:

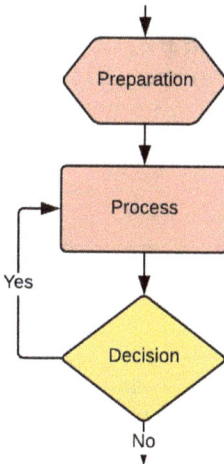

Figure 11.9: Program loop

1. Preparation: Set initial values for a) variables to be modified during each pass of the loop and b) variables, like counters, that will determine when to exit the loop.
2. A process to be repeated multiple times: Examples include adding numbers to a running total, searching a table for a particular value, and calling the same set of functions multiple times.
3. Decision when to exit the loop: Most loops have an exit objective such as all of the desired numbers have been added, the entire table has been searched, the desired value has been located, etc.

We are familiar with loops from the Verilog *for* construct which builds a physical subcircuit for each iteration through the loop. Now we want a loop such that each "time" we pass through it, the product is multiplied by the next lower integer. The following one line of C++ coding would calculate the factorial.

- for (N = FACT - 1; N >= 2; N = N - 1)   FACT = FACT * N;

The loop we are creating has the following parts:

1. Preparation: The running product will be in a variable named FACT. Variable N will be the count-down variable to determine when to exit the loop
2. Process to be repeated multiple times: Each time through the loop, the running product will be multiplied by the next lower integer value. The value will also be decremented for the next pass.
3. Decision when to exit the loop: I've chosen to exit the loop after multiplying by two. From a computational viewpoint, there is no advantage of multiplying by 1.

Branch, Jump, Goto

There's not much point in having status codes if we can't use their values to change program flow. Almost all CPUs have instructions that alter the PC to skip over some code or to go to a different part of the program. In the ARM, these are referred to as branch instructions, while in many CPUs they are called jumps. In many higher level programming languages, they're called Gotos.

I like spaghetti, but only on my dinner plate, not in program code. Many programming languages, such as C and Java, do not even have a Goto statement because many novice programmers used too many in their programs. Their programs were as tangled as a plate of spaghetti and were almost impossible to maintain.

```
1. //      Factorial of 6 = 6 ! = 6*5*4*3*2*1
2.         FACT = 6                 // Factorial of 6 = 6 ! = 6*5*4*3*2*1
3.         N = FACT - 1             // Initilize loop counter (multiplier)
4. FLoop:  FACT = FACT * N          // On each pass, multiply by one lower
5.         N = N - 1                // Decrement multiplier for next pass
6.         if (N >= 2) go to FLoop  // Go back until multiplier = 2

1. //      Factorial of 6 = 6 ! = 6*5*4*3*2*1
2.         R0 = 6                   // R0 will hold "working" factorial
3.         R1 = R0 - 1              // Initilize loop counter (multiplier)
4. FLoop:  R0 = R0 * R1             // On each pass, multiply by one lower
5.         R1 = R1 - 1              // Decrement multiplier for next pass
6.         if (R1 >= 2) go to FLoop // Go back until multiplier = 2
```

Figure 11.10: Basic program to generate factorial

I've unfolded the above C++ *for* statement for calculating the factorial to the five line "basic" program in Figure 11.10. I used the Basic programming language for the example because it has a "go to" statement which is very similar to the branch instruction in assembly language and machine code.

There are two copies of the same program which differ only by their variable names. The top one uses names appropriate from an applications viewpoint, while the bottom copy uses variables that match the register names in the ARM assembly language in Figure 11.11.

1. @		Calculate factorials using ARM assembly language coding
2.		
3.	MOV	R0,#6	@ Factorial of 6 = 6 ! = 6*5*4*3*2*1
4.	SUB	R1,R0,#1	@ Initialize R1 as multiplier (loop counter)
5. FLoop:	MUL	R0,R1	@ On each pass, multiply by one lower
6.	SUB	R1,#1	@ Decrement multiplier for next pass
7.	CMP	R1,#2	@ Compare multiplier to loop ending condition
8.	BGE	FLoop	@ Go back until multiplier = 2

Figure 11.11: ARM assembly language to calculate factorial

The above ARM assembly language code will calculate factorials for any integer between 2 and 12, and we will use it "as is," but not yet because of the following two missing "ingredients":

1. No multiply instruction: We won't get the multiply instruction on line 5 until Chapter 13. Since this example is about multiplying, that's a somewhat important missing piece. Not having a multiply instruction has been a problem solved by many programmers in the past. We will just make another loop: one that calculates multiplication by repeated addition.
2. No branch instruction: We won't get the branch instruction on line 8 until Chapter 15. In the 32-bit ARM processor, register R15 is the PC. From a programming style perspective, I don't like manipulating the PC with MOV instructions, but for now we have no alternative.

Nested Loops

Actually, multiplication is a quick form of multiple additions. A product is generated by a number (known as the multiplicand) that is to be added to itself a certain number of times (the multiplier). In an example such as 6 × 5, the multiplicand can be 6 and the multiplier is 5 resulting in 6 + 6 + 6 + 6 + 6 = 30.

A very common programming technique is one loop nested within another. Each loop will have its own exit condition.

- The outer loop will be similar to the previous code in Figure 11.11: It will calculate the factorial of 6 by doing $6 \times 5 \times 4 \times 3 \times 2$.
- The inner loop will actually do the multiplication by repeated addition: For example, $6 \times 5 = 6 + 6 + 6 + 6 + 6$.

In assembly language, a single register typically controls each loop. In this example, ARM registers R1 and R3 control each pass through the loops:

- The outer loop: Register R1 counts down by 1 on each pass through the loop calculating the factorial.
- The inner loop: Register R3 counts down by 1 on each pass through the loop calculating a product.

Although nested loops are a powerful technique, it's very easy to write large nested loops with confusing code where one loop's data and counters interfere with that of the other.

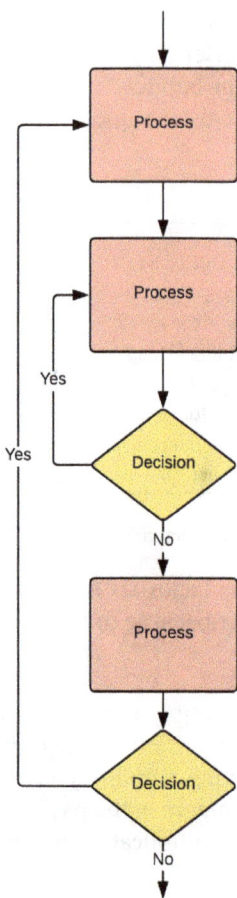

Figure 11.12: Nested loops

Figure 11.13 shows the ARM assembly language that will calculate the factorial using nested addition loops. The branches on lines 9 and 12 will be converted to MOV instructions for now.

Listing 11.6 contains the ARM factorial program as well as a new user interface that allows full speed execution, breakpoints, and the ability to examine all 32 bits of any general purpose register.

1. **CPU_UI:** The User Interface has been changed to allow dumping of any 32-bit R register and the setting of a breakpoint address.
2. **CPU:** The CPU execution cycle is modified to allow breakpoints to pause a program's execution.

Computer Architecture Tutorial Using an FPGA

3. **DataProcIns:** Same as previous example in Listing 11.5.
4. **Macros:** Same as previous example in Listing 11.5.
5. **ProgMod:** Contains the factorial program shown in Listing 11.6.

1. @		Program to calculate factorial using nested iterative summation loops.
2.		
3.	MOV	R0,#6	@ Factorial of 6 = 6 ! = 6*5*4*3*2*1
4.	SUB	R1,R0,#1	@ Initialize R1 as multiplier (loop counter)
5. FLoop:	MOV	R2,R0	@ On each pass, multiply by one lower.
6.	SUB	R3,R1,#1	@ Initialize counter for summation loop.
7. MLoop:	ADD	R0,R2	@ On each pass, add multiplicand to total.
8.	SUBS	R3,#1	@ Decrement number of additions to do.
9.	BGT	MLoop	@ Go back until "multiply" is completed.
10.	SUB	R1,#1	@ Decrement multiplier for next pass.
11.	CMP	R1,#2	@ Compare multiplier to loop ending condition.
12.	BGE	FLoop	@ Go back until multiplier = 2

Figure 11.13: Nested loop in ARM assembly language

Breakpoints

The argument list to the CPU module has been expanded to include a breakpoint address (PCbp), a register ID number (dmpID), and a 32-bit variable (hexDisp) to display the contents of any one of the 16 general purpose R registers. Line 66 in Listing 11.6 A shows how the register contents are obtained.

```
53.  CPU (KEY, reset, CLOCK_50, SW[8:4], SW[3:0], hexDisp, LEDR);
54.  endmodule
55.
56.  //
57.  //--------------- "ARM" CPU imitation ----------------
58.  //
59.  module CPU (KEY, reset, clk, PCbp, dmpID, hexDisp, LEDR);
60.     input [1:0] KEY;
61.     input reset, clk;
62.     input [4:0] PCbp;
63.     input [3:0] dmpID;
64.     output [9:0] LEDR;
65.     output [31:0] hexDisp;
66.     assign hexDisp = R[dmpID];
67.     assign LEDR[5:0] = PCir[5:0];          // Address of current instruction
68.     assign LEDR[9:6] = CPSR[31:28];        // N, Z, C, V status bits
```

Listing 11.6 A: Module CPU argument list change for dumping a selected R register

The decode state has been enhanced on line 151 in Listing 11.6 B to run at full speed and only pause if the instruction fetch address is the same as the breakpoint address.

The && is a logical "and" and the || is a logical "or." They generate a value of either True or False, rather than & and | which do a bit-by-bit comparison of bit-array variables.

```
145.        decode:                          // Disassemble the instruction
146.          begin
147.            RnVal <= R[RnID];             // Source data register
148.            RmVal <= R[RmID];             // 2nd op data register
149.            if (~KEY[1])                  // Instruction cycle interlock
150.              run <= 0;
151.            if (~run &&
                  (~KEY[0] || PCir! = PCbp))  // Only stop at breakpoint
152.              begin
153.                run <= 1;
```

Listing 11.6 B: Breakpoint coding within the decode state of CPU instruction cycle

The user interface now operates as follows:

1. The ARM imitation will run at full speed until it comes to the breakpoint address.
2. While paused at the breakpoint address, any of the registers can be viewed by setting SW[3:0] to 0 to 15. A new breakpoint address can even be set.
3. Program execution can be continued simply by pushing KEY[0] followed by KEY[1] which will run the program until the breakpoint address is reached.
4. SW[9] selects whether the upper or lower 16 bits is displayed.

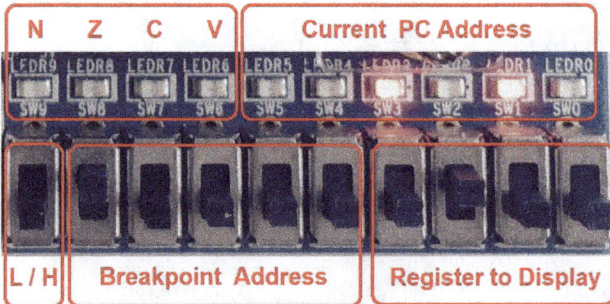

Figure 11.14: User interface with breakpoint address and 32-bit register display

```
386.
387.    // Program to calculate factorial using nested iterative summation loops.
388.
389.    `MOV    R0,0    `_3        // 0: Reset "boot" address
390.    `MOV    R0,12   `_3        // 1: 6 ! = 6*5*4*3*2*1 = 720
391.    `SUB    R1,R0,1  `_4       // 2: Initialize R1 as multiplier (loop counter).
392. FLoop = IP;                   // Outer loop: Calculate the factorial
393.    `MOV    R2,R0    `_3        // 3: On each pass, multiply by one lower.
394.    `SUB    R3,R1,1  `_4        // 4: Initialize counter for summation loop.
395. MLoop = IP;                    // Inner loop: Multiply using addition
396.    `ADD    R0,R2    `_3        // 5: On each pass, add multiplicand to total.
397.    `SUB`S  R3,1     `_4        // 6: Decrement number of additions to do.
398.    `MOV`GT  R15,MLoop  `_4  // 7: Go back until "multiply" is completed.
399.    `SUB    R1,1     `_3        // 8: Decrement multiplier for next pass.
400.    `CMP    R1,2     `_3        // 9: Compare multiplier to end condition.
401.    `MOV`GE  R15,FLoop  `_4   // 10: Go back until multiplier = 2.
402. ILoop = IP;                    // Infinite loop (stay in one place)
403.    `MOV    R15,ILoop  `_3      // 11: Keep setting PC to this location.
404.    progMem[30] = 0;
```

Listing 11.6 C: Program FACTORIAL listing

The code in Listing 11.6 will work with both the DE10-Lite and the DE2-115. The Listing 11.7 file is included on GitHub, and it has the user interface presenting all eight hexadecimal digits at the same time for DE2-115 users. It also has the initial value of R0 changed from 6 to 12 to calculate 12 !, which is 32'h1C8CFC00 (decimal value of 479,001,600).

Trace With Breakpoints

Using breakpoints to pause a program's execution to examine register contents is a common technique for testing and debugging software. Basically, set the breakpoint address, then allow the program to continue executing, and when it stops at the next breakpoint address, examine any or all register contents.

The following three examples provide ideas on how to use breakpoints to diagnose a running program.

Example 1: Just let the program run until it ends.

1. Set SW[8] high, and then start the program by pushing KEY[0] followed by KEY[1].
2. Since SW[8] is high, the breakpoint address will be at least 16, and the program will never stop because its highest address is 11.
3. The factorial will appear on the hex displays, and the "CPU" seems like it has stopped. However, I deliberately "stuck" it in an infinite loop, continuously branching back to the same address.

Example 2: Trace the top of the outer loop:

1. Set SW[8:4] to 00011 (address 3), which is the address of the FLoop label. Start the program by pushing KEY[0] followed by KEY[1].
2. When the program stops, examine the registers noted below, and then push KEY[0] followed by KEY[1] to run to the next pass through the loop.

What to look for on each pass through the outer loop:

1. R0: Set SW[3:0] to 0000. It contains $12 \times 11 \times 10 \times \ldots$
2. R1: Set SW[3:0] to 0001. Current multiplier and count down register changes on each pass: 12, 11, 10, ...
3. R15: Compare the PC address in R15 to the address on the LEDs. This difference will be discussed in Chapter 15.

Example 3: Trace the bottom of the inner loop:

1. Set SW[8:4] to 00111 (address 7), which is line 398 in Listing 11.6 C. Start the program by pushing KEY[0] followed by KEY[1].
2. When the program stops, examine the registers noted below, and then push KEY[0] followed by KEY[1] to step to the next pass through the loop.

What to look for on each pass at the bottom of the inner loop:

1. R0: This is the product as it's being built.
2. R3: The multiplier is decremented on each pass.
3. LEDR[9:6]: Status bits N, Z, C, V

Any Other Status Bits?

Sure. The status register has 32 bits. Status bits N, Z, C, and V appear in bit positions 31, 30, 29, and 28, respectively. The exact bit positions generally are not important to programmers because they use the mnemonics such as EQ and GT to test the bit status. The following status bits are also present in the CPSR, but they will not be discussed at this time.

- Q (bit 27): Underflow
- J (bit 24): Jazelle
- E (bit 9): Endian (Chapter 14)
- I (bit 7): Interrupt inhibit (Chapter 17)
- F (bit 6): Fast interrupt inhibit (Chapter 17)
- T (bit 5): Thumb (Chapter 19)
- M (bits 4 - 0): Privilege mode & register set

The two instructions MSR and MRS can be used to load and save the CPSR register contents to another general purpose R register, but they will not be described in this book.

Highlights and Comparisons

In this chapter, the ARM instruction format has been expanded to its full 32 bits, seven more data processing instructions have been included, and status bits are now being set and examined.

Chapter 11 contains seven listing files of Verilog code

1. User interface which dumps 32-bit numbers, four hex digits at a time
2. User interface which dumps 32-bit numbers, eight hex digits at a time
3. Run 32-bit version of XOR program
4. Update asdp to accommodate building instructions with status bit processing
5. Run XOR program in 32-bit format
6. Application program to calculate factorials
7. Factorial appplication for the DE2-115 having eight 7-sevgment displays.

The ARM architecture is very unique in which instructions modify and check the condition flags of the CPSR:

1. Each data processing instruction has the option of whether to change the values of the NZCF flags or leave them as they were. On almost every other CPU design, arithmetic and logic functions (add, sub, and, or, ...) always change the NZCV flags, and load/store instructions (mov, ldr, ...) never change the NZCV flags.
2. Almost every ARM instruction can be conditionally executed depending on the value in the NZCF flags. Almost all other CPUs only have branch (also known as jump) instructions that examine the flags.

Review Questions

1. *Sequential circuits described in a Verilog *always* block should use non-blocking assignments (i.e., use <= instead of =). Why does line 108 of Listing 11.3 have "IP = 0" instead of "IP <= 0"?
2. *The factorial program in Listing 11.6 C won't work on an actual ARM processor, exactly as it is written. What is wrong with how it substitutes MOV PC instructions for real branch instructions?
3. The ADC and ADDS instructions are used to perform double precision. When would the ADCS instruction be necessary?
4. Both the TEQ and CMP instruction can test whether two values are

equal. Which one sets the status bits for a following instruction to branch using GE or GT status?

5. Nested loops in a Verilog hardware description consume a large number of the gates available on the FPGA. Likewise, nested loops in software consume a lot of time. Why is the approach currently taken to calculate the factorial so much slower when it gets to larger numbers? Why is that not so much of a problem when using the real multiply instruction? See problem 5 below and also Chapter 13.

Exercises

1. Rewrite the asdp1 function so that the S and condition fields are reversed. For example, allow `ADD`S`EQ instead of `ADD`EQ`S. See Table 11.5, Figure 11.7, Figure 11.8, and Listing 11.4 D.

2. As in the above exercise, rewrite the asdp1 function to allow both orders of S and EQ. However, this probably is not a good idea. Too much flexibility can lead to problems as witnessed on the Internet with HTML where too much forgiveness led to very sloppy application code.

3. Many applications of factorials have one factorial divided by another. Not only do we not have a multiply instruction in the ARM imitation at the moment, there is not even a divide instruction in the 32-bit ARM architecture. This is easily solved as in the example 7 ! / 4 ! = 7 × 6 × 5. Modify the factorial assembly language program so that a different ending condition (other than 2) is used. Notes: Larger values such as 15 ! / 13 ! can also be accommodated. Don't just change the ending with a "hard coded" constant in the compare instruction, but put the ending value in a register at the top of the program.

4. The factorial stops at 12 ! because that is the largest value that will fit in 32 bits. However, by using the carry status bit, we can go much higher. Modify the factorial assembly language program to use double precision. You will need the following modifications:
   a. Two registers will be needed to hold the running total.
   b. The ADDS instruction will be needed to add the lower 32 bits of the double register.
   c. The ADC instruction will be used to add the upper 32 bits.

5. Review question 5 (above) states that this nested loop approach consumes much more time as we get to larger numbers than if we used a multiply instruction. Quickly modify the code to include another register that counts how many times we pass through the inner loop. Run the program for a few values such as 4 !, 6 !, 8 !, 10!, and 12!, and then plot a curve. You may want to do a logarithmic plot.

— 12 —
Shift Instructions

Shifting bits within a register is used extensively in digital applications and has been available on almost every CPU including the first microprocessor, the Intel 4004. Serial to parallel conversion is commonly done by shifting, and multiplication (or division) by a power of two is done by shifting bits within a register. Sometimes bits simply need to be repositioned.

Shifting is included in the set of ARM data processing instructions. Actually, it is included in every ARM data processing instruction. Most CPUs have individual instructions that "just shift," but not the ARM, which combines shifting with ADD, XOR, TST, etc.

Chapter 12 provides a thorough description of the second operand of ARM data processing instructions. It is composed of either two registers, two constants, or a register with a constant, but always contains a shift operation.

Introductions

The following four shift "instructions" are included in each of the ARM data processing instructions. When they appear "by themselves" in assembly language, they are actually part of a MOV instruction.

- **LSL:** Logical Shift Left
- **LSR:** Logical Shift Right
- **ASR:** Algebraic Shift Right
- **ROR:** Rotate (circular) shift Right

Bit Shift Operations

Almost all CPU architectures support three types of shifts:

- **Logical:** Bits shifted out from either end of the register are discarded, and new zero bits fill in on the opposite end.
- **Circular (also referred to as rotate):** Bits shifted out one end of the register come back in on the other end.
- **Arithmetic:** Similar to a logical right shift, except the arithmetic shift brings in copies of the sign bit instead of zero.

The number of bits to shift can be indicated either from the contents of a register or an immediate value in the instruction.

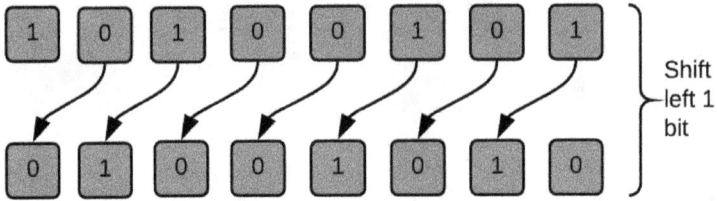

Figure 12.1: Logical left shift loses bit on left and brings in zero on right.

Logical shifts have many applications. Two common applications are for converting between serial and parallel, and for multiplying an integer by a power of two. For example, a one bit shift to the left is multiplying by two, while a two bit shift is multiplying by four.

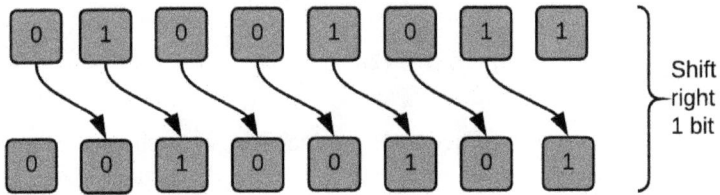

Figure 12.2: Logical right shift moves out bit on right and brings in zero on left.

In a circular shift, also referred to as a rotate, no "1 bits" are lost. Bits shifted out one end come back in on the opposite end. The ARM only provides a rotate to the right, but a rotate to the left can be done by a right rotate. For example, a left rotate of 5 bit positions is identical to rotating right 27 bit positions (register size of 32 minus 5 bit shift).

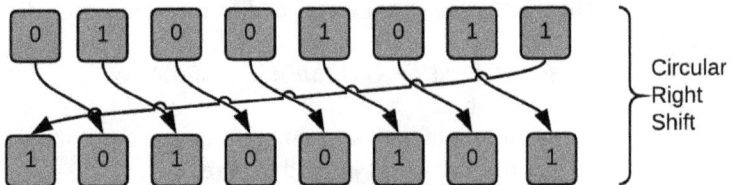

Figure 12.3: Rotate one bit position to the right — no bit lost

A shift to the right is like dividing by a power of two, but be aware of two basic problems. Division can have a remainder which will get truncated, not rounded. Also, a logical right shift will bring in a zero in bit 31, thereby converting a negative number to an inappropriate positive number. The arithmetic shift will solve the negative problem, but a rounding error is still present. Please see Appendix C if you need an explanation why the high-order bit (bit 31) is a "1" for negative numbers.

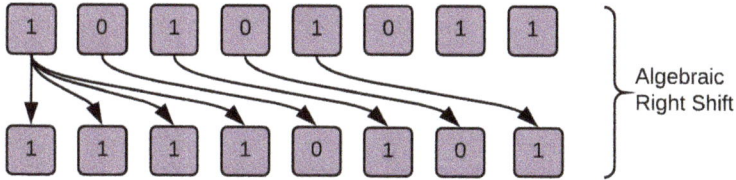

Figure 12.4: Algebraic shift copies the sign bit.

Shift Demonstration

Before building the shift into the ARM instruction format, let's build a test circuit demonstrating the behavior of the four shifts provided within the ARM instruction set.

- **LSL (2'b00):** Logical Shift Left
- **LSR (2'b01):** Logical Shift Right
- **ASR (2'b10):** Algebraic Shift Right
- **ROR (2'b11):** Rotate (circular) shift Right

The demonstration will shift the bit pattern located in the switches and display it on the LEDs. Note: The test code will shift an 8-bit register, while the code for the ARM will require a 32-bit register to be shifted. A parameter (REGSIZE) will enable the code to be easily changed from one register size to another. The following identifiers will be used in the Verilog coding of the ARM shift register operations. They correspond to the fields within the ARM instruction used to evaluate the second operand.

- shiftType (Switches 12, 11): Type of shift to be performed (2 bits)
- shiftCount (Switches 10...8): Number of bit positions to shift (3 bits, range is 0 to 7)
- RmVal (Switches 7...0): Original binary data pattern (8 bits)
- op2Val (LEDs 7...0): Bit pattern after shift (8 bits)
- clk (KEY[0]): Bits are shifted when the clock "ticks"

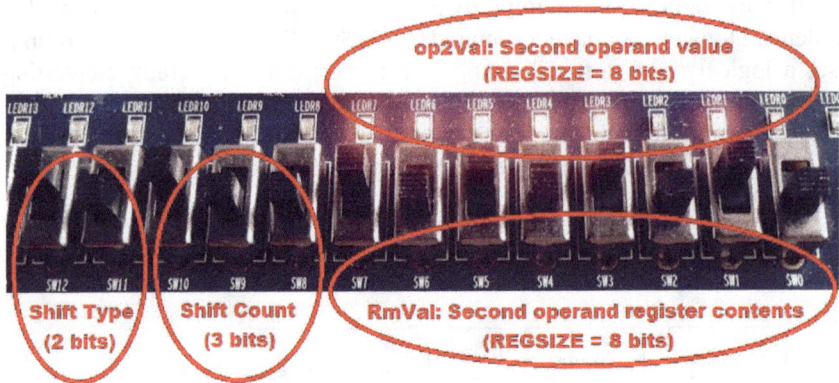

Figure 12.5: Test layout on the DE2-115 FPGA development board

```
1.   // Switches 12, 11: Type of shift to be performed
2.   // Switches 10...8: Number of bit positions to shift
3.   // Switches 7...0: Original binary data pattern
4.   // LEDs 7...0: Bit pattern after shift
5.
6.   module TopLevel (SW, KEY, LEDR);
7.      input [17:0] SW;
8.      input [1:0] KEY;
9.      output [17:0] LEDR;
10.     parameter REGSIZE = 8;              // Size of ARM 32-bit register
11.
12.     parameter LSL = 2'b00;              // Logical Shift Left
13.     parameter LSR = 2'b01;              // Logical Shift Right
14.     parameter ASR = 2'b10;              // Algebraic Shift
15.     parameter ROR = 2'b11;              // Rotate (circular) Right
16.
17.     reg [REGSIZE-1:0] op2Val;
18.     wire [REGSIZE-1:0] RmVal;
19.     wire [1:0] shiftType;
20.     wire [2:0] shiftCount;
21.     wire clk;
22.
23.     assign clk = ~KEY[0];
24.     assign shiftType = SW[12:11];
25.     assign shiftCount = SW[10:8];
26.     assign RmVal = SW[REGSIZE-1:0];
27.     assign LEDR[REGSIZE-1:0] = op2Val;
28.
```

Listing 12.1 A: Test setup for demonstrating bit shifting operations

Notes regarding Listing 12.1:

- Lines 1 - 4: Comments describing switch assignments for test
- Line 10: This code tests 8-bit registers. This parameter will be set to 32 when it is used for the "real" ARM 32-bit registers.
- Lines 12 - 15: ARM codes for each of its shifts
- Lines 24 - 27: Variables used in shifting code
- Line 29: Integer type (loop variable "i") is only used at "compile time." It does not represent a physical signal or register.
- Line 30: This is sequential code (i.e., Key[0] must be pushed to perform the shift and to update the LED display).
- Lines 31 - 47: Generate one block of code for each bit in the register (i.e., the total number of gates generated is proportional to the register size).
- Lines 42 - 46: There are three right shifts: logical (zero fills in on the left), algebraic (sign bit copied in on the left), and rotate (bit "lost" from right end fills in on the left).

```
29.    integer i;         // Loop counter
30.    always @ (posedge(clk))
31.    for (i=0;i<REGSIZE;i=i+1)
32.      begin
33.        if (shiftType= =LSL)
34.          if (i >= shiftCount)
35.            op2Val[i] <= RmVal[i-shiftCount];
36.          else
37.            op2Val[i] <= 0;
38.        else           // The three right shifts (LSR, ASR, ROR)
39.          if (i < REGSIZE-shiftCount)
40.            op2Val[i] <= RmVal[i+shiftCount];
41.          else
42.            case (shiftType)
43.              LSR: op2Val[i] <= 0;
44.              ASR: op2Val[i] <= RmVal[REGSIZE-1];
45.              default: op2Val[i] <= RmVal[i+shiftCount-REGSIZE];
46.            endcase
47.      end
48.  endmodule
```

Listing 12.1-B: Code to generate LSL, LSR, ASR, and ROR bit shifts

Although the DE10-Lite only has 10 switches and 10 LEDs, it can support a reduced version of the above test program. Basically, the shift register is reduced from 8 bits to 6, and the maximum shift count is reduced from 7 to 3. The shift codes identifying the four types of shifts will remain the same.

- Switches 9, 8 (shiftType): Type of shift to be performed (2 bits)
- Switches 7, 6 (shiftCount): Number of bit positions to shift (2 bits)
- Switches 5...0 (RmVal):Original binary data pattern (6 bits)
- LEDs 5...0 (op2Val): Bit pattern after shift (6 bits)

Note that the REGSIZE parameter was changed on line 10 from 8 to 6 to accommodate the DE10-Lite. Similar "register shift" code will be placed within the ARM imitation, and its REGSIZE will then be changed to 32.

```
1.  // Switches 9, 8: Type of shift to be performed
2.  // Switches 7, 6: Number of bit positions to shift (max of 3)
3.  // Switches 5...0: Original binary data pattern
4.  // LEDs 5...0: Bit pattern after shift
5.
6.  module TopLevel (SW, KEY, LEDR);
7.      input [9:0] SW;
8.      input [1:0] KEY;
9.      output [9:0] LEDR;
10.     parameter REGSIZE = 6;          // Size of ARM 32-bit register
11.
12.     parameter LSL = 2'b00;          // Logical Shift Left
13.     parameter LSR = 2'b01;          // Logical Shift Right
14.     parameter ASR = 2'b10;          // Algebraic Shift
15.     parameter ROR = 2'b11;          // Rotate (circular) Right
16.
17.     reg [REGSIZE-1:0] op2Val;
18.     wire [REGSIZE-1:0] RmVal;
19.     wire [1:0] shiftType;
20.     wire [1:0] shiftCount;
21.     wire clk;
22.
23.     assign clk = ~KEY[0];
24.     assign shiftType = SW[9:8];
25.     assign shiftCount = SW[7:6];
26.     assign RmVal = SW[REGSIZE-1:0];
27.     assign LEDR[REGSIZE-1:0] = op2Val;
28.
```

Listing 12.2: Modifications to shift test for use with DE10-Lite

Second Operand Details

The ARM data processing instruction machine code format is provided in the copy of Figure 11.5. The twelve bits of the second operand are actually somewhat more complicated than how we have been using them.

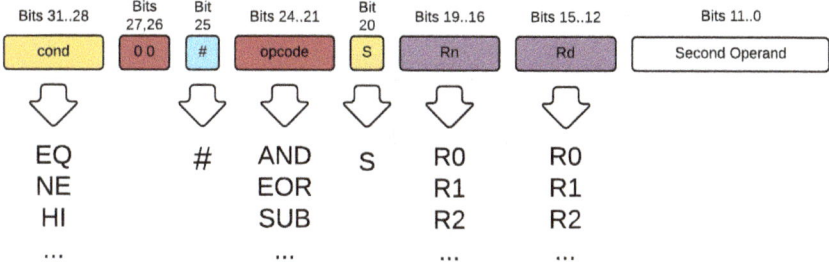

Bits 31..28	Bits 27,26	Bit 25	Bits 24..21	Bit 20	Bits 19..16	Bits 15..12	Bits 11..0
cond	0 0	#	opcode	S	Rn	Rd	Second Operand

EQ	#	AND	S	R0	R0
NE		EOR		R1	R1
HI		SUB		R2	R2
...		...		...	...

Copy of Figure 11.5: Positions in machine code

Algebraically, the general format of a subtraction instruction is $D = N - M{\times}2^S$, where D and N represent register contents, and M and S can be either register contents or constants. This format is true of every one of the sixteen data processing instructions:

$D = N - M{\times}2^S$ where 2^S is not really an exponent, but a register shift.

- **D**: Destination register to hold the 32-bit result
- **N**: First operand register: In subtraction, it is the minuend (quantity from which another quantity is to be subtracted)
- **M×2^S**: Second operand: In subtraction, it is the subtrahend (quantity to subtract). There are three possible formats for M and S:
  1. **M** and **S** are both constants: This is somewhat like scientific notation where M is a constant (0 through 255) and S is a shift count (0 through 30, even integers only).
  2. **M** is a register and **S** is a constant: The contents of register M are shifted (logical, algebraic, or circular) by a constant (range of 0 through 31).
  3. **M** and **S** are both in registers: The contents of register M are shifted (logical, algebraic, or circular) by the value in register S.

ARM data processing instructions have three general formats, depending on two "immediate" flag bits at bit positions 25 and 4.

- **SUB R1, R2, #4:** The second operand is an immediate constant composed of two numbers: a base value and a rotate amount. If only one number is present, then the rotate amount is assumed to be zero.
- **SUB R1, R2, R3, LSR #17:** The second operand is a register that is shifted a constant amount.
- **SUB R1, R2, R3, LSR R6:** The second operand is a register that is shifted by the amount specified in another register.

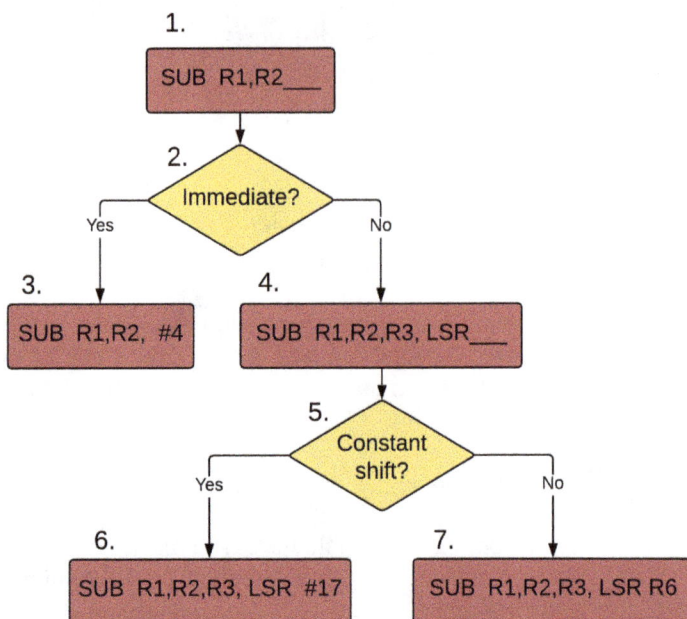

Figure 12.6: Three formats of the second operand in every data processing instruction

Figure 12.6 shows how the immediate flag bits determine the format of the second operand.

1. SUB R1,R2 _____: All formats start with Rd and Rn.
2. Is the second operand an immediate constant (bit 25)?
3. SUB R1,R2, #4: Instruction with immediate constant
4. SUB R1,R2,R3, LSR __: Second operand is a register being shifted.
5. Is the shift count a constant (bit 4)?
6. SUB R1,R2,R3, LSR #17: Instruction with constant shift count
7. SUB R1,R2,R3, LSR R6: Instruction with shift count in register

Shifting and Execution Speed

Here's a prelude to a couple of questions at the end of the chapter.

1. Do you think it takes the ARM processor longer to shift a register ten bit positions or five bit positions?
2. The ARM includes a shift (most often a shift of zero) with every data processing instruction. Would the ARM run faster if the shift was not done at all?

Computer Architecture Tutorial Using an FPGA

I won't give you the answers, but here's a clue. What is the *for* loop doing on line 31 in Listing 12.1? Remember: Verilog is a hardware description language, not a programming language.

Shifting and Assembly Language

We really don't want to enter the data processing instruction format in binary or even hexadecimal since it now has as many as 12 pieces (look at figures 12.7 through 12.11). First we will update the assembly language coding and examine what it produces using Listing 12.3. This entails a new task (ash) and new functions (ash1, ash2, and asdp4) along with a modification for the asdp3 function to handle all the variations of the second operand.

In preceding chapters, the second operand was either a constant or the Rm register. Now, even the constant has two parts, and the Rm register can be accompanied by a second register (Rs), a shift, and another constant.

```
185.  // Function asdp3 constructs the instruction's operand fields for function asdp2.
186.  // `SUB R1,7 Format 1: Rd = Rd - constant
187.  // `SUB R1,R2 Format 2: Rd = Rd - Rm
188.  // `SUB R1,R2,7 Format 3: Rd = Rn - constant
189.  // `SUB R1,R2,R3 Format 4: Rd = Rn - Rm
190.  // `SUB R1,R2,R3,SHR R5 Format 5: Rd = Rn - (Rm shifted)
191.
192.    function [31:0] asdp3 ();
193.      input [15:0] opCode, Q1, Q2, Q3, Q4, Q5;
194.      if (Q3>0)
195.        asdp3 = opCode<<21 | asdp4(Q1, Q2, Q3, Q4, Q5);
196.      else                    // Rn is same as Rd
197.        asdp3 = opCode<<21 | asdp4(Q1, Q1, Q2, Q3, Q4);
198.    endfunction
199.
200.  // Function asdp4 fills in the Rd, Rn, and 12-bit second operand field Rm/constant.
201.
202.    function [31:0] asdp4 ();
203.      input [15:0] Rd,Rn,Rm,Sh,Rs;
204.      if (Rm<R0)              // Rm field is a constant?
205.        asdp4 = 1<<25 | Rd[3:0]<<12 | Rn[3:0]<<16 | Rm[7:0] | Sh[4:1]<<8;
206.      else
207.        if (Rs<R0)            // Shift value is constant?
208.          asdp4 = Rd[3:0]<<12 | Rn[3:0]<<16 | {Rs[4:0],Sh[1:0],1'b0,Rm[3:0]};
209.        else                  // Shift value is in register Rs
210.          asdp4 = Rd[3:0]<<12 | Rn[3:0]<<16 |
                    {Rs[3:0],1'b0,Sh[1:0],1'b1,Rm[3:0]};
211.    endfunction
```

Listing 12.3 A: Modifications to the asdp3 function and a new asdp4 function

12: Shift Instructions   215

Second Operand is an Immediate Constant

Figure 12.7 shows the immediate constant format (point 3 in Figure 12.6). It has the correct bit pattern for the second operand having a value of 4, but it is oversimplified. Rather than being a single 12 bit binary number having a decimal range of 0 through 4095, the ARM implementation enables a 32-bit number to be generated by combining an 8-bit base with a 4-bit shift. It sacrifices precision in order to provide a greater range.

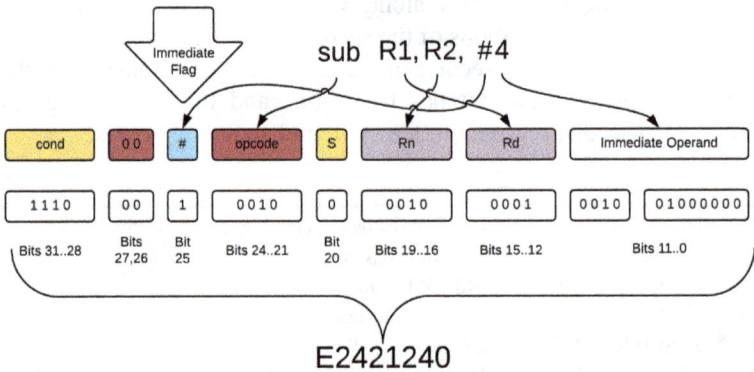

Figure 12.7: "Assemble" SUB with immediate value to ARM machine code 0xE2421240

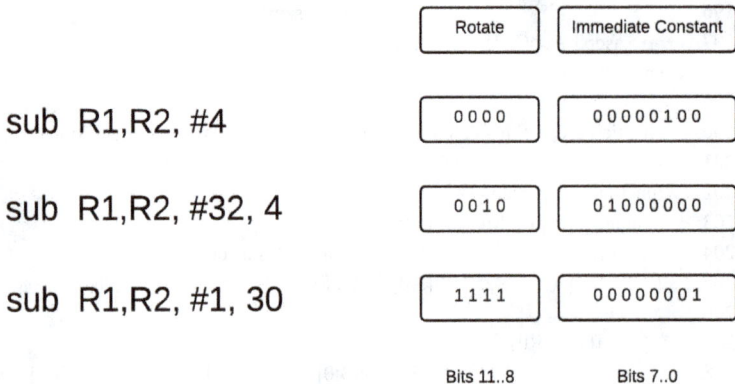

Figure 12.8: Three different ways to subtract the constant 4.

Basically, any 8-bit pattern can be moved to any even bit-position. Figure 12.8 shows three examples of an immediate constant being converted into an 8-bit value along with a rotate value. Each of the examples produces exactly the same result (an immediate value of 4). Note that the rotate is only to the right, and its range of 0 through 15 (4-bit field) must be multiplied by 2 in order to make the full 30 bit rotation possible.

Figure 12.9 shows five different hexadecimal constants, and how the base and shift count are used to produce a single integer. Obviously, 12 bits cannot be used to represent every possible 32 bit number, and #0x101 is not possible but would have been if a straight 12-bit format was used.

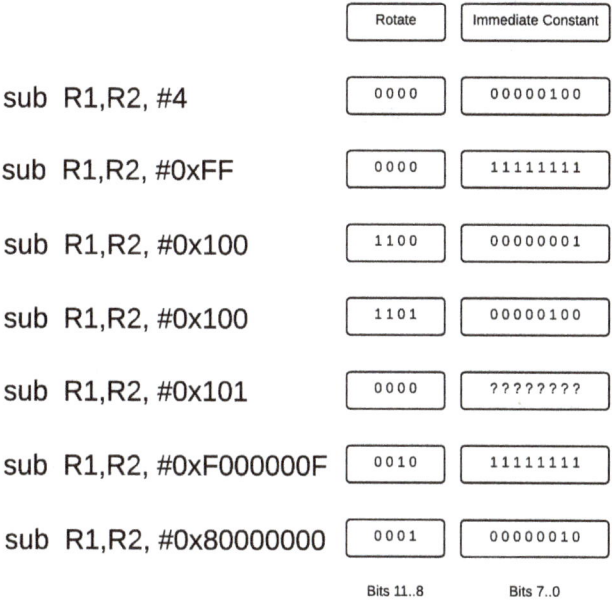

	Rotate	Immediate Constant
sub R1,R2, #4	0000	00000100
sub R1,R2, #0xFF	0000	11111111
sub R1,R2, #0x100	1100	00000001
sub R1,R2, #0x100	1101	00000100
sub R1,R2, #0x101	0000	????????
sub R1,R2, #0xF000000F	0010	11111111
sub R1,R2, #0x80000000	0001	00000010
	Bits 11..8	Bits 7..0

Figure 12.9: Six immediate value examples and one impossible value (0x101)

- 0x101 (binary 100000001) cannot work because it is more than an 8-bit pattern.
- 0x102 (binary 100000010) cannot work because even though it is an 8-bit pattern, it cannot be shifted to an odd bit position.
- 0x103 (binary 100000011) cannot work because it is more than an 8-bit pattern.
- 0x104 (binary 100000100) works. Not only is it a 7-bit pattern, but it shifts to an even bit position.

Second Operand is a Register with a Shift

We are now at point 4 in Figure 12.6. Bits 0 to 3 of the instruction contain the ID of the register to be shifted, and bits 5 and 6 provide the type of shift. Bit 4 determines whether the shift count is a constant or is in register Rs.

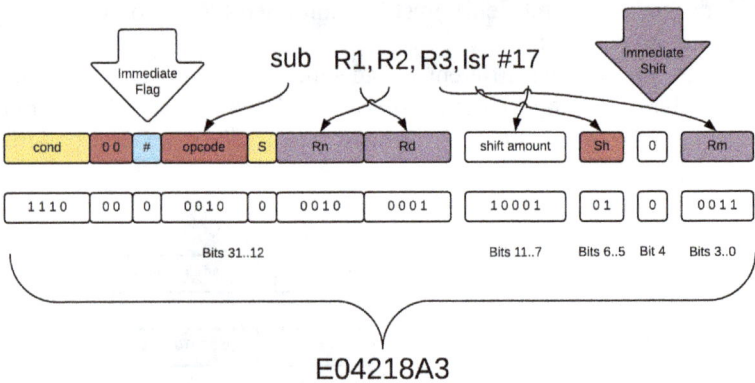

Figure 12.10: Second operand is a register shifted by a constant value.

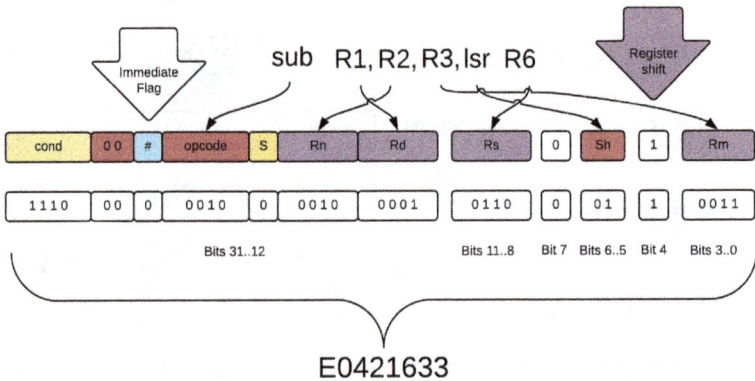

Figure 12.11: Second operand is a register shifted by the amount contained in another register.

Shift Instructions

When you look at a typical ARM assembly language program, you will probably see what appears to be "stand alone" shift instructions. These are really MOV instructions with a shift in the second operand. Listing 12.3 B shows what looks like opcodes being defined, and Listing 12.3 C shows how these opcodes are converted into MOV instructions using the newly added ash task, along with ash1 and ash2 functions. All they are doing is rearranging the arguments from the "shift instructions" to the proper positions for a MOV instruction.

Computer Architecture Tutorial Using an FPGA

70.	`define	LSL ash (16'h1000,	// Logical Shift Left
71.	`define	LSR ash (16'h1001,	// Logical Shift Right
72.	`define	ASR ash (16'h1002,	// Algebraic Shift Right
73.	`define	ROR ash (16'h1003,	// Rotate (circular) Right

Listing 12.3 B: Assembly language for what looks like "shift instructions"

If we did not allow for defaults and dropouts (such as using LSL R3,10 instead of LSL R3,R3,10), the ash1 and ash2 functions would be simpler.

```
212.
213.    // Task ash is called by bit-shift opcode macros `LSL, `LSR, `ASR, `ROR
214.    // The number of parameters will vary between three and six.
215.        // `LSL`EQ`S R1,R2,R3 `_6 // General format with 6 parameters.
216.        // `LSL R1,R2 `_3 // Default values chosen for 3 parameters.
217.
218.    task ash ();
219.        input [15:0] Shft,P1,P2,P3,P4,P5,P6,P7;
220.        progMem[IP] <= ash1(Shft,P1,P2,P3,P4,P5);
221.        IP = IP + 1;
222.    endtask
223.
224.    // Function ash1 is called by task ash to construct a MOV instruction (4'b1101).
225.
226.    function [31:0] ash1 ();
227.        input [15:0] Shft,P1,P2,P3,P4,P5;
228.        if (P5 > 0)
229.            ash1 = asdp1 (4'b1101,P1,P2,P3,P4,Shft,P5,0);
230.        else
231.            if (P4 > 0)
232.                if (P2 < R0)
233.                    ash1 = asdp1 (4'b1101,P1,P2,P3,P3,Shft,P4,0);
234.                else
235.                    ash1 = asdp1 (4'b1101,P1,P2,P2,Shft,P3,0,0);
236.            else
237.                if (P3 > 0)
238.                    if (P1 < R0)
239.                        ash1 = asdp1 (4'b1101,P1,P2,P2,Shft,P3,0,0);
240.                    else
241.                        ash1 = asdp1 (4'b1101,P1,P2,Shft,P3,0,0,0);
242.                else
243.                    ash1 = asdp1 (4'b1101,P1,P1,Shft,P2,0,0,0);
244.    endfunction
```

Listing 12.3 C: Task and functions for "shift instructions"

Now that we have completed the ARM data processing format, go ahead and compile, download, and test the "assembly language" produced in Listing 12.3.

12: Shift Instructions   219

Verify that each text line of assembly code produces the correct machine language code (like you did for Listing 11.3). Simply set the address of 0 through 31 (binary 00000 through 11111) on switches SW[4:0] and then push KEY[1]. Those of you with a DE2-115 may prefer to include the user interface code from lines 39 through 46 of Listing 11.2 to see all eight hex digits at once.

```
254.   IP = 0;
255.   `MOV   R1,'b0101   `_3      // 0: E3A01005 Move 0b0101 into R1
256.   `MOV   R2,0   `_3           // 1: E3A02000 Move 0 into R2
257.   `MOV   R5,25,30   `_4       // 2: E3A05F19 Move 'x19 rotated 30 bits
258.   `MOV   R5,R6   `_3          // 3: E1A05006 Rd = Rm
259.   `MOV   R1,R1,LSR,10   `_5   // 4: E1A01521 Move with a constant shift
260.   `MOV   R2,R3,ASR,R4   `_5   // 5: E1A02453 Shift count in register
261.   `MVN   R1,2,4   `_4         // 6: E3E01202 Move neg. of 2 rot. 4
262.   `ADD   R5,R6,25   `_4       // 7: E2865019 Rd = Rn + constant
263.   `ADD   R5,R6,R9   `_4       // 8: E0865009 Rd = Rn + Rm
264.   `ADC   R11,R1   `_3         // 9: E0ABB001 Add with carry
265.   `SUB   R5,R6,7   `_4        // 10: E2465007 Subtract, but no status
266.   `SUB`S   R5,R6,7   `_5      // 11: E2565007 Subtract, update status
267.   `SUB`EQ   R5,R6,7   `_5     // 12: 02465007 If Z-flag, subtract
268.   `SUB`EQ`S   R5,R6,7   `_6   // 13: 02565007 Test Z-flag and update status
269.   `SBC`EQ   R12,R2   `_4      // 14: 00CCC002 Subtract with carry
270.   `SUB   R1,R2,R3,LSR,17   `_6   // 15: E04218A3 Constant shift
271.   `SUB   R1,R2,R3,LSR,R6   `_6   // 16: E0421633 Shift in register
272.   `ORR   R3,R11,R12   `_4     // 17: E18B300C (R11) | (R12) => R3
273.   `EOR   R4,R1,R2   `_4       // 18: E0214002 Exclusive OR instruction
274.   `BIC   R3,5,4   `_4         // 19: E3C33205 Clear bits 30 and 28
275.   `BIC   R12,R2,R1   `_4      // 20: E1C2C001 Not(R1) & (R2) => R12
276.   `TST   R1,'b0101   `_3      // 21: E3110005 Test 0b0101 in R1
277.   `CMP   R2,R6   `_3          // 22: E1520006 Compare two registers
278.   `LSR   R1,10   `_3          // 23: E1A01521 MOV R1,R1,LSR 10
279.   `LSR   R1,R3   `_3          // 24: E1A01331 MOV R1,R1,LSR R3
280.   `ASR   R2,R3,3   `_4        // 25: E1A021C3 MOV R2,R3,ASR 3
281.   `ROR   R4,R5,R6   `_4       // 26: E1A04675 MOV R4,R5,ROR R6
282.   `LSR`GT   R4,1   `_4        // 27: C1A040A4 MOVGT R4,R4,LS4 1
283.   `LSR`S   R4,2   `_4         // 28: E1B04124 MOVS R4,R4,LSR 2
284.   `LSR`NE`S   R4,30   `_5     // 29: 11B04F24 MOVNES R4,R4,LSR 30
285.   progMem[30] = 0;
```

Listing 12.3 D: Variety of formats to test ARM assembly language to machine code conversion

Try your own combination of opcodes, operands, and conditions. Compile, download, and verify your own examples of converting assembly language to machine code.

Conditional Operator for Bit Shifts

Now that we have the assembly language that generates shifts for the machine code instructions, its time to upgrade the ARM imitation to support the execution of the shifting operations. The Verilog code in Listing 12.1 for demonstrating the various bit shifting features of the ARM processor is a sequential (i.e., clocked) circuit.

Clocked circuits are usually desired in computer architecture, but I don't want to "waste" one clock cycle in every data processing instruction just doing a register shift. I can, however, develop a dataflow circuit using the *assign* statement that has the shifting done continuously as the data changes.

A new ShiftMod module to execute the shift operations will be constructed. The shifting code from Listing 12.1 will be converted to a dataflow approach in two stages, so as to minimize confusion.

I will use the Verilog conditional operator (a.k.a., ternary operator) which is a shorthand version of combining *if* with *else* on one command line. It has the following three parts:

- Condition   ?   Do if True   :   Do if False   ;

```
34.          if (i >= shiftCount)
35.              op2Val[i] <= RmVal[i-shiftCount];
36.          else
37.              op2Val[i] <= 0;

          --- Replaced by ---

34.          op2Val[i] <= (i >= shiftCount) ? RmVal[i-shiftCount] : 0;
```

Figure 12.12: Conditional operator replaces *if* and *else*.

In Figure 12.12, lines 34 through 37 of Listing 12.1 are replaced by a single line (34 in upcoming Listing 12.4). The conditional operator can also be nested. For example, multiple conditions can be processed as indicated below:

- Condition 1   ?   Do if condition 1 is True   :
- Condition 2   ?   Do if condition 2 is True   :
- . . .
- Condition N   ?   Do if condition N is True   :
- Do if all preceding conditions are False   ;

This is somewhat like a Verilog case statement where it processes the first true case, but not every true case. If none of the conditions is true, then the final "else" acts as the default case and is processed.

```
39.        if (i < REGSIZE-shiftCount)
40.            op2Val[i] <= RmVal[i+shiftCount];
41.        else
42.          case (shiftType)
43.             LSR: op2Val[i] <= 0;
44.             ASR: op2Val[i] <= RmVal[REGSIZE-1];
45.             default: op2Val[i] <= RmVal[i+shiftCount-REGSIZE];
46.          endcase

        --- Replaced by ---

36.        op2Val[i] <= (i < REGSIZE-shiftCount) ? RmVal[i+shiftCount] :
37.           (shiftType==LSR) ? 0 :
38.           (shiftType==ASR) ? RmVal[REGSIZE-1] :
39.           RmVal[i+shiftCount-REGSIZE];
```

Figure 12.13: Conditional operator replaces *case* statement.

In Figure 12.13, lines 39 through 46 of Listing 12.1 are replaced by a single conditional operator. Listing 12.4 incorporates the above two changes and will run exactly the same as Listing 12.1 for demonstrating the shifting operations. For those of you with DE10-Lite units, the same change can be made to Listing 12.2.

```
29.    integer i;   // Loop counter
30.    always @ (posedge(clk))
31.    for (i=0;i<REGSIZE;i=i+1)
32.      begin
33.        if (shiftType==LSL)
34.            op2Val[i] <= (i >= shiftCount) ? RmVal[i-shiftCount] : 0;
35.        else  // The three right shifts (LSR, ASR, ROR)
36.            op2Val[i] = (i < REGSIZE-shiftCount) ? RmVal[i+shiftCount] :
37.              (shiftType==LSR) ? 0 :
38.              (shiftType==ASR) ? RmVal[REGSIZE-1] :
39.              RmVal[i+shiftCount-REGSIZE];
40.      end
41. endmodule
```

Listing 12.4: Conditional operator updates to Listing 12.1.

Listing 12.4 still uses the clocked sequential behavioral approach. I've basically replaced the *if*, *else*, and *case* statements with the ? : conditional operator which is also available for use on dataflow *assign* statements.

The two conditional operators in Listing 12.4 can also be combined to form one statement. This one *assign* statement is then placed within a *generate* block in Listing 12.5 to form an asynchronous dataflow circuit. The same switches and

LEDs are being used as before, but the clock from KEY[0] is no longer needed. The shifting operation has been placed into a new module named ShiftMod that will also be incorporated into the next ARM imitation.

```
6.  module TopLevel (SW, LEDR);
7.    input [17:0] SW;
8.    output [17:0] LEDR;
9.    parameter REGSIZE = 8;              // Size of ARM 32-bit register
10.
11.   wire [REGSIZE-1:0] RmVal;
12.   wire [1:0] shiftType;
13.   wire [2:0] shiftCount;
14.   wire clk;
15.
16.   assign shiftType = SW[12:11];
17.   assign shiftCount = SW[10:8];
18.   assign RmVal = SW[REGSIZE-1:0];
19.   ShiftMod (RmVal, shiftType, shiftCount, LEDR[REGSIZE-1:0]);
20. endmodule
21.
22. module ShiftMod (RmVal, shiftType, shiftCount, op2Val);
23.   parameter REGSIZE = 8;              // Size of ARM 32-bit register
24.
25.   parameter LSL = 2'b00;             // Logical Shift Left
26.   parameter LSR = 2'b01;             // Logical Shift Right
27.   parameter ASR = 2'b10;             // Algebraic Shift Right
28.   parameter ROR = 2'b11;             // Rotate (circular) Right
29.
30.   output [REGSIZE-1:0] op2Val;
31.   input [REGSIZE-1:0] RmVal;
32.   input [1:0] shiftType;
33.   input [2:0] shiftCount;
34.   genvar i;
35.   generate
36.   for (i=0;i<REGSIZE;i=i+1)
37.     begin:blkname
38.       assign op2Val[i] = (shiftType==LSL) &&
39.         (i >= shiftCount) ? RmVal[i-shiftCount] :
40.         (shiftType==LSL) ? 0 :
41.         (i < REGSIZE-shiftCount) ? RmVal[i+shiftCount] :
42.         (shiftType==LSR) ? 0 :
43.         (shiftType==ASR) ? RmVal[REGSIZE-1] :
44.         RmVal[i+shiftCount-REGSIZE];
45.     end
46.   endgenerate
47. endmodule
```

Listing 12.5: Module ShiftMod performs the LSL, LSR, ASR, and ROR shifts for the ARM imitation.

12: Shift Instructions   223

The code in Listing 12.5 is somewhat concise. If this was a real application, and not a book demonstration, I would have included some global comments describing what the code is doing.

The above code is not totally obscure due to the choice of variable names and the use of parameters. As an example of how poorly documented the exact same code can be, take at look at Figure 12.14.

```
1.  module TopLevel (SW, KEY, LEDR);
2.     input [17:0] SW;
3.     input [1:0] KEY;
4.     output [17:0] LEDR;
5.     ShiftMod (SW[7:0], SW[12:11], SW[10:8], LEDR[REGSIZE-1:0]);
6.  endmodule
7.  module ShiftMod (W, X, Y, Z);
8.     input [7:0] W;
9.     input [1:0] X;
10.    input [2:0] Y;
11.    output [7:0] Z;
12.    genvar U;
13.    generate
14.       for (U=0;U<8;U=U+1)
15.          begin:blkname
16.             assign Z[U] = (X==0) &&
17.                (U >= Y) ? W[U-Y] :
18.                (X==0) ? 0 :
19.                (U < 8-Y) ? W[U+Y] :
20.                (X==1) ? 0 :
21.                (X==2) ? W[7] :
22.                W[U+Y-8];
23.          end
24.    endgenerate
25. endmodule
```

Figure 12.14: Code example with a poor choice of variable names

Whether it be hardware or software, there is more to coding than getting the code to work. Not only do the compilers and assemblers have to read hardware and software descriptions, but we humans must do so as well. If you insist on writing obscure code like that in Figure 12.14, do not put your name anywhere near it.

Why all the emphasis on "maintenance"? Maintenance is partially "bug" fixes to correct errors, but more importantly, it involves the enhancements to incorporate features in the future that were not anticipated when the code was first designed.

Multiply by Shifting and Adding

As a demonstration, we will again perform the factorial example. However, its performance will be improved by doing the multiplication through shifting and adding instead of numerous repeated additions.

$$
\begin{array}{r}
1100 \\
\times\, 1011 \\
\hline
1100 \\
1100 \\
0000 \\
+\,1100 \\
\hline
10000100
\end{array}
$$

We can multiply in binary using the same procedure that we as humans multiply in base ten. Figure 12.15 shows the multiplication of 12 by 11 in binary.

1. The first row of the product is 1×1100, the second is 10×1100, the third is $0 \times 100 \times 1100$, and the fourth is 1000×1100.
2. All of the shifted multiplicands are added together giving 10000100.

Figure 12.15: Multiplication by shifting and adding

Something to keep in mind for the next chapter which describes multiplication instructions: As can be seen above, multiplying two 4-bit binary numbers can result in an 8-bit product. Likewise, multiplying two 32-bit numbers can result in a 64-bit product. This is one of the reasons why multiplication is not included in the set of ARM data processing instructions.

Factorial Example Using Shifting and Adding

Twelve factorial is $12 \times 11 \times 10 \times 9 \ldots$ Using the nested-addition iteration loop approach, 12×11 is calculated by adding 12 to itself 11 times. As a more efficient alternative, consider doing only three additions, one for each of the "1" bits in 1011.

The "iterative addition" loop is compared to the "shifting and addition" loop by the two series shown below. In the first step of calculating 12 !, both techniques calculate $12 \times 11 = 132$, but the second is much faster because it performs fewer calculations.

- $12 \times 11 = 12 + 12 + 12 + 12 + 12 + 12 + 12 + 12 + 12 + 12 + 12 = 132$
- $12 \times 11 = (1 \times 12) + (2 \times 12) + (8 \times 12) = 12 + 24 + 96 = 132$

The number 11_{10} is expressed in binary as 1011 which is $1 \times 2^3 + 0 \times 2^2 + 1 \times 2^1 + 1 \times 2^0$ which as $8 + 2 + 1$. Shifting left one bit position is like multiplying by 2.

The second step of calculating 12 ! is 132×10. The number 10_{10} is expressed in binary as 1010 which is $1\times2^3 + 0\times2^2 + 1\times2^1 + 0\times2^0$. A comparison of the two approaches on the second pass through the outer loop is shown below:

- $132 \times 10 = 132 + 132 + 132 + 132 + 132 + 132 + 132 + 132 + 132 + 132 = 1320$
- $132 \times 10 = (2 \times 132) + (8 \times 132) = 264 + 1056 = 1320$

Figure 12.16 contains an ARM assembly language program that calculates factorials using the shifting and adding approach. With the exception of the two branch instructions on lines 12 and 15, it is the identical code that will be used to demonstrate features of the ARM imitation (upcoming Listing 12.6).

As in the factorial example in Chapter 11, this program uses two loops, one nested within the other.

- Outer Loop: Count down from 12 to 2. This is identical to the outer loop used in Chapter 11. On each pass, the previous accumulative product in register R0 (the multiplicand) will be multiplied by the current value of the count down variable contained in register R1 (the multiplier).
- Inner Loop: Multiply using the shifting and adding approach. The multiplicand will be shifted and added for each "1" bit appearing in the multiplier. When all the "1" bits are processed, the loop will exit.

1. @			Program to calculate factorial using nested shift and add loops.
2.			
3.	MOV	R0,#12	@ Factorial of 12 = 12 ! = 12*11*10*9*8*7*6*5*4*3*2*1
4.	SUB	R1,R0,#1	@ Initialize R1 as multiplier (loop counter)
5. FLoop:	MOV	R2,R1	@ Copy current value of multiplier into R2
6.	MOV	R3,R0	@ "Cumulative product" completed thus far
7.	MOV	R0,0	@ Initialize sum of "shift and add" to zero
8. MLoop:	RORS	R2,#1	@ Shift bit 0 into sign position (Z-flag)
9.	ADDMI	R0,R3	@ Here's the "add" of shift and add
10.	LSL	R3,#1	@ Here's the "shift" of shift and add
11.	BICS	R2,2,2	@ Clear bit 31 and set Z-flag if done
12.	BNE	MLoop	@ Go back until "multiply" is completed.
13.	SUB	R1,#1	@ Decrement multiplier for next pass.
14.	CMP	R1,#2	@ Factorial stops when 2 is reached
15.	BGE	FLoop	@ Go back until multiplier = 2

Figure 12.16: Nested "shift and add" loop in ARM assembly language

- Line 5: Top of outer loop: R0 will contain 12, then 12×11 (132), then $12 \times 11 \times 10$ (1320), ...
- Line 8: Top of inner loop: R3 will contain the current product from

the outer loop (from R0), that will then be shifted left one bit position on each pass through this inner loop.

- Line 12: Bottom of inner loop: until all the "1" bits in the multiplier have been processed, control will go back to the top of the inner loop.
- Line 15: Bottom of outer loop: Until the multiplier is decremented to 2, control will branch back to the top of the outer loop.

Shifting in the Execute State

Two new modules will be added to the ARM imitation: **Op2Mod** which calculates all the possible formats for the second operand and **ShiftMod** from Listing 12.5 which calculates all the possible register shifts.

Figure 12.17 shows the improvement in support for the second operand here in Chapter 12 compared to that in Chapter 11. Previously, the second operand was considered to be either a 12-bit immediate constant or the contents of register Rm (Listing 11.6 lines 214 and 215). Now, lines 241 through 255 of Listing 12.6 show how the ShiftMod module is instantiated to finish the features of the ARM data processing instructions.

```
214.  wire [31:0] op2;              // Calculated value of second operand
215.  assign op2 = (iFlag) ? op2raw : Rm;

            --- Replaced by ---

241.  module Op2Mod (RmVal, RsVal, op2raw, iFlag, op2Val);
242.    input [31:0] RmVal;          // Possible 2nd operand data register contents
243.    input [31:0] RsVal;          // Possible 2nd operand shift register contents
244.    input [11:0] op2raw;         // Second operand "as is" in instruction
245.    input iFlag;                 // Immediate value flag
246.    output [31:0] op2Val;        // Calculated value of second operand
247.
248.    wire [31:0] op2imVal;        // Value if 2nd op. is immediate
249.    wire [31:0] shiftCount;      // Calculated shift count (fixed or Rs)
250.    wire [31:0] op2shifted;      // Value if 2nd op. is in Rm
251.    assign op2imVal = {op2raw[7:0],24'b0,op2raw[7:0]} >> {op2raw[11:8],1'b0};
252.    assign shiftCount = (op2raw[4]) ? RsVal : op2raw[11:7];
253.    ShiftMod (RmVal, op2raw[6:5], shiftCount, op2shifted);
254.    assign op2Val = (iFlag) ? op2imVal : op2shifted;
255.  endmodule
```

Figure 12.17: Module Op2Mod replaces simple processing of second operand.

- Line 251: Immediate value is 8-bit base with a 4-bit shift count
- Line 252: The shift count is either in a register or is a constant
- Line 253: Shifting is done by module ShiftMod
- Line 254: Second operand value is either shifted register or immediate

Listing 12.6 is the final version of the data processing instruction format. It contains both the assembly language conversion (already shown in Listing 12.3) and the machine code execution.

1. **CPU_UI:** The User Interface is the same as in Listings 10.2 and 11.6.
2. **CPU:** Breakpoints and displays are also the same as in Listing 11.6
3. **DataProcIns:** The calculation of the second operand of the data processing instructions has been moved to new module Op2Mod.
4. **Op2Mod:** Calculates the second operand, whether it is only an immediate constant or data in one register shifted by either a constant value or the amount in another register.
5. **ShiftMod:** New module that continuously calculates all shifts (taken from Listing 12.5).
6. **Macros:** This set of macros in Listing 12.3 and here in Listing 12.6 supports assembly language for all 16 data processing instructions, including shifting.
7. **ProgMod:** The factorial program using shifting and adding as described in Figure 12.16.

```
486.    always @ (posedge(reset))
487.       begin
488.          integer FLoop, MLoop, ILoop;
489.          IP = 0;
490.
491.  // Program to calculate factorial using nested Shift and add loops.
492.
493.    `MOV   R0,0   `_3          // 0: Reset "boot" address
494.    `MOV   R0,12  `_3          // 1: 12 ! = 12*11*...2*1 = 32'h1C8CFC00
495.    `SUB   R1,R0,1  `_4        // 2: Initialize R1 as multiplier.
496.  FLoop  =  IP;                // Outer loop: Calculate the factorial
497.    `MOV   R2,R1  `_3          // 3: Copy current multiplier into R2.
499.    `MOV   R3,R0  `_3          // 4: Cumulative product done thus far
499.    `MOV   R0,0   `_3          // 5: Initialize sum to 0 for "shift and add"
500.  MLoop  =  IP;                // Inner loop: Multiply using shift & add
501.    `ROR`S  R2,1  `_4          // 6: Shift bit from bit 0 to sign
502.    `ADD`MI  R0,R3  `_4        // 7: Here's the "add" of "shift and add."
503.    `LSL   R3,1   `_3          // 8: Here's the "shift" of "shift and add."
504.    `BIC`S  R2,2,2  `_5        // 9: Clear bit 31 and set Z-flag if done.
505.    `MOV`NE  R15,MLoop  `_4    // 10: Go back until "multiply" is completed.
506.    `SUB   R1,1   `_3          // 11: Decrement multiplier for next pass.
507.    `CMP   R1,2   `_3          // 12: Compare multiplier for exit condition.
508.    `MOV`GE  R15,FLoop  `_4    // 13: Go back until multiplier = 2.
509.  ILoop  =  IP;                // Infinite loop (stay in one place)
510.    `MOV   R15,ILoop  `_3      // 14: Keep setting PC to this location.
511.    progMem[30] = 0;
```

Listing 12.6: Shifting and adding FACTORIAL program listing

Trace With Breakpoints

Using breakpoints to pause a program's execution to examine register contents will be done here just like it was done at the end of Chapter 11. Basically, set the breakpoint address, then allow the program to continue executing, and when it stops at the next breakpoint address, examine any or all register contents.

When paused, any of the 16 registers can be examined by changing SW[3:0]. No clock pulse is needed. SW[9] can select either the upper 16 bits or lower 16 bits of the register to be displayed. DE10-115 users may choose to download and execute Listing 12.7, which is the same as 12.6, except it displays all 32 bits of a register at one time.

The following three examples provide ideas on how to use breakpoints to diagnose a running program.

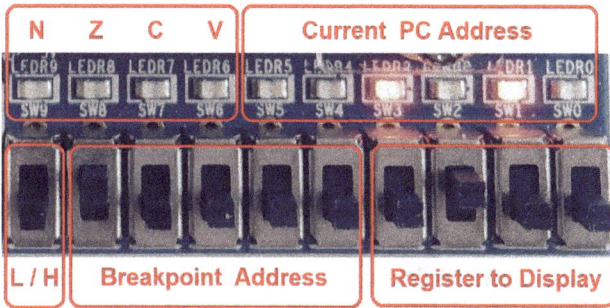

Copy of Figure 11.14: User interface with breakpoint address and display

Example 1: Just let the program run until it ends.

1. Set SW[8] high, and then start the program by pushing KEY[0] followed by KEY[1].
2. The factorial will appear on the hex displays, and the LEDs will show that the program is "stuck" in an infinite loop at address 14 ('b01110). Since, the breakpoint address was never reached, the program never halted.

Example 2: Trace near the top of the inner loop:

1. Set SW[8:4] to 00111 (address 7), which is line 502 in Listing 12.6. Start the program by pushing KEY[0] followed by KEY[1].
2. When the program stops, examine the registers noted below, and then push KEY[0] followed by KEY[1] to run to the next pass through the loop.
   a. R0: The factorial as it is being constructed on each pass through the loops

b. R1: Current multiplier and count down register changes on each pass of outer loop: 12, 11, 10, ...
c. R2: The previous instruction (line 501) had just rotated bit 0 into the sign bit position.
d. LEDs: The sign flag shown in LEDR[9] will be set if bit 0 had a "1" before the rotate.

Example 3: Trace the bottom of the inner loop:

1. Set SW[8:4] to 01010 (address 10). This instruction checks to see if all the "1" bits of the multiplier have been used to "shift and add" the current multiplicand. Start the program by pushing KEY[0] followed by KEY[1].
2. When the program stops, examine the registers noted below, and then push KEY[0] followed by KEY[1] to step to the next pass through the loop.
   a. R2: "1" bits in multiplier yet to be processed
   b. R15: Address of next instruction if "fall out of loop"
   c. LEDR[9:6]: Status bits, where the Z-flag, LEDR[8], indicates if loop should continue.

Highlights and Comparisons

In Chapter 12, the set of ARM data processing instructions has been completed. The evaluation of the second operand, with its multiple shifting capabilities, is included with every instruction described thus far. Most of the remaining instructions do not share this capability, but the memory load and store instructions described in Chapter 16 calculate memory addresses using an almost identical shifting process.

Algebraically, the general format of the 16 data processing instructions is $D = N - M \times 2^S$, where D and N represent register contents, and M and S can be either register contents or constants. This format is true of every one of the sixteen data processing instructions:

$D = N - M \times 2^S$ where 2^S is not really an exponent, but a register shift.

- **D**: Destination register to hold the 32-bit result
- **N**: First operand register: In subtraction, it is the minuend (quantity from which another quantity is to be subtracted)
- **M×2^S**: Second operand: In subtraction, it is the subtrahend (quantity to subtract). There are three possible formats for M and S:
  1. **M** and **S** are both constants: This is somewhat like scientific notation where M is a constant (0 through 255) and S is a shift count (0 through 30, even integers only).
  2. **M** is a register and **S** is a constant: The contents of register

M are shifted (logical, algebraic, or circular) by a constant (range of 0 through 31).

3. **M** and **S** are both in registers: The contents of register M are shifted (logical, algebraic, or circular) by the value in register S.

There are basically three general formats, depending on the two "immediate" flag bits at bit positions 4 and 25. If bit 25 is set, then the whole second operand is a constant. It's not simply a single 12-bit value having a range of 0 through 4095, but is an 8-bit value that is rotated to the right by two times the value in bits 8 through 11.

If immediate flag bit 25 is zero, then the second operand value is calculated by shifting a register (Rm) by either a constant value (if bit 4 = 0) or the amount in a register (Rs).

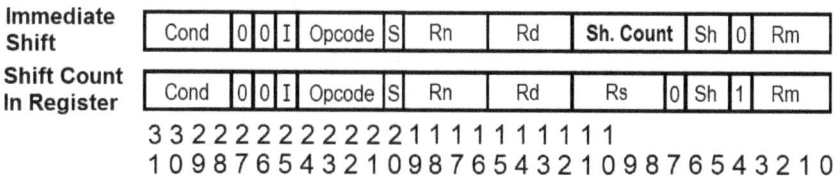

Immediate Shift

| Cond | 0 | 0 | I | Opcode | S | Rn | Rd | Sh. Count | Sh | 0 | Rm |

Shift Count In Register

| Cond | 0 | 0 | I | Opcode | S | Rn | Rd | Rs | 0 | Sh | 1 | Rm |

```
3 3 2 2 2 2 2 2 2 2 2 2 1 1 1 1 1 1 1 1 1 1
1 0 9 8 7 6 5 4 3 2 1 0 9 8 7 6 5 4 3 2 1 0 9 8 7 6 5 4 3 2 1 0
```

Figure 12.18: Second operand located in fields in the lower 12 bits of data processing instruction.

What's still missing in the ARM instruction set?

1. More instructions: Common instructions, like multiplication, will be described in Chapter 13.
2. We need a better way to change program flow, which will be described in Chapter 15.
3. Moving data between registers and main memory has not been described. Chapters 14, 16, and 18 will describe instructions that access memory.
4. Moving data to and from external devices is described in Chapter 17.
5. Using the Thumb 16-bit instruction set is described in Chapter 19.
6. Parallel processing using the NEON coprocessor is described in Chapters 20 and 21.
7. Chapter 22 describes floating point format, which is available within the NEON coprocessor.

Review Questions

1. *How can the decimal value 5120 be loaded into a 32-bit register with a single MOV instruction? Clue: 5120 = 4096+1024.
2. Do you think it takes the ARM processor longer to shift a register ten bit positions or five bit positions?
3. The ARM includes a shift (most often a shift of zero) with every data processing instruction. Would the ARM run faster if the shift was not done at all?
4. The second operand is 12 bits, but not simply a 12-bit number with a range of 0 through 4095. Instead, it enables a 32-bit number to be generated by combining an 8-bit base with a 4-bit shift. What does it mean when we say this format "sacrifices precision in order to provide a greater range?" See Figure 12.8 for examples.
5. A normal 12-bit field can represent 4096 unique numbers. How many unique numbers can be represented in the ARM data processing format that divides the immediate constant into an 8-bit base and 4-bit shift count? Note: There are many numbers that can be generated by more than one combination of base value with a shift.

Exercises

1. Many computers have double shift instructions, where two registers can shift and share bits as though they are one. Write a string of assembly language statements to perform a double register shift by the number of bits in a third register. Most likely you will need a combination of LSL, LSR, ORR, and SUB instructions.
2. Many applications of factorials have one factorial divided by another. Not only do we not have a multiply instruction in the ARM imitation at the moment, there is not even a divide instruction in the 32-bit ARM architecture. This is easily solved as in the example 7 ! / 4 ! = 7 × 6 × 5. Modify the factorial assembly language program so that a different ending condition (other than 2) is used. Notes: Larger values such as 15 ! / 13 ! can also be accommodated. Don't just change the ending with a "hard coded" constant in the compare instruction, but put the ending value in a register at the top of the program. Note: This is very similar to Exercise 11.3.

— 13 —
Multiplication Instructions

To someone using a calculator, multiplication and division appear to be no more complex than addition and subtraction. However, from our elementary school education, we know that is not the case. Both multiplication and division involve a process consisting of multiple steps. Division even has an "educated guess and check" phase to it.

Chapter 13 implements the ARM multiplication instructions in the following three stages:

1. **Assembler:** Macros, tasks, and functions will be provided to convert six ARM multiplication instructions in "assembly language" to 32-bit machine code that is identical with that for a real ARM processor.
2. **Execute:** A new Verilog module will be created to execute the multiplication machine code instructions. It will be integrated into the previous system from Chapter 12 so that programs consisting of both data processing and multiplication instructions can be run.
3. **Application Demonstrations:** Three ARM assembly language programs will be run, complete with breakpoints and register dumps.
   a. **Factorial:** Finally, we get the real multiply instruction in the factorial programming example.
   b. **Double Precision:** The factorial program will be expanded to support a 64-bit product (needed for 19 !).
   c. **Division:** Division through multiplying by a reciprocal is demonstrated. The reciprocals will be calculated by iterative shifting and subtraction.

Introductions

The ARM architecture provides the following six multiplication instructions. They are similar to the data processing instructions, except they only work with registers (i.e., you cannot multiply a register by an immediate constant), and they may generate either a 64-bit or a 32-bit product.

- **MUL:** 32-bit product from operands in two 32-bit registers
- **MLA:** 32-bit product added to a running total
- **SMULL:** Signed 64-bit product in two 32-bit registers
- **SMLAL:** Signed 64-bit product added to a running total
- **UMULL:** Unsigned 64-bit product in two 32-bit registers
- **UMLAL:** Unsigned 64-bit product added to a running total

Multiplication Instruction Formats

In the ARM architecture, the multiplication instructions have a different structure than the other arithmetic and logic instructions. It was noted in Chapter 12 (Figure 12.15) that multiplying two 4-bit numbers results in an 8-bit product. Likewise, multiplying two 32-bit numbers results in a 64-bit product. Within the ARM architecture, four of the multiplication instructions produce a 64-bit product, while two of the instructions only give the lower 32 bits (assuming the upper 32 bits are zeroes).

- **MUL:**   Rd   <=   (Rs × Rm) [31 : 0]
- **MLA:**   Rd   <=   (Rs × Rm + Rn) [31 : 0]
- **SMULL:** {RdH, RdL}   <=   Rs × Rm
- **SMLAL:** {RdH, RdL}   <=   Rs × Rm + {RdH,RdL}
- **UMULL:** {RdH, RdL}   <=   Rs × Rm
- **UMLAL:** {RdH, RdL}   <=   Rs × Rm + {RdH,RdL}

Figure 13.1 compares the format of the multiply instructions to the data processing instructions. Notice that the multiplies do not have an immediate flag bit (i.e., no constant possible), and no register shift capability. The Rs register field contains the multiplicand, and the field that indicated shift type is a constant 4'b1001. The destination register, Rd, has switched positions with the Rn register for the MUL and MLA instructions.

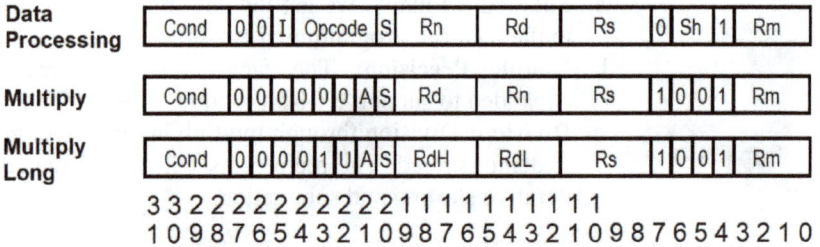

Data Processing: Cond | 0 | 0 | I | Opcode | S | Rn | Rd | Rs | 0 | Sh | 1 | Rm

Multiply: Cond | 0 | 0 | 0 | 0 | 0 | 0 | A | S | Rd | Rn | Rs | 1 | 0 | 0 | 1 | Rm

Multiply Long: Cond | 0 | 0 | 0 | 0 | 1 | U | A | S | RdH | RdL | Rs | 1 | 0 | 0 | 1 | Rm

```
3 3 2 2 2 2 2 2 2 2 2 2 1 1 1 1 1 1 1 1 1 1
1 0 9 8 7 6 5 4 3 2 1 0 9 8 7 6 5 4 3 2 1 0 9 8 7 6 5 4 3 2 1 0
```

Figure 13.1: Compare bit positions in multiply instructions to data processing instructions

- Bits 31 - 28: Multiplication is only executed for matching CPSR value
- Bits 27 - 24: 4b'0000 for all multiplication instructions
- Bit 23: Long flag: Set for UMULL, SMULL... (64-bit result)
- Bit 22: Unsigned flag: Set for UMULL, UMLAL (unsigned result)
- Bit 21: Accumulate flag: Set for MLA, SMLAL, UMLAL
- Bit 20: S flag: Set for updating CPSR status bits N and Z
- Bits 19 - 16: Rd, 32-bit product for MUL and MLA
- Bits 19 - 16: RdH, register for long product upper 32-bits
- Bits 15 - 12: Rn for MLA instruction
- Bits 15 - 12: RdL, register for long product lower 32-bits

- Bits 11 - 8: Rs (multiplicand register)
- Bits 7 - 4: 4b'1001 for all multiplication instructions
- Bits 3 - 0: Rm (multiplier register)

Assembly of Multiplication Instructions

Listing 13.1 contains the macros, tasks, and functions that will perform the assembly language to machine code conversion for the six multiplication instructions. Listing 13.2 is identical to Listing 13.1 except it provides eight hexadecimal digits in the user interface display for DE2-115 users.

```
1.  // Listing 13.1 introduces assembly of multiplication instructions
2.  // 1) Assembles 32-bit ARM multiplication instructions
3.  // 2) Dumps 32-bit words from memory in hexadecimal
4.  //
5.  // Modules and macros contained in this file:
6.  // 1) CPU_UI: User Interface that dumps 32-bit words, 16 bits at a time
7.  // 2) Macros for only assembling ARM multiplication instructions
8.  // 3) ProgMod: Memory containing "ARM" program
```

Listing 13.1 A: File to test the assembly of multiplication instructions

Listing 13.1 file is 183 lines and contains only enough code to assemble and test the various formats of the multiplication instructions. Its macros, tasks, and functions will be integrated into the larger code that will execute both the data processing instructions (from Listing 12.6) and the newly added multiplication instructions.

```
50.  //
51.  //---------------- Macro definitions for assembly language ----------------
52.  //

61.  `define MUL asmul (3'b000,      // Multiply giving 32-bit product
62.  `define MLA asmul (3'b001,      // Multiply, accumulate 32-bit product
63.  `define UMULL asmul (3'b100,    // Unsigned 64-bit product
64.  `define UMLAL asmul (3'b101,    // Unsigned 64-bit product, accumulate
65.  `define SMULL asmul (3'b110,    // Signed 64-bit product
66.  `define SMLAL asmul (3'b111,    // Signed 64-bit product, accumulate
```

Listing 13.1 B: Multiplication instruction macros for assembly language conversion

The assembly of the six multiplication formats is very similar to that done for the data processing instructions. The asmul task is shown in 13.1 C, and of course, the full listing containing the functions it calls is provided in the downloaded file.

Verify that Listing 13.1 generates the correct ARM machine code by compiling, downloading, and testing it as was done in previous chapters. Listing 13.1-D shows examples of multiplication instructions being assembled.

```
116.   // Task asmul is called by multiplication macros `MUL, `MLA, `UMULL, ...
117.   // The number of parameters will vary between three and seven.
118.   // `MLA`EQ`S R1,R2,R3,R4 `_7 // General format with 7 parameters
119.   // `MUL R1,R2 `_3 // Some parameters are optional
120.
121.   task asmul ();
122.       input [15:0] P0,P1,P2,P3,P4,P5,P6,P7;
123.       progMem[IP] <= asmul1(P0,P1,P2,P3,P4,P5,P6);
124.       IP = IP + 1;
125.   endtask
```

Listing 13.1 C: Task asmul assembles multiply instructions

Copy of Figure 11.4: Controls for running instruction dump

The copy of Figure 11.4 shows the test setup for viewing 32-bit numbers on the 7-segment displays.

1. Push KEY[0] to initialize.
2. Set the switches to desired memory address (0 up to 9).
3. Push KEY[1] to update and show machine code instruction
4. Repeat steps 2 and 3 to see the contents of other addresses.

```
171.   IP   =   0;
172.   `MUL    R5, R6, R7  `_4              // 0: E0050796
173.   `MUL    R4, R5  `_3                  // 1: E0040495
174.   `MUL`S  R4, R5  `_4                  // 2: E0140495
175.   `MUL`NE  R1, R2  `_4                 // 3: 10010192
176.   `MUL`GT`S  R3, R4  `_5               // 4: C0130394
177.   `MLA   R2, R3, R4, R5  `_5           // 5: E0225493
178.   `UMULL  R1, R2, R3, R4  `_5          // 6: E0821493
179.   `UMLAL  R1, R2, R3, R4  `_5          // 7: E0A21493
180.   `SMULL  R1, R2, R3, R4  `_5          // 8: E0C21493
181.   `SMLAL  R1, R2, R3, R4  `_5          // 9: E0E21493
182.   progMem[25]  <=  0;
```

Listing 13.1 D: Assembly of assortment of multiply instructions

Execution of Multiplication Instructions

Now that we can build the ARM machine code for the multiply instructions, how do we execute it in a running program? How can we integrate multiplication with the existing data processing instructions? Basically, let's follow the pattern done in developing the data processing instructions:

1. Describe a new module that "does the work" of producing a 64-bit product and accumulated product. It will be given the name MultiplyIns.
2. Instantiate the new MultiplyIns module. This not only creates it, but its argument list connects its data lines to those of the main CPU module.
   a. Input data is from Rm, Rs, Rd, and Rn register contents.
   b. A clock pulse indicates when the multiplication is to take place.
   c. Output is in a 64-bit product (Rs × Rm).
3. Modify the CPU instruction cycle to include multiplication and distinguish it from the data processing instructions.
   a. The decode circuit will set a signal, SelMul, when the multiplication opcodes are present.
   b. The writeBack state will store the results of the multiplication into the appropriate Rd and Rn destination register(s).
   c. The CPSR will be updated.

```
1.  // Listing 13.3 Full-speed factorial using ARM 32-bit instructions
2.  // 1) Assembles ARM data processing and multiplication instructions
3.  // 2) Executes 32-bit Factorial program
4.  // 3) Dumps any "one" 32-bit register at a time.
5.  // 4) Breakpoint address can be set to "stall" assembler program
6.  //
7.  // Modules and macros contained in this file:
8.  // 1) CPU_UI: User Interface that dumps 32-bit words, 16 bits at a time
9.  // 2) CPU: 32-bit CPU with data processing and multiplication instructions
10. // 3) DataProcIns: 16 "ARM like" DP instructions using 32-bit registers
11. // 4) Op2Mod: Fill in second operand including all shift possibilities
12. // 5) ShiftMod: Calculate all shift possibilities for Op2Mod
13. // 6) MultiplyIns: Module to calculate multiplication
14. // 7) Macros for assembling ARM data processing and multiplication instructions
15. // 8) ProgMod: Memory containing 12! assembly language program
```

Listing 13.3 A: Directory header in Listing_13-3.txt

```
130.   // Instantiate program memory and ALU processing
131.
132.   wire SelDP, SelMul, SelM64;
133.   assign SelDP = ok2exe && IR[27:26]==0 && IR[7:4]!=4'b1001;
134.   assign SelMul = ok2exe && IR[27:26]==0 && IR[7:4]==4'b1001;
135.   assign SelM64 = SelMul && IR[23];
136.   ProgMod (R[PC], CPU_state==fetch, reset, IR);
137.   Op2Mod (RmVal, RsVal, op2raw, iFlag, op2Val);
138.   DataProcIns (opCode, RnVal, op2Val, CPU_state==execute && SelDP, CPSR,
       RdVal_DP, RdVal_DPU, CPSR_DP);
139.   MultiplyIns (opCode, RdVal, RmVal, RnVal, RsVal, CPU_state==execute &&
       SelMul, CPSR, RdV64_M, CPSR_M);
```

Listing 13.3 B: Instantiate ALU modules for instruction execution, including multiplication.

The new MultiplyIns module is instantiated on line 139 shown in Listing 13.3 B. It has the following nine arguments:

1. opCode: Field indicating which instruction: MUL, MLA, UMULL, ...
2. RdVal: 32-bit Rd register value needed for UMLAL and SMLAL
3. RmVal: 32-bit Rm register value (multiplier)
4. RnVal: 32-bit Rn register value needed for all three multiply and accumulate instructions (MLA, UMLAL and SMLAL)
5. RsVal: 32-bit Rs register value (multiplicand value)
6. CPU_state==execute && SelMul: Clock "tick" that says when the multiply operation will take place.
7. CPSR: Current value of CPSR (i.e., before the multiplication)
8. RdV64_M: Calculated 64-bit product (or accumulated sum)
9. CPSR_M: Value of CPSR updated by multiplication instruction

```
190.   writeBack:               // Update specific registers
191.     begin
192.       {R[RnID],R[RdID]} <=
193.         (RdVal_DPU & SelDP) ? {R[RnID],RdVal_DP} :
194.         (SelM64) ? RdV64_M :
195.         (SelMul) ? {RdV64_M[31:0],R[RdID]} :
196.           {R[RnID],R[RdID]};
197.       if (SU) CPSR <=
198.         SelDP ? CPSR_DP :
199.         SelMul ? CPSR_M :
200.           CPSR;
201.       CPU_state <= fetch;
202.     end
```

Listing 13.3 C: The writeBack state selects which data updates Rd and Rn.

The writeBack state of the instruction cycle has two registers to update: Rn and Rd (Listing 13.3 C line 192). The Verilog conditional operator selects the updated values based upon the opcode value of the current instruction (flags SelDP, SelMul, SelM64). On line 197, the CPSR will be updated based upon whether the "S" bit is set in the instruction and which opcode is currently being executed.

```
302. //---------------- ALU for multiplication instructions ----------------
303. //
304. module MultiplyIns (opCode, Rn, Rm, Rd, Rs, clk, CPSR, Rd64, CPSRMul);
305.     input [3:0] opCode;
306.     input [31:0] Rd,Rm,Rn,Rs;                 // Note: Rn and Rd switched
307.     input clk;
308.     input [31:0] CPSR;
309.     output reg [63:0] Rd64;                    // 64-bit product
310.     output reg [31:0] CPSRMul;                 // Updated N and Z
311.     always @ (posedge(clk))
312.       begin
313.         case (opCode)
314.           0: Rd64 <= Rs * Rm;                  // MUL
315.           1: Rd64 <= Rs * Rm + Rn;             // MLA
316.           4: Rd64 <= Rs * Rm;                  // SMULL
317.           5: Rd64 <= Rs * Rm + {Rd,Rn};        // SMLAL
318.           6: Rd64 <= Rs * Rm;                  // UMULL
319.           7: Rd64 <= Rs * Rm + {Rd,Rn};        // UMLAL
320.         endcase
321.         CPSRMul <=
322.           {opCode[1]==0 & {Rs[31]^Rm[31],Rs==0 | Rm==0},
323.               CPSR[1:0]};
324.       end
325. endmodule
```

Listing 13.3 D: ALU module for multiplication instructions

Listing 13.3 D shows MultiplyIns, the ALU module for calculating the product for the ARM multiply instructions. Its arguments are described above where it is instantiated. Notice that the Rn and Rd values have switched positions compared to where they are located in the data processing instructions. This is due to the way the MUL and MLA instructions are defined in the ARM processor itself. See Figure 13.1.

There are two outputs: the 64-bit product and the CPSR flags (only N and Z are updated). The carry and overflow CPSR flags really don't have meaning for multiplication like they do for the data processing instructions and are not set in the real ARM hardware either.

Multiplication Applications

Finally, we get to put the multiply instruction into the factorial program. We will also perform a double precision version where the factorial of 19 can be calculated, which will require a 64-bit value.

The first 585 lines of files Listing_13-4.txt and Listing_13-5.txt are the same as in Listing_13-3.txt. The ARM assembly language programs at the end of each file are different in order to demonstrate the following applications:

1. Single precision factorial
2. Double precision factorial
3. Division by shifting an subtracting
4. Division by multiplying by the inverse

Trace With Breakpoints

Any of the 16 registers can be examined by changing SW[3:0]. See the examples at the end of the previous two chapters.

Specific breakpoint addresses can be set, and then reset to new values as each breakpoint is reached. However, if you are only interested in the final register values, that can be achieved by setting SW[8:4] to the address of the infinite loop at the end of each program or a high address that cannot be reached.

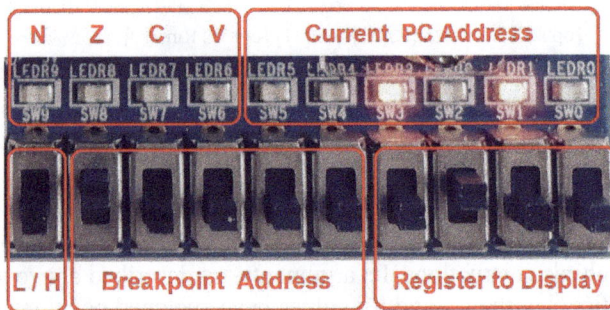

Copy of Figure 11.14: User interface with breakpoint address and display

File Listing_13-3.txt contains the Verilog code to assemble the multiply instructions, execute them as part of the ARM imitation, and test them by calculating 12!. Go ahead and compile, download, and run this factorial program. If SW[8:0] is set to 010000010, the answer for 12! will be present in register R2 as 32'h1C8CFC00. If you are interested in watching the factorial loop count down from 12 to 1, then set the breakpoint address to 6.

Computer Architecture Tutorial Using an FPGA

```
585.
586.   // Program to calculate single precision factorial in R2
587.   // 12! = 479,001,600 = 1C8C,FC00 H
588.
589.           `MOV   R0, 0     `_3        // 0: Reset "boot" address
590.           `MOV   R2, 12    `_3        // 1: 12 ! = 12*11*...2*1 = 32'h1C8CFC00
591.           `SUB   R1, R2, 1   `_4      // 2: Initialize R1 as multiplier/counter
592.   FLoop  =   IP;                      // Loop: Calculate the factorial
593.           `MUL   R2, R1    `_3        // 3: On each pass, multiply by one lower
594.           `SUB   R1, 1     `_3        // 4: Decrement multiplier for next pass.
595.           `CMP   R1, 2     `_3        // 5: Compare to loop ending condition.
596.           `MOV`GE  R15, FLoop  `_4    // 6: Go back until multiplier = 2.
597.   ILoop  =   IP;                      // Infinite loop (stay in one place)
598.           `MOV   R15, ILoop  `_3      // 7: Keep setting PC to this location.
599.
600.           progMem[30]  =  0;
601.    end
```

Listing 13.3 E: Single precision multiply for 12 factorial

Double Precision Factorial

Factorials greater than 12 produce binary numbers that will not fit in 32 bits. The 64 bits of double precision enable us to go somewhat higher, but even 64 bits are insufficient once we pass 20. Calculating factorials between 13 and 19 does give us a nice demonstration of using the 64-bit UMULL instruction.

The MUL instruction produces a 32-bit product from two 32-bit factors, and the UMULL instruction produces a 64-bit product also from two 32-bit factors. However, there is not an ARM instruction that works with two 64-bit factors or at least one factor being 64 bits and the other being 32 bits. What this means is once we "cross over" into double precision in our factorial calculation, we are stuck.

For numbers greater than 12, we side step the problem by dividing it into two parts and doing only one double precision multiply at the end. For example 15! = (15 × 14 ×13) × 12!, where both (15 × 14 ×13) and 12! fit within 32 bits. The multiplication loop in the example in Listing 13.4 "reinitializes itself" when it passes through 12.

The double precision factorial loop has the following changes made to the factorial loop just demonstrated:

1. When the R1 loop counter is at 13, it saves the current running product into register R3. The running multiply in R2 is then reinitialized to a value of 1 to continue the loop to calculate 12!.
2. When the loop is exited when R1=2, the final double precision factorial is calculated by multiplying R3 times R2.

```
585.
586.   // Program to calculate double precision factorial in R9, R8
587.
588.   // 12! =    479,001,600 =   1C8C,FC00 H
589.   // 17! =    355,687,428,096,000 =   1,437E,EECD,8000 H
590.   // 18! =    6,402,373,705,728,000 =   16,BEEC,CA73,0000 H
591.   // 19! = 121,645,100,408,832,000 =   1B0,2B93,0689,0000 H
592.
593.          `MOV   R0, 0   `_3              // 0: Reset "boot" address
594.          `MOV   R2, 19  `_3              // 1: 19 ! = 19*18*...*2*1
595.          `MOV   R3, 1   `_3              // 2: Initialize for factorial < 13 !
596.          `SUB   R1, R2,1  `_4            // 3: Initialize R1 as multiplier/counter
597.  FLoop  =  IP;                           // Loop: Calculate the factorial
598.          `MUL   R2, R1  `_3              // 4: On each pass, multiply (n-1)
599.          `SUB   R1, 1   `_3              // 5: Decrement multiplier for next pass.
600.          `CMP   R1, 12  `_3              // 6: Divide calculation at 12!
601.          `MOV`EQ  R3, R2   `_4           // 7: Save first product until 12! done.
602.          `MOV`EQ  R2, 1    `_4           // 8: Restart lower part of factorial.
603.          `CMP   R1,2  `_3               // 9: Compare to loop ending condition.
604.          `MOV`GE  R15, FLoop   `_4       // A: Go back until multiplier = 2.
605.          `UMULL   R8, R9, R2, R3 `_5     // B: Combine upper and lower factorials.
606.  ILoop  =  IP;                           // Infinite loop (stay in one place)
607.          `MOV   R15, ILoop   `_3         // C: Keep setting PC to this location.
608.
609.          progMem[30] = 0;
```

Listing 13.4: 64-bit double precision factorial (19!)

- Line 605: Double precision final answer in {R9,R8} = R2 × R3
- Line 600: Position in loop when two factors are separated
- Line 601: First factor is saved in register R3
- Line 602: Initialize for second factor
- Line 595: Initialize R3 to 1 (used with factorials less than 13)
- Line 594: Put factorial to be calculated here (2 to 19)

Go ahead and compile, download, and run the double precision factorial program in Listing 13.4. The answer will appear in registers R8 and R9. Try a few other values besides 19!. Yes, the Verilog compilation does take a couple of minutes, but it is doing a lot more work than just converting a few lines of assembly language to machine code.

Fractions in Binary

Subtraction is the inverse of addition. It gives rise to negative integers. Division is the inverse of multiplication. It gives rise to fractions.

Appendix C explains several ways to represent the negative of a number in binary, but how do we represent a fraction in binary? With whole numbers in binary, we start with the rightmost bit, and count 0, 1, 10, 11, 100, etc. gradually using more bits to the left as we count higher. For fractions, we start with the leftmost high-order bit, and gradually use more bits to the right as we represent smaller "fractions" of numbers.

A decimal point separates a decimal whole number from a decimal fraction. Likewise, a binary point separates a binary whole number from a binary fraction. In the following example, it seems reasonable to make the first bit to the right of the binary point equal to 2^{-1} which is ½. The second bit to the right of the binary point is 2^{-2} which is $½^2$ which is ¼.

- $111.1_2 = 1 \times 2^2 + 1 \times 2^1 + 1 \times 2^0 + 1 \times 2^{-1} = 4 + 2 + 1 + ½$
- $11.11_2 = 1 \times 2^1 + 1 \times 2^0 + 1 \times 2^{-1} + 1 \times 2^{-2} = 2 + 1 + ½ + ¼$

Division by Shifting and Subtracting

The ARM processor does not have a division instruction. The NEON coprocessor that often accompanies ARM processors has a divide instruction. See Chapter 20 for details. However, relying on the availability of an alternate processor might not always be the best option.

In Chapter 12, we were able to multiply by shifting and adding as demonstrated in Figure 12.15 and Listing 12.6. The ARM processor does have a multiply instruction, so that technique is seldomly used in practice except in cases that multiply by a power of 2.

We can divide by just shifting bits to the right, but "only shifting" only works if we are dividing by a number that is a power of 2. Following the "long division" example of 1/10 (one divided by decimal ten) shown in Figure 13.2 leads to a technique for calculating a quotient by shifting and subtracting. The quotient will not be exact because 1/10 cannot be represented exactly in base 2 (for the same reason 1/3 and 1/7 cannot be represented exactly in base 10).

$$\frac{1}{10_{10}} = \frac{1}{1010_2} = 1010 \overline{\smash{\big)}\,\begin{array}{l} 0.000110011 \\ 1.000000000 \\ -\underline{1010} \\ 1100 \\ -\underline{1010} \\ 10 \cdots \end{array}}$$

Figure 13.2: Binary "long division"

Division Example Using Reciprocals

Almost all CPUs perform subtraction by "adding the complement." They convert the subtraction "A - B" into the "A + -B" equivalent. Likewise, we can convert division into a multiplication using the reciprocal of the multiplier: "A / B" is changed to the "A × 1/B" equivalent.

Listing 13.5 has an example where 25 is divided by 10 to give 2.5. First it generates the reciprocal of 10 using the "long division" approach from Figure 13.2, and then it uses the 64-bit UMULL multiplication to get 2 in one register and 0.5 in another.

```
585.
586.  // Program to calculate 1/10, then multiply by 25 to get 2.5
587.
588.  // 1/10 = .0001100110011001..b = .1999..h
589.
590.      `MOV    R0, 0      `_3        // 0: Reset (i.e., "boot" address)
591.      `MOV    R5, 10     `_3        // Divisor (value to find reciprocal)
592.      `MOV    R6, 31     `_3        // Loop count (size of register - 1)
593.      `MOV    R1, 1      `_3        // Single bit that will be shifted
594.      `MOV    R2, 1      `_3        // Dividend of 1 for reciprocal
595.      `MOV    R9, 0      `_3        // Reciprocal will be built in R9.
596.  DLoop  =   IP;                    // Step through each of the 32 bits
597.      `LSL    R2, 1      `_3        // Multiply current dividend by 2.
598.      `CMP    R2, R5     `_3        // Test if bit needed in quotient.
599.      `ORR`GE  R9, R9, R1, LSL, R6  `_7  // Include bit in quotient.
600.      `SUB`GE  R2, R5      `_4      // Subtract for bit just included.
601.      `SUB`NE`S R6, 1      `_5      // Decrement for this pass in loop.
602.      `MOV`GE  R15, DLoop  `_4      // Exit if exact value or 31 bits.
603.
604.      `MOV    R8, 25     `_3        // C: 25 will be divided by 10
605.      `UMULL   R10, R11, R8, R9  `_5  // D: 2 in R11 and 0.5 in R10
606.
607.      `MOV    R15, IP  `_3          // Infinite loop
```

Listing 13.5: "Long division" calculation and reciprocal

- Lines 591,594: 1/10 will be calculated, R2 = 1, R5 = 10
- Line 595: The quotient (1/10) will be built in R9
- Lines 596 - 602: Loop of shifting and subtracting. Each pass through the loop generates one binary column of the quotient in Figure 13.2.
- Line 602: If the fraction is exact, such as if we were calculating 3/4, then the loop will finish early
- Line 604: Set up for dividing 25 by 10.
- Line 605: 64-bit product (whole number in R10, fraction in R11)

Go ahead and compile, download, and run the division program in Listing 13.5. I recommend setting at least the following two breakpoints:

1. Address C; The reciprocal of ten has been calculated and appears in register R9.
2. Address E; The value of 2 ½ is in registers R10 and R11. Note that ½ is $0.011111..._2$ instead of 0.1_2 because the reciprocal of 10 is slightly less that 1/10 because 1/10 cannot be represented exactly in binary.

Another interesting breakpoint location is address 6 which is the top of the division loop. You can watch the 1/10 (binary .00010011001...) being built in register R9 as the number of passes through the loop decreases from 31 to 0.

Highlights and Comparisons

The sum of products instructions (MLA, UMULAL, SMULAL) are popular in many computer architectures, not just that of the ARM. The NEON coprocessor to be introduced in Chapter 20, also supports multiply with accumulate. These instructions assist in calculating dot and scalar products in science, engineering, and computer graphics applications.

Working with fractions in binary, such as those in Listing 13.5, is a good start at getting comfortable with understanding the floating point format to be presented in Chapter 22.

Multiplication instructions are used extensively in computer programs. We have found that shifting and adding is a much faster method for calculating a product than simply adding the minuend to itself multiple times. There are even faster algorithms, albeit more complicated, that are actually used in real CPUs.

Review Questions

1. * "By hand, without a computer," convert the following decimal fractions into binary and provide the answers in hexadecimal.
   a. $0.5_{10} = ½ =$
   b. $0.625_{10} =$
   c. $0.25_{10} =$
   d. $0.03125_{10} =$
   e. $0.0078125_{10} =$
2. * "By hand, without a computer," convert the following binary fractions from hexadecimal back into real numbers in base 10.
   a. $.C0000000_{16} = 0.11_2 =$
   b. $.E0000000_{16} =$
   c. $.10000000_{16} =$
   d. $.50000000_{16} =$

3. Search the Internet to find the subtle difference between SMULL and UMULL.
4. * How can you load ¾ into a register using a MOV instruction?
5. * Using a calculator for division by powers of 2, convert the following binary fractions from hexadecimal into the decimal fraction that each is "approaching."

    a. $.33333333_{16}$
      $= 0.0011001100110011001100110011001100110011_2 => 0.2_{10}$
    b. $.66666666_{16} =>$
    c. $.CCCCCCCC_{16} =>$
    d. $.E6666666_{16} =>$

6. * Why will multiplying by 0.1 always result in a loss of precision in binary computers?

Exercises

1. Write and test an assembly language program using the MLA instruction to calculate $10 \times 4 + 3 \times 5 + 6 \times 7$. You can use the first 585 lines of listings 13.3, 13.4, or 13.5 for the assembler and ARM execution processor.
2. Modify Listing 13.5 to square ½ and show that its answer is ¼. Also show that $3 \times ¼ = ¾$.

— 14 —
Load & Store Instructions

Addressing memory is more complicated than you might expect. Four bits within an instruction can identify one of sixteen general purpose registers. However, 30 bits are required to select one byte from a gigabyte of memory. How do we get these large memory addresses into a machine code instruction?

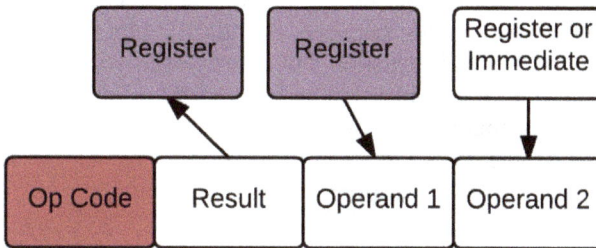

Copy of Figure 9.2: Instruction format in a RISC architecture

Most CISC architectures support multiple instruction lengths to accommodate memory addresses within their machine code formats. However, two common characteristics of RISC architectures, including that of the ARM, are all instructions are the same size, and none of the arithmetic and logic operations access data directly from memory.

The ARM has several instructions dedicated to moving data between its 16 general purpose registers and main memory. Chapter 14 implements the ARM load and store instructions in the following three stages:

1. **Assembler:** Macros, tasks, and functions will convert four ARM load and store instructions from assembler format to machine code. Three pseudo ops will also be implemented for initializing data in memory.
2. **Execute:** A new Verilog module will be created to hold read/write memory and move data between it and the general purpose R registers. It will be integrated into the previous system so that an ARM program can use memory storage as part of its calculations.
3. **Application Demonstrations:** A short program that mixes byte stores with word loads will demonstrate the little endian feature of memory present in most of today's computers.

Introductions

The following four load and store instructions move data between any one of the 16 general purpose R registers and main memory.

- **STR:** Store contents of a 32-bit R register into memory
- **STRB:** Store contents of lower 8 bits of an R register into memory
- **LDR:** Load all 32 bits of an R register from memory
- **LDRB:** Load lower 8 bits of an R register from memory

Bits, Bytes, Words

How many bits does one machine code instruction move between the registers and memory? In the ARM processor, that number depends on the type of load/store instruction and varies between 8 and 512 bits. Note: This is the number of bits moved by one instruction, but not necessarily in one clock cycle.

Chapter 14 introduces the instructions that move a single word or byte between a register and memory. Chapter 16 will add increased capability and complexity to these same load and store instructions. Chapters 18 and 21 will describe moving data between multiple registers and memory.

Byte or Word Addressable

Computer applications have traditionally been divided into two major categories: text processing or number "crunching." Text processing typically involves the manipulation of strings of ASCII characters, where each character is stored in one byte. Engineering and scientific number crunching applications typically work with integers and floating point numbers, where each number is stored in a single word.

Can text strings, integers, and floating point be stored in the same physical memory? They don't have to be, but in the vast majority of computer architectures, bytes and words are stored together.

So if four bytes are in one word, do we address the common memory space as words or bytes? Today, almost all computers are byte addressable, where a word is composed of four adjacent bytes. In the mainframe days, many computer architectures were word addressable, where each word would contain between two and ten bytes, depending on the particular computer architecture.

Indirect Memory Addressing

The ARM architecture has an addressing mode which is referred to as "indirect addressing" where any of the 16 general purpose R registers can point to a byte in memory. Having 32 bits, these registers can address any of 4.29 billion bytes.

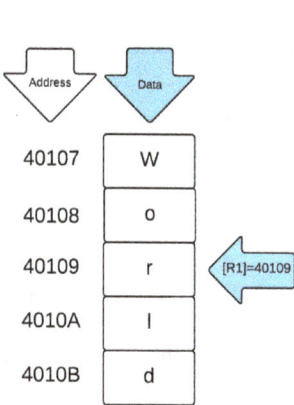

Figure 14.1 illustrates the indirect mode where register R1 contains the memory address 40109. This "base" address can be modified with an offset that will be described in Chapter 16.

The ARM architecture has even another "base register" that is used by the operating system for relocation of user programs, but it will not be described in this book.

Figure 14.1: Indirect addressing points to a memory location.

Load and Store Instruction Format

Figure 14.2 shows the assembly language and the corresponding machine code of an instruction which stores the full 32 bits of register R2 into four consecutive bytes of memory beginning at the byte address contained in register R4. The parentheses around register R4 imply that R4 does not contain the data but rather points to it in memory (i.e., indirect). In most ARM assemblers, square brackets are used instead of parentheses, but that could not be done with the Verilog macro syntax used in this book.

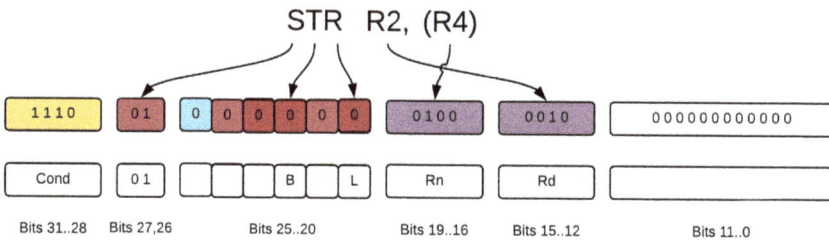

Figure 14.2: Store full 32-bit register into memory.

These LDR/STR instructions are conditionally executed, the same as with the data processing and multiply instructions. The machine code for the STR, STRB, LDR, and LDRB instructions differ only by the Byte and Load bits as shown in the figures. Other control bits and an offset provide additional features which will be described in Chapter 16.

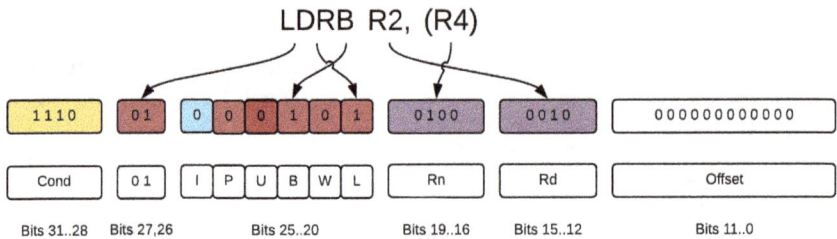

Figure 14.3: Load lower 8 bits of register from byte in memory.

- Cond (bits 31 - 28): Status indicating whether the instruction should be executed (1110 indicates always execute).
- 0 1 (bits 27, 26): Opcode indicating that this is a LDR/STR instruction.
- IPUBWL (bits 25..20): Bits indicating specifically what is to be done by the LDR/STR instruction. All bits will be explained in Chapter 16.
  - B (bit 22): "1" indicates Byte (LDRB), "0" indicates word
  - L (bit 20): "1" indicates Load (LDR), "0" indicates store

- Rn (bits 19 - 16): Base address register, R0 through R15
- Rd (bits 15 - 12): Data register to provide/receive data
- Offset (bits 11 - 0): Offset to address in Rn

Alignment

Most byte-oriented CPU architectures prefer multi-byte memory accesses to be on aligned boundaries. A word should be loaded from an address that is a multiple of 4 bytes. Likewise a double-word and half-word should be on addresses that are multiples of 8 and 2, respectively. Words that are not aligned can lead to performance degradation or even a memory failure interrupt, depending on the processor version and manufacturer.

The assemblers for most CPU architectures have an ALIGN pseudo op that relieves programmers of the responsibility of counting bytes to get the correct alignment. I have included an ALIGN task that duplicates that feature.

Pseudo Instructions

Assembly language programs consist of lines of text that are converted into machine code instructions. Each text line has the following four columns:

1. Label: Name associated with instruction's memory address
2. Op-code: The operation being performed (add, sub, shift, ...)
3. Operands: Location of the data (usually a register or constant)
4. Comment: Describes why the instruction is being used

Computer Architecture Tutorial Using an FPGA

Assemblers have other "instructions" that do not correspond to machine code instructions even though they look the same. These pseudo instructions sometimes generate code, and other times just set parameters used by the assembler to create the code. I have added the following three pseudo "ops," two generate code and one adjusts the address to hold the next instruction.

- **WORD:** Initialize 32-bit word in memory
- **BYTE:** Initialize 8-bit byte in memory
- **ALIGN:** Set IP address to a multiple of bytes

Assembly of Load and Store Instructions

Listing 14.1 contains the macros, tasks, and functions that will perform the assembly language to machine code conversion for the memory access instructions and pseudo ops. Listing 14.2 is identical except it provides eight hexadecimal digits in the user interface display for DE2-115 users.

```
1.  // Listing 14.1 introduces assembly of load/store instructions
2.  // 1) Assembles 32-bit ARM multiplication instructions
3.  // 2) Dumps 32-bit words from memory in hexadecimal
4.  //
5.  // Modules and macros contained in this file:
6.  // 1) CPU_UI: User Interface that dumps 32-bit words, 16 bits at a time
7.  // 2) Macros for only assembling ARM LDR/STR/WORD... instructions
8.  // 3) ProgMod: Memory containing "ARM" program
```

Listing 14.1 A: File to test the assembly of load and store instructions

```
50.  //
51.  //--------------- Macro definitions for assembly language ----------------
52.  //

61.  `define STR asmem (3'b000,      // Store full 32-bit R register
62.  `define STRB asmem (3'b100,     // Store low-order byte in register
63.  `define LDR asmem (3'b001,      // Load full 32-bit R register
64.  `define LDRB asmem (3'b101,     // Load lower 8 bits and zero fill
65.  `define WORD asdat (4,          // Initialize 32-bit word in memory
66.  `define BYTE asdat (1,          // Initialize 8-bit byte in memory
67.  `define ALIGN asdat (0,         // Align IP to multiple of bytes
```

Listing 14.1 B: LDR, STR, WORD, BYTE, ALIGN macros for assembly language conversion

Listing 14.1 file has 187 lines and contains only enough code to assemble and test the various formats of the load and store instructions, including the word, byte, and align pseudo-ops. Its macros, tasks, and functions will be integrated into the larger code that will execute both the data processing instructions (from Listing 12.6) and the newly added multiplication instructions in Chapter 13.

```
117.  // Task asmem is called by load/store opcode macros `LDR, `STR, ...
118.  // The number of parameters will vary between three and eight.
119.  // `LDR`EQ R1,(R2) `_4 // General format (for now) is 4 parameters
120.  // `LDR R1,(R2) `_3 // Conditional parameter is optional
121.
122.  task asmem ();
123.    input [15:0] P0,P1,P2,P3,P4,P5,P6,P7;
124.    {progMem[IP+3],progMem[IP+2],progMem[IP+1],progMem[IP]} <=
          asmem1(P0,P1,P2,P3,P4,P5,P6,P7);
125.    IP = IP + 4;
126.  endtask
```

Listing 14.1 C: Task asmem assembles load/store instructions

The assembly of these new instructions and pseudo ops is very similar to that done for the data processing and multiplication instructions. The asmem task is shown in 14.1 C, and of course, the full listing containing the functions it calls is provided in the downloaded file. Take special note to lines 124 and 125 which accommodate the 32-bit instruction being placed into four consecutive bytes.

```
147.  // Task asdat is called by WORD, BYTE, ALIGN macros
148.
149.  task asdat ();
150.    input [31:0] P0,P1,P2,P3,P4,P5,P6,P7;
151.    if (P0==0)                                          // align
152.      IP = (IP+P1-1) & ~(P1-1);
153.    else
154.      if (P0==1)                                        // byte
155.        begin
156.          progMem[IP] <= P1;
157.          IP = IP + 1;
158.        end
159.      else
160.        if (P0==4)                                      // word
161.          begin
162.            {progMem[IP+3],progMem[IP+2],progMem[IP+1],progMem[IP]}
                  <= P1;
163.            IP = IP + 4;
164.          end
165.  endtask
```

Listing 14.1 D: Task asdat assembles memory pseudo-ops

The three pseudo ops are implemented with just a single task, asdat, and do not call any functions for assistance. Line 152 in Figure 14.1 D shows how the ALIGN pseudo op aligns the next IP address to be used to a multiple of 2, 4, 8, 16, ... bytes. For example, if the current IP address is 13, an "ALIGN   8" command will change the IP to 16 (double word boundary). Most often, an "ALIGN   4" command is used to set IP to a multiple of 4 for word alignment.

Copy of Figure 11.4: Controls for running instruction dump

The copy of Figure 11.4 shows the test setup for viewing 32-bit numbers on the 7-segment displays.

1. Push KEY[0] to initialize.
2. Set the switches to desired memory address (0 up to 31).
3. Push KEY[1] to update and show machine code instruction
4. Repeat steps 2 and 3 to see the contents of other addresses.

Go ahead and compile, download, and execute the Verilog code in Listing 14.1. Follow the same procedures as in previous chapters to verify that the proper machine code instructions are generated. Notice that the instructions shown in Listing 14.1 E now begin on byte addresses.

I deliberately provided data to the WORD pseudo op on line 184 that did not fill the entire 32-bits. When you load address 18 (hexadecimal), the value is right-justified in the 32 bit register. Also, try addresses 19, 1A, 1B, and 1C (i.e., you will be starting in the middle of the word). Any surprises?

```
174. begin
175. IP           =        0              ;
176.             `LDR     R4,(R5)    `_3   // 0: E4154000
177.             `LDR`GT  R4,(R5)    `_4   // 4: C4154000
178.             `LDRB    R4,(R5)    `_3   // 8: E4554000
179.             `STR     R1,(R2)    `_3   // C: E4021000
180.             `STRB    R1,(R2)    `_3   // 10: E4421000
181.             `BYTE    8'H98      `_2   // 14: 8'h98
182.             `BYTE    25         `_2   // 15: 8'h19
183.             `ALIGN   4          `_2   // Set for word boundary
184.             `WORD    'HABCDEF   `_2   // 18: 32;h00ABCDEF
185. progMem[99]  <=      0              ;
186. end
```

Listing 14.1 E: Assembly of assortment of memory access instructions and pseudo-ops

14: Load & Store Instructions   253

Big and Little Endian

We usually load words using the LDR instruction from an array written by the STR instruction or generated by the WORD assembler pseudo op. What happens if we load words that were written by the STRB instruction or generated by the BYTE pseudo op?

At first glance it appears that when a "whole" word is loaded from a string of bytes in memory, the bytes are in reverse order. It's as if the bytes were loaded into the wrong "end" of the register.

Today, almost all computers load bytes starting from the "little end" (i.e., bit 0) of a register. In the past, some byte-addressable computers loaded their registers starting with the "big end" (bit 31).

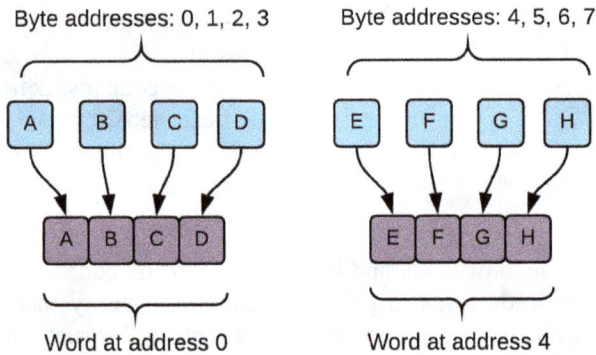

Figure 14.4: Big endian: The first byte goes to the "Big End" of the register.

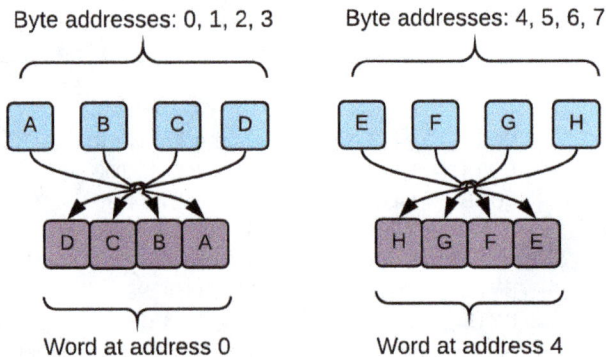

Figure 14.5: Little endian: The first byte goes to the "Little End" of the register.

Figure 14.4 illustrates the string "ABCDEFGH" being loaded as words in big endian format while Figure 14.5 shows the same string being loaded in little endian format. Note: The ARM 7 processor defaults to little endian format but can be switched to big endian by setting CPSR bit 9.

Big endian format seems very natural to many people, while little endian seems so backward. Why would anyone prefer little endian? One advantage is in casting (i.e., conversion) among types of variables. In the example in Figure 14.6, memory address 1620 can represent an integer of various sizes, all with the same value of 25.

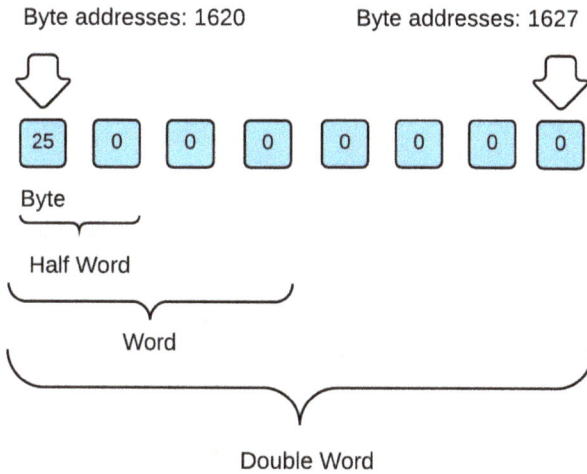

Figure 14.6: "Automatic" casting of a constant into different variable types

For the Load Byte ARM instruction, the 8 bit byte is loaded into the lower 8 bits, while the upper 24 bits will be set to zero. There is a special LDRSB instruction described in the Highlights and Comparisons section that sign fills the upper 24 bits.

Harvard and Von Neumann

The early computer designs had the instructions located in a physical medium that was independent of the data storage. Babbage's analytical engine had the instructions on punched cards modeled after the Jacquard loom of the 1800s. Some computers in the 1940s had their programs entered and stored in plug boards. This design, having the instructions and data in separate media or at least on separate I/O buses or data channels, is referred to as a Harvard architecture.

The Harvard architecture is still very popular today, but mostly with micro-controllers used in embedded systems, rather than microprocessors used for general computing. A micro-controller is a small System On a Chip (SOC) containing a CPU, some memory, and specialized I/O data lines. Some advantages of the Harvard architecture follow:

- Improved performance because instructions and data can be fetched from the two separate buses at exactly the same time.
- The instruction size does not depend on the byte and word sizes.
- Programs cannot be accidentally written over as if they were data.

Figure 14.7: Harvard architecture has independent instructions bus and data bus.

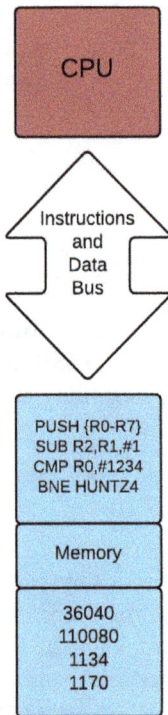

Figure 14.8: Von Neumann architecture

A different approach is attributed to John Von Neumann in which the instructions and data are in the same memory or at least on the same memory bus. Most general purpose computers, including the mainframes from the 1960s, minicomputers from the 1970s, and microcomputers from the 1980s have the Von Neumann architecture. Some advantages of the Von Neumann design follow:

- Because a program is simply data stored on the computer's disk, this architecture enables a large variety of programs to be selected to be run.
- Multiple applications can be independently run on the same computer at the same time, and an operating system can easily share the computing resources among these applications.
- Only one memory bus is needed, thereby reducing costs.
- Programs can be developed on the same machine in which they will eventually run.

Computer Architecture Tutorial Using an FPGA

Execution of Load and Store Instructions

Now that we can build the ARM machine code for the load and store instructions, how do we execute it in a running program? How can we integrate memory access with the existing data processing and multiplication instructions? Basically, let's follow the pattern done in developing the previous instructions:

1. Describe a new module that contains RAM memory and "does the work" of loading and storing single bytes or words. It will be given the name DataMod.
2. Instantiate the new DataMod module. This not only creates it, but its argument list connects its data lines to those of the main CPU module.
   a. Input data is from the Rd and Rn register contents.
   b. A clock pulse indicates when the load or store is to take place.
   c. Output is a 32-bit value for LDR and LDRB.
3. Modify the CPU instruction cycle to include loading and storing data from memory and distinguish it from the other ARM instructions.
   a. The decode circuit will set a signal, SelRAM, when the load and store opcodes are present.
   b. The writeBack state will store the results of the LDR and LDRB instructions into the appropriate Rd destination register.
   c. The CPSR will be <u>not</u> be updated for load and store instructions.

```
1.  // Listing 14.3: Demonstrate RAM memory with LDR and STR instructions
2.  // 1) Assembles most ARM instructions for computation and memory access
3.  // 2) Executes demonstration of little endian memory access
4.  // 3) Dumps any "one" 32-bit register at a time.
5.  // 4) Breakpoint address can be set to "stall" assembler program
6.  //
7.  // Modules and macros contained in this file:
8.  // 1) CPU_UI: User Interface that dumps 32-bit words, 16 bits at a time
9.  // 2) CPU: 32-bit CPU with data processing and multiplication instructions
10. // 3) DataProcIns: 16 "ARM like" DP instructions using 32-bit registers
11. // 4) Op2Mod: Fill in second operand including all shift possibilities
12. // 5) ShiftMod: Calculate all shift possibilities for Op2Mod
13. // 6) MultiplyIns: Module to calculate multiplication
14. // 7) DataMod: RAM memory and load/store instructions
15. // 8) Macros for assembling ARM data processing and multiplication instructions
16. // 9) ProgMod: Program to demonstrate little endian feature of memory
```

Listing 14.3 A: Directory header for Listing_14.3.txt file

```
132.   // Instantiate program memory and ALU processing
133.
134.   wire SelDP, SelMul, Sel64, SelRAM;
135.   assign SelDP = ok2exe && IR[27:26]==0 && IR[7:4]!=4'b1001;
136.   assign SelMul = ok2exe && IR[27:24]==0 && IR[7:4]==4'b1001;
137.   assign SelM64 = SelMul && IR[23];
138.   assign SelRAM = ok2exe && IR[27:26]==2'b01;
139.   ProgMod (R[PC], CPU_state==fetch, reset, IR);
140.   Op2Mod (RmVal, RsVal, op2raw, iFlag, op2Val);
141.   DataProcIns (opCode, RnVal, op2Val, CPU_state==execute && SelDP, CPSR,
       RdVal_DP, RdVal_DPU, CPSR_DP);
142.   MultiplyIns (opCode, RdVal, RmVal, RnVal, RsVal, CPU_state==execute &&
       SelMul, CPSR, RdV64_M, CPSR_M);
143.   DataMod (IR[25:20], R[RdID], RnVal, CPU_state==execute && SelRAM,
       RdVal_RAM);
```

Listing 14.3 B: Instantiate ALU modules for instruction execution, including load/store.

The new DataMod module is instantiated on line 143 shown in Listing 14.3 B. It has the following five arguments:

1. IR[25:20]: IPUBWL command bits (B => Byte, L=> Load)
2. R[RdID]: Data value needed for STR and STRB
3. RnVal: Memory byte-address from Rn register
4. CPU_state==execute && SelRAM: Clock "tick" that says when the memory transfer will take place.
5. RdVal_RAM: Data loaded from memory by LDR and LDRB

```
195.   writeBack:              // Update specific register
196.     begin
197.       {R[RnID],R[RdID]} <= (RdVal_DPU & SelDP) ? {R[RnID],RdVal_DP} :
198.       (SelM64) ? RdV64_M :
199.       (SelMul) ? {RdV64_M[31:0],R[RdID]} :
200.       (SelRAM) ? {RnVal,RdVal_RAM} :
201.         {R[RnID],R[RdID]};
202.       if (SU) CPSR <=
203.         SelDP ? CPSR_DP :
204.         SelMul ? CPSR_M :
205.           CPSR;
206.       CPU_state <= fetch;
207.     end
```

Listing 14.3 C: The writeBack state selects which data updates Rd and Rn.

The writeBack state of the instruction cycle has two registers to update: Rn and Rd (Listing 14.3 C line 200). Currently, the DataMod module will not alter Rn, but that will change in Chapter 16 when additional features are added to the load and store instructions. The Verilog conditional operator selects the updated values based upon the opcode value of the current instruction (flags SelDP, SelMul, SelM64, SelRAM).

On line 202, the CPSR will be updated based upon whether the "S" bit is set in the instruction and which opcode is currently being executed. The load and store instructions do not contain an "S" bit and do not update the CPSR.

```
333. //---------------- ALU for load/store instructions and RAM memory ----------------
334. //
335. module DataMod (IPUBWL, Rd, Rn, clk, Rd32);
336.    input [5:0] IPUBWL;
337.    input [31:0] Rd,Rn;
338.    input clk;
339.    output reg [31:0] Rd32;          //
340.    reg [7:0] RAM[0:99];             // Read/Write Random Access Memory
341.
342.    always @ (posedge(clk))
343.      if (IPUBWL[0])                 // Load
344.        if (IPUBWL[2])               // byte
345.          Rd32 <= {24'b0,RAM[Rn]};
346.        else                         // Load word
347.          Rd32 <= {RAM[Rn+3],RAM[Rn+2],RAM[Rn+1],RAM[Rn]};
348.      else                           // Store
349.        begin
350.          if (IPUBWL[2])             // byte
351.            RAM[Rn] <= Rd[7:0];
352.          else                       // Store word
353.            begin
354.              RAM[Rn+3] <= Rd[31:24];
355.              RAM[Rn+2] <= Rd[23:16];
356.              RAM[Rn+1] <= Rd[15:8];
357.              RAM[Rn] <= Rd[7:0];
358.            end
359.          Rd32 <= Rd;
360.        end
361. endmodule
```

Listing 14.3 D: ALU module for RAM memory and load/store instructions

The DataMod module in Listing 14.3 D contains the read/write RAM memory and the load/store instruction execution. The argument list was previously described in Listing 14.3 B. It's operation is pretty straightforward and based on the two command bits: B in IPUBWL[2] and L in IPUBWL[0]. The output data,

Rd32, will be "updated" even for the STR and STRB instructions (line 359) because the CPU instruction cycle writeBack state always assumes a new value.

- Line 340: 100 bytes of read/write RAM memory (This is part of the FPGA, not the RAM memory that is included on the DE10-Lite).
- Lines 344 - 347: Load from memory
- Lines 350 - 359: Store into memory
- Line 359: Even though it's a store instruction, the writeBack state of the CPU instruction doesn't need to know

Little Endian Application Demonstration

The load and store instructions will be demonstrated in a simple program that stores 20 bytes into memory and then reads back the first 8 bytes as two words. In the process, the little endian characteristic of byte-addressable memory can be observed.

684. //	Program to demonstrate little endian		
685.			
686.	`MOV	R0, 0	`_3 // 0: Reset (i.e., "boot" address)
687.	`MOV	R2, 0	`_3 // 1: R2 will be loop counter
688. SLoop	=	IP	;
689.	`STRB	R2, (R2)	`_3 // 2: Store 0, 1, 2, ... into bytes
690.	`ADD	R2, 1	`_3 // 3: Increment loop index
691.	`CMP	R2, 20	`_3 // 4: Store 0 up to 20 in bytes
692.	`MOV`NE	R15, SLoop	`_4 // 5: "Branch" back until 20
693.			
694.	`MOV	R3, 0	`_3 // 6: Pointer for word loads
695.	`LDR	R4, (R3)	`_3 // 7: Load first 4 bytes
696.	`ADD	R3, 4	`_3 // 8: Point to second word
697.	`LDR	R5, (R3)	`_3 // 9: Load second 4 bytes
698.			
699.	`MOV	R15, IP	`_3 // A: Infinite loop
700.			

Listing 14.3 E: ARM program demonstrating little endian memory fetches

- Lines 687 - 692: Loop to initialize 20 bytes of memory to 0, 1, 2, ..., 19 at RAM address in register R2
- Line 695: Load R4 with 32'h03020100 from memory
- Line 697: Load R5 with 32'h07060504 from memory
- Line 699: Infinite loop (program stalls)

Go ahead and compile, download, and run the example program in Listing 14.3. Just set the breakpoint address to binary 1010 or above to check that the registers were loaded in the little endian sequence.

Chapter 14 introduced load and store instructions for single bytes and words in memory. Chapter 16 will elaborate on these same instructions giving them indexing and index updating capabilities. Chapter 18 will describe instructions that store and load multiple registers at one time. There are also a few other instructions that were later squeezed into the original ARM instruction format (STRH, LDRH, LDRSB, LDRSH).

Instruction	Number of bits stored	Bit position in registers
STR	32 bits (1 word)	Bits 0 to 31 (whole register)
STRB	8 bits (1 byte)	Bits 0 to 7 (right side)
STRH	16 bits (1 half word)	Bits 0 to 15 (right half register)
STM	Between 32 and 512 bits (1 to 16 words)	All 32 bits of each of the selected registers

Table 14.1: Operations that store data into memory from registers

All 16 ARM general purpose registers, R0 through R15, have 32 bits. Table 14.1 indicates that the instructions that store a partial register (STRB, STRH) copy the low order bits to memory. The high order bits of the register (bits 31-8 for STRB and bits 31-16 for STRH) are not copied into memory.

Instruction	Number of bits loaded	Bit position in registers
LDR	32 bits (1 word)	Bits 0 to 31 (whole register)
LDRB	8 bits (1 byte)	Bits 0 to 7 (right side), zero fill bits 8 to 31
LDRSB	8 bits (1 byte)	Bits 0 to 7, sign extend bits 8 to 31
LDRH	16 bits (1 half word)	Bits 0 to 15 (right half register), zero fill bits 16 to 31
LDRSH	16 bits (1 half word)	Bits 0 to 15, sign extend bits 16 to 31
LDM	Between 32 and 512 bits (1 to 16 words)	All 32 bits of each of the selected registers

Table 14.2: ARM instructions that load data from memory into registers

In the partial register store instructions, the high order bits of the register are ignored. Are they also ignored in the corresponding partial register load instructions? No. The high order bits are either set to zero (LDRB, LDRH), or they are filled with the sign bit. For LDRSB, bit 7 of the byte in memory will not only be copied into bit 7 of the register, but also bits 8 through 31 of the register. For the half-word load (LDRHS), the same bit loaded into bit 15 of the register will also be copied into bits 16 through 31. In other words, the sign of the byte or half-word in memory will be the sign of the data loaded into the register.

The DE10-Lite and DE2-115 development boards have separate RAM memory chips that can be used by user code. In this book, I did not use that memory in order to simplify the examples and focus on the Verilog coding. However, using that memory instead of allocating resources from the FPGA has many advantages.

The term microprocessor is usually reserved for integrated circuit chips that contain only the CPU, and it usually has only digital I/O lines. Like most microprocessors, the ARM 7 has a Von Neumann architecture. The ARM 9 is unique among the family of ARM processors in that it has a separate address bus and data bus, and it therefore falls within the general characteristics of a Harvard architecture. I hope you noticed that I implemented the memory in this chapter as though it was a Harvard architecture.

Review Questions

1. *The addressing of computer memory has been a nightmare for decades. A large part of the problem has ironically been due to a good factor: the tremendous drop in memory prices. Please explain?
2. *Compare the merits of a byte-addressable computer architecture to one that is word addressable.
3. In Chapter 14, the ARM imitation was implemented as a Harvard architecture. What characteristics of the Verilog code make it a Harvard, rather than Von Neumann, architecture? How could this be changed to make it more compatible with most ARM designs?

Exercise

1. The LDRSB, LDRH, LDRSH, and STRH instructions have been squeezed into the original ARM instruction format. Check the Internet for their machine codes, and then modify Listing 14.1 with the macros and task needed to assemble them.

— 15 —
Branch Instructions

What's wrong with directly modifying the PC register to change program flow? Isn't that what the branch instructions do anyway? Actually, the vast majority of computer architectures keep the PC as an internal register that user programs can only change through branch or "jump" instructions. Even the 64-bit version of the ARM processor does not permit its PC to be modified directly.

Chapter 15 implements and demonstrates the ARM branch instructions in the following three stages:

1. **Assembler:** Macros, task, and functions will be provided to convert the ARM branch instructions to the 32-bit machine code that is identical to that which a real ARM would execute.
2. **Execute:** The branch instruction will be integrated into the CPU instruction cycle of fetch/decode/execute.
3. **Application Demonstrations:** The factorial program will be coded three ways to demonstrate the use of branch instructions.
   a. **Factorial:** Same as before, but using the real branch instruction instead of a MOV to the PC.
   b. **Factorial Subroutine:** Typical approach to calling a subroutine.
   c. **Recursive Factorial Subroutine:** The factorial subroutine will call itself to calculate a smaller factorial.

Introductions

The following four ARM instructions are available to change the PC so that a different section of a program can be run:

- **B:** Branch; Change the PC by adding or subtracting an offset from its current value.
- **BL:** Branch with Link; The PC is changed and a return address is saved in the Link Register (LR is R14).
- **BX:** Branch and Exchange; Branch to the location contained in one of the general purpose registers and possibly change to Thumb mode.
- **SWI:** Software Interrupt; Branch to a hardwired address within the operating system for a specific software service.

Program Flow

The data processing and multiplication instructions do all of the calculations within a computer program. Load and store instructions move data between main memory and the CPU when needed for processing. The order in which these operations take place is controlled by the branch instructions, which modify the PC for the following reasons:

1. Special data requires special processing
2. Loops
3. Calling subroutines

B: Branch Instruction Format

At first glance, it appears the ARM processor has many different branch instructions. However, as you might suspect, instructions like BGE, BGT, BEQ, etc. are really just the same "B" branch instruction that relies on those four high-order NZCV condition bits to indicate whether the branch is to be taken. Figure 15.1 compares the format of a branch instruction to that of the data processing instructions.

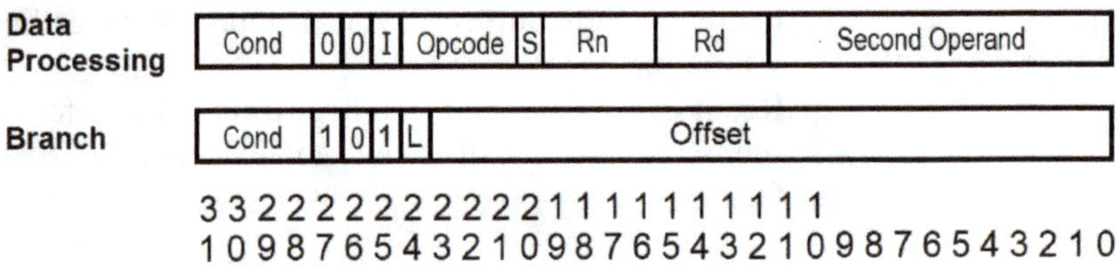

Figure 15.1: Branch instruction format

BL: Branch with Link Instruction

The BL instruction is simply the above branch instruction with the L bit set. Notice that there is not a 4-bit "register field" in the machine code instruction. It is assumed that the LR register (general purpose register R14) will be loaded with the address of the instruction immediately following the BL instruction. This will be the "return address" that a called subroutine will use when it is finished with its assigned job.

A subroutine is a section of code that is "called" to perform a specific job. Depending upon the programming language and application, subroutines are also known as functions, modules, procedures, and methods. Examples of jobs a subroutine can perform:

- Calculate the factorial of a number
- Display a number to the user
- Get input from a specific device such as a temperature sensor
- Change the speed of a motor
- Perform a particular type of analysis such as a least-squares fit of data

BX: Branch and Exchange Instruction

A branch and exchange instruction copies the contents of any one of the general purpose registers to the PC, and it can switch the CPU instruction decode circuitry to expect 16-bit format Thumb instructions. Figure 15.2 compares its format to that of the branch instruction and the data processing instructions. Notice that it is conditionally executed, just like all of the ARM instructions, and it has one 4-bit register field.

The BX instruction with its Rm field set to 14 (i.e., the LR register) is commonly used to return from a subroutine that has been called with the BL instruction. At the same time, the CPU can be set to Thumb mode. See Chapter 19 for more information on this alternate, and abridged, instruction format for the ARM processor.

Since ARM instructions are all 32 bits in length, all instructions fall on a 4-byte alignment in memory. The CPU thereby assumes bits 0 and 1 of the address in the Rm register are zero. This assumption enables the CPU to re-purpose bit 0 of the Rm register as a flag for switching into Thumb mode.

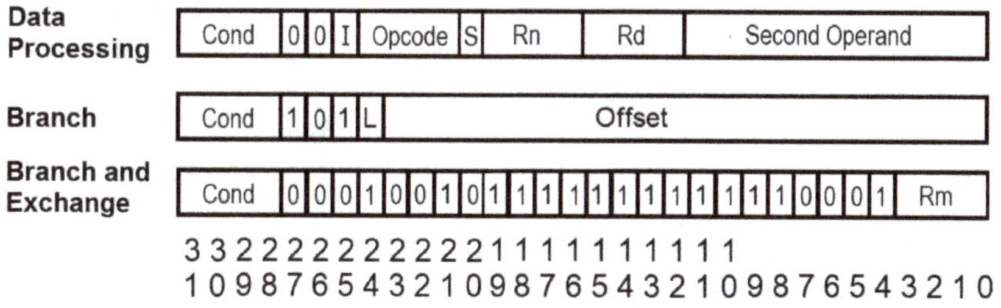

Figure 15.2: Branch and exchange instruction format

SWI: Software Interrupt Instruction

Computer operating systems, such as Windows and Linux, provide a variety of services to user programs through "subroutines" that are called using a "software interrupt." The SWI is similar to the BL instruction, but it always branches to the same location (address 8 in the ARM), and the processor is switched into a special supervisory mode.

Many ARM assemblers provide both the SWI and SVC (service call) mnemonics for this instruction. I will not be implementing the software interrupt in this book's ARM imitation, but only mentioning it because of its importance with implementing multi-user operating systems.

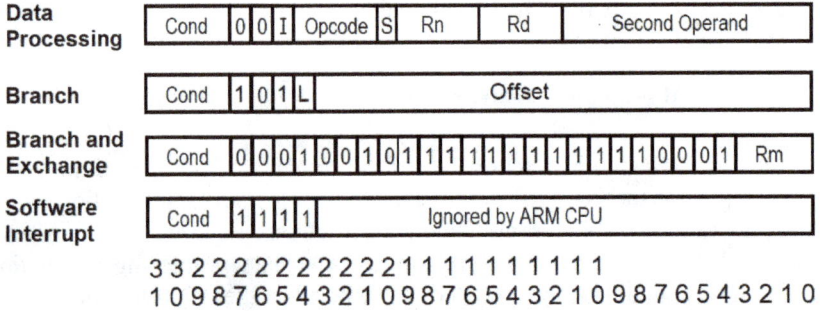

```
Data
Processing   | Cond | 0 0 I | Opcode | S | Rn | Rd |   Second Operand

Branch       | Cond | 1 0 1 L |                Offset

Branch and
Exchange     | Cond | 0 0 0 1 0 0 1 0 1 1 1 1 1 1 1 1 1 1 1 1 0 0 0 1 | Rm

Software
Interrupt    | Cond | 1 1 1 1 |          Ignored by ARM CPU

              3 3 2 2 2 2 2 2 2 2 2 2 1 1 1 1 1 1 1 1 1 1
              1 0 9 8 7 6 5 4 3 2 1 0 9 8 7 6 5 4 3 2 1 0 9 8 7 6 5 4 3 2 1 0
```

Figure 15.3: Software interrupt

Multiprogramming and Relative Addresses

Multiprogramming is a software technique where multiple independent programs are "running" within a single CPU by sharing the CPU time and other resources such as memory and input/output channels. It requires an operating system, such as Windows or Linux, to manage the sharing of resources and the switching of the CPU among the several different programs. Hardware features such as memory protection and privileged instructions are also required.

When a program is compiled (or assembly program assembled), it is not known where in memory it will be actually be loaded. A dozen different programs may be running at any one "time," and the operating systems will load them wherever it sees fit. CPUs designed for multiprogramming have additional memory mapping registers, sometimes called "base," bank, or translation registers, that accommodate this need. Branch instructions with relative offsets also assist in this program relocation requirement.

Today's computer systems can have gigabytes of memory, which require an address space of at least 30 bits. Although a program may be very large, most of its branches are very close. Relative branching is available as an instruction format in most CPU architectures where a value is added to or subtracted from, the current PC address. The ARM B and BL instructions contain relative "branch to" locations, while the BX uses an absolute byte address contained in the LR register. The SWI instruction always branches to absolute byte address 8.

Pipelining

Fetch, decode, and execute are the three fundamental states of almost every CPU design. As in the design presented in this book, each of these states is performed in its own independent circuitry. This allows all three states to be operating at the same time. This capability leads to a theoretical performance boost of up to 300% for program execution.

Obviously, for a given instruction, it cannot be executing while it is being decoded or before it is even loaded from memory. Pipelining is a common technique where the CPU is in the process of executing three different instructions at the same time: one is being fetched from memory, while a different one is being decoded, and a third one is actually performing the desired operation.

Pipelining works well, but it requires some assumptions, and it complicates the CPU operation and design. Generally, it is assumed that instructions are being executed in the order they appear in memory. What happens when that assumption is wrong, such as if the instruction is a branch? Basically, the pipeline is flushed, and the two instructions following the branch will have their memory fetches and instruction decode ignored. Branches hurt performance.

A second complication of pipelining, and the reason I'm describing it here, has to do with calculating relative addresses for branch instructions. Is the "branch to" address relative to the immediately available value of the PC register, or is it relative to the memory address of the branch instruction itself? Due to pipelining in the ARM, the relative address is eight bytes (i.e., two instructions) greater than the byte address of the branch instruction.

Assembly of Branch Instructions

The Listing 15.1 file has 176 lines and contains only enough code to assemble and test the various formats of the branch instructions. Its macros, task, and functions will be integrated into the ARM imitation in Listing 15.3 for running test and demonstration programs.

```
50.  //
51.  //--------------- Macro definitions for assembly language ----------------
52.  //
62.     `define B asbr (4'b1010,              // Branch
63.     `define BL asbr (4'b1011,             // Branch with Link
64.     `define BX asbr (4'b0001,             // Branch and Exchange
```

Listing 15.1 A: B, BL, and BX macros for assembly language conversion

```
110.  // ----- Tasks and Functions that implement the "assembler" -----
111.
112.    integer IP;
113.
114.  // Task asbr is called by branch opcode macros `B, `BL, and `BX
115.  // The number of parameters will be either two or three.
116.    // `B`EQ    TopAdr    `_3 // General format is 3 parameters
117.    // `BX    R14    `_2 // Conditional parameter is optional
118.
119.    task asbr ();
120.      input [15:0] P0,P1,P2,P3,P4,P5,P6,P7;
121.      {progMem[IP+3],progMem[IP+2],progMem[IP+1],progMem[IP]} <=
          asbr1(P0,P1,P2);
122.      IP = IP + 4;
123.    endtask
124.
125.  // Function asbr1 is called by task asbr to construct the condition fields.
126.    // `BL Format 1: Always branch (no status change possible)
127.    // `BL`EQ Format 2: Branch only if Z-flag set
128.
129.    function [31:0] asbr1 ();
130.      input [15:0] P0,P1,P2;
131.      if (P2 == 0)                          // Format 1
132.        asbr1 = 'hE<<28 | asbr2(P0,P1);
133.      else
134.        asbr1 = P1<<28 | asbr2(P0,P2);
135.    endfunction
136.
137.  // Function asbr2 fills the branch's register or address field.
138.
139.    function [31:0] asbr2 ();
140.      input [15:0] Q0, Q1;
141.      integer IPoff;
142.      IPoff = Q1 - IP - 8;
143.      if (Q0 == 1)                          // BX
144.        asbr2 = {24'h12FFF1, Q1[3:0]};
145.      else                                  // B or BL
146.        asbr2 = {Q0, IPoff[25:2]};
147.    endfunction
148.
```

Listing 15.1 B: Task and functions to assemble branch instructions

Compile, download, and examine the instructions created in Listing 15.1. Notice the following in the offset field for the branch instruction:

- Line 116: The argument for the B and BL instructions is the byte address containing the next instruction to be executed (represented by label TopAdr).

- Line 117: The argument for the BX instruction is a register ID number. The register is assumed to contain the byte address of the next instruction to be executed
- Line 120: The asbr task is called with eight arguments to be compatible with the `_2 and `_3 ending macros. No more than three are ever used by the branch instructions.
- Line 129: Function asbr1 handles the conditions for execution (GT, EQ, LT, etc.) and calls function asbr2.
- Line 142: The branch offset value is calculated based upon the PC value while the CPU is in the execute state. By that time, two more instructions (8 bytes) have been fetched and are in the pipeline.
- Line 146: Notice how the lower two relative address bits are not present, but assumed to be zero (i.e., this is a word address offset). This increases the relative branch range by a factor of 4 to be $\pm 2^{22}$ instructions ($\pm 2^{24}$ bytes).
- Line 160: Branch and Exchange example
- Line 163: Branching "backward" to a lower memory address gives a negative offset.
- Lines 164, 165, 168: Calling the same Sub1 subroutine. Note that the offsets are different in each call.
- Line 170: Typical return from subroutine using the LR (i.e., register R14)

```
159.
160.          `BX`GT    R4      `_3 // 0: C12FFF14
161. Lab1     =         IP;        //
162.          `BX       R5      `_2 // 4: E12FFF15
163.          `B`EQ     Lab1    `_3 // 8: 0AFFFFFD
164.          `BL       Sub1    `_2 // C: EB000001
165.          `BL`EQ    Sub1    `_3 // 10: 0B000000
166.          `BX       R14     `_2 // 14: E12FFF1E
167. Sub1     =         IP;        //
168.          `BL       Sub1    `_2 // 18: EBFFFFFE
169.          `BL`EQ    Sub2    `_3 // 1C: 0B000000
170.          `BX       R14     `_2 // 20: E12FFF1E
171. Sub2     =         IP;        //
172.          `BX       R0      `_2 // 24: E12FFF10
173.
```

Listing 15.1 C: Assembly of assortment of branch instructions

Execution of Branch Instructions

Now that we can build the ARM machine code for the branch instructions, how do we execute it in a running program? The implementation will be different from what was done for the data processing, multiplication, and data load/store instructions. No data registers are being modified and the CPSR is unchanged. Since the PC is being altered, I have placed the modifications directly into the CPU instruction cycle:

1. The decode circuit will set signals, SelB, SelBL, SelBX, when the branch opcodes are present.
2. The writeBack state will update changes to the PC and LR.
3. The CPSR will <u>not</u> be updated for branch instructions

```
1.   // Listing 15.3 Demonstrate ARM program branching and subroutines
2.   // 1) Assembles most ARM instructions including branch instructions
3.   // 2) Executes demonstration of factorial calculation
4.   // 3) Dumps any "one" 32-bit register at a time.
5.   // 4) Breakpoint address can be set to "stall" assembler program
6.   //
7.   // Modules and macros contained in this file:
8.   // 1) CPU_UI: User Interface that dumps 32-bit words, 16 bits at a time
9.   // 2) CPU: 32-bit CPU with data processing and multiplication instructions
10.  // 3) DataProcIns: 16 "ARM like" DP instructions using 32-bit registers
11.  // 4) Op2Mod: Fill in second operand including all shift possibilities
12.  // 5) ShiftMod: Calculate all shift possibilities for Op2Mod
13.  // 6) MultiplyIns: Module to calculate multiplication
14.  // 7) DataMod: RAM memory and load/store instructions
15.  // 8) Macros for assembling ARM data processing and multiplication instructions
16.  // 9) ProgMod: Program to demonstrate factorial calculated in a loop
```

Listing 15.3 A: Directory header for Listing_15.3.txt file

```
146.
147.    assign SelB = ok2exe && IR[27:24]==4'b1010;
148.    assign SelBL = ok2exe && IR[27:24]==4'b1011;
149.    assign SelBX = ok2exe && IR[27:4]==24'h12FFF1;
150.    assign BrOff = IR[23] ? {6'h3F, IR[23:0], 2'b00} : IR[23:0]<<2;
151.
```

Listing 15.3 B: Command signal flags determined by branch instruction opcodes

Text lines 147 through 149 of Listing 15.3 B show the branch command signals being set based upon their unique opcode bit patterns appearing in the ARM instruction. Line 150 calculates the branch offset:

1. It is a two's complement number in bit positions 0 through 23 (see Figure 15.3). For negative numbers, the sign bit (IR[23]) must be extended to a full 32 bits to match the size of the PC register.
2. The value must be multiplied by 4 (i.e., shifted left 2 places) because the PC is in bytes, while the "branch to" location is in words.

201.	writeBack:	// Update specific register
202.	begin	
203.	{R[RnID],R[RdID]} <= (RdVal_DPU & SelDP) ? {R[RnID],RdVal_DP} :	
204.	(SelM64) ? RdV64_M :	
205.	(SelMul) ? {RdV64_M[31:0],R[RdID]} :	
206.	(SelRAM) ? {RnVal,RdVal_RAM} :	
207.	{R[RnID],R[RdID]};	
208.	if (SU) CPSR <=	
209.	SelDP ? CPSR_DP :	
210.	SelMul ? CPSR_M :	
211.	CPSR;	
212.	{R[LR],R[PC]} <= (SelBX) ? {R[LR], RmVal} :	
213.	(SelBL) ? {R[PC], PCir+8+BrOff}:	
214.	(SelB) ? {R[LR], PCir+8+BrOff} :	
215.	{R[LR], R[PC]};	
216.	CPU_state <= fetch;	
217.	end	

Listing 15.3 C: The writeBack state can now also modify the PC and LR.

The writeBack state will now update two more registers (lines 212 through 215) to complete the execution of the branch instructions. Note: There is no conversion to Thumb format by this BX instruction code, but that could easily be added later.

- **SelBX:** Copy byte address in Rm register to the PC.
- **SelB:** Add the offset to current PC value.
  - The PC is in bytes and is 8 more than the address of the branch instruction.
  - The offset, BrOff, can be positive or negative and has already been converted from words to bytes.

- **SelBL:** Like SelB, an offset is added to current PC value. Also, the LR is loaded with the address of the instruction following the BL instruction.

Branch Application Examples

Finally, we get to put the real branch instruction into the factorial program. The factorial program will be coded three ways to demonstrate the use of branch instructions:

1. **Factorial:** Same as before, but using the real branch instruction instead of a MOV to the PC.
2. **Factorial Subroutine:** Typical approach to calling a subroutine.
3. **Recursive Factorial Subroutine:** The factorial subroutine will call itself to calculate a smaller factorial.

Trace with Breakpoints

File Listing_15-3.txt contains the Verilog code to assemble the branch instructions, execute them as part of the ARM imitation, and test them by calculating 12!. Go ahead and compile, download, and run this factorial program.

```
734.
735. //        Program to calculate single precision factorial in R2
736. //        12! = 479,001,600 = 1C8C,FC00 H
737.
738.          `MOV    R0, 0      `_3 // 0: Reset "boot" address
739.          `MOV    R2, 12     `_3 // 1: 12 ! = 12*11*...2*1 = 32'h1C8CFC00
740.          `SUB    R1, R2, 1  `_4 // 2: Initialize R1 as multiplier (loop counter).
741. FLoop    =      IP;               // Loop: Calculate the factorial
742.          `MUL    R2, R1     `_3 // 3: On each pass, multiply by one lower
743.          `SUB    R1, 1      `_3 // 4: Decrement multiplier for next pass.
744.          `CMP    R1, 2      `_3 // 5: Compare to loop ending condition.
745.          `B`GE   FLoop      `_3 // 6: Go back until multiplier = 2.
746. ILoop    =      IP;               // Infinite loop (stay in one place)
747.          `B      ILoop      `_4 // 7: Keep branching to this location.
748.
```

Listing 15.3 D: Single precision multiply for 12 factorial

Any of the 16 registers can be examined by changing SW[3:0]. See the examples at the end of the previous chapters. If SW[8:0] is set to 010000010, the answer for 12! will be present in register R2 as 32'h1C8CFC00. If you are interested in watching the factorial loop count down from 12 to 1, then set the breakpoint address to 6 (SW[8:0] is 001100010).

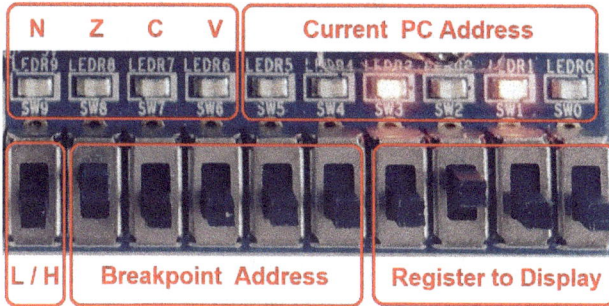

Copy of Figure 11.14: User interface with breakpoint address and display

Subroutines

One of the principal hallmarks of the industrial revolution was the use of interchangeable parts in the manufacturing process. In a similar manner, subroutines, macros, and operating system services are building blocks for developing large sophisticated software applications. These three building blocks are predefined program segments that can be used over and over again by calling them from the application program.

- Service: A section of operating system code that is called (SWI or SVC) to perform a common task for all user programs.
- Subroutine: A section of user-written code that is "jumped to" or "branched to" to perform a common task within a user program.
- Macro: A section of user-written code that is essentially "copied and pasted" into multiple locations within the program source code.

The advantages of using subroutines are many:

- Subroutines help organize the construction of the program.
- The code only takes up memory space once.
- It's less work to modify or correct one area of common code rather than many copies of almost identical code.
- "Information hiding" occurs because one part of the program is unable to directly access data in another part of the program and accidentally change it.
- Division of programming assignments among different programmers is easier.

The disadvantages of subroutines are few.

- There is a slight performance degradation compared to "in-line code"

15: Branch Instructions 273

due to the overhead of the call and return.

- It can lead to too much of a good thing: Too many tiny subroutines can lead to confusion.

Figure 15.4 illustrates a factorial subroutine and a main program that calls it. Listing 15.4 implements this example. The first 734 lines of files Listing_15-4.txt and Listing_15-5.txt are the same as in Listing_15-3.txt.

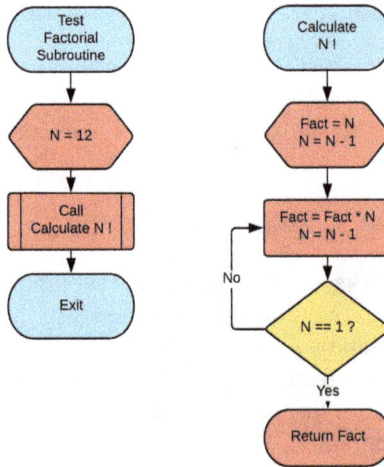

Figure 15.4: Main program is calling a subroutine to calculate the factorial.

```
734.
735. //      Program to calculate single precision factorial in R2
736. //      12! = 479,001,600 = 1C8C,FC00 H
737.
738.        `MOV   R0, 0       `_3 // 0: Reset "boot" address
739.        `MOV   R2, 12      `_3 // 1: 12 ! = 12*11*...2*1 = 32'h1C8CFC00
740.        `BL    FactN       `_2 // 2: Call factorial function
741.        `B     IP          `_2 // 3: Infinite loop ends program
742.
743. FactN  =     IP;               // Subroutine entry address
744.        `SUB   R1, R2, 1   `_4 // 4: Initialize R1 as multiplier (loop counter).
745. FLoop  =     IP;               // Loop: Calculate the factorial
746.        `MUL   R2, R1      `_3 // 5: On each pass, multiply by one lower
747.        `SUB   R1, 1       `_3 // 6: Decrement multiplier for next pass.
748.        `CMP   R1, 2       `_3 // 7: Compare to loop ending condition.
749.        `B`GE  FLoop       `_3 // 8: Go back until multiplier < 2.
750.        `BX    LR          `_2 // 9: Return with factorial in R2.
751.
```

Listing 15.4: Factorial subroutine

- Line 739: The argument N is loaded into register R2.
- Line 740: Function (or subroutine) call
- Line 741: Infinite loop to do nothing. If this was a real application, we would do more coding here or return control to the operating system.
- Line 744: Function entry point (i.e., target of the BL call).
- Line 750: "Return" to the calling program using the address saved in the LR.

Recursive Factorial Function

From a programming viewpoint, a function is a subroutine that returns a single value. For example, $A = Factorial\ (N)$ represents a function that has one input argument and returns a single value. A recursive function is one that actually calls itself, such as $Factorial\ (N) = N \times Factorial\ (N - 1)$. Here 12 ! = 12 × 11 ! and 11 ! = 11 × 10 !, and so on until 1 ! is reached.

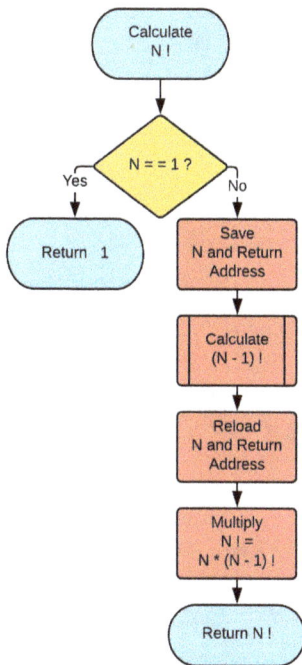

Figure 15.5 illustrates a flowchart for a factorial function that uses recursion to calculate its "answer." As far as the main program is concerned this recursive function is identical to the one previously shown in Figure 15.4 and implemented in Listing 15.4. Internally, the two factorial functions are quite different.

There is still a loop, of course: 12, 11, 10, etc., but it is built within the function calls. In the example of 12!, we have twelve copies of the FactN function "open" at one time. How and where do we keep all those intermediate calculations?

The input argument N and the return address are saved in a data structure known as a stack. This is an area in read/write memory that is commonly used for temporary storage of a variable amount of data.

Figure 15.5: Recursive function

Recursive subroutines have to be careful about the following three factors:

1. Data variables must be unique for each call. This "re-entrant" capability means that each time the function is called, it must have its

own set of variables and not destroy the contents of another call to the same function.

2. The return address must be saved in a location that is unique to each call. Functions and subroutines in the ARM environment use the BL instruction that loads the return address into register R14. This value must be saved before another call is made, or else the "bridges are burnt" and there will be "no way home."

3. There must be an ending condition, where the function is not called again. Otherwise, the program will go into an infinite loop and never return. For the factorial calculation, 1 ! is the ending condition.

735. //	Program to calculate single precision factorial in R2
736. //	12! = 479,001,600 = 1C8C,FC00 H
737.	
738.	`MOV   R0, 0      `_3 // 0: Reset "boot" address
739.	`MOV   SP, 120    `_3 // 1: Initialize SP Stack Pointer
740.	`MOV   R2, 12     `_3 // 2: 12 ! = 12*11*...2*1 = 32'h1C8CFC00
741.	`BL    FactN      `_2 // 3: Call factorial function
742.	`B     IP         `_2 // 4: Infinite loop ends program
743.	
744. //	FactN: Function that calculates factorials, N! = FactN (N).
745. //	Input: R2: Value of N. Range = [1, 12]
746. //	LR: Return address
747. //	SP: Stack must have at least 120 bytes available
748. //	Output: R2: Calculated value of N. Range = [1, 479001600]
749. //	R1,R3: Used as scratch, i.e., not saved
750.	
751. FactN	= IP; // Function FactN entry address
752.	`CMP   R2, 1     `_3 // 5: Test for special, 1! = 1
753.	`BX`EQ LR `_3 // 6: Return with factorial in R2
754.	`SUB   SP, 4     `_3 // 7: Open up 1 word of storage on stack
755.	`STR   R2, (SP)  `_3 // 8: Save the input value of N
756.	`SUB   SP, 4     `_3 // 9: Open up another word of storage
757.	`STR   LR, (SP)  `_3 // A: Save return address
758.	`SUB   R2, 1     `_3 // B: Prepare argument of (N-1)
759.	`BL    FactN     `_2 // C: Get value for (N-1)! in R2
760.	`LDR   R3, (SP)  `_3 // D: Restore return address into R3
761.	`ADD   SP, 4     `_3 // E: Return storage space on stack
762.	`LDR   R1, (SP)  `_3 // F: Restore input value of N into R1
763.	`ADD   SP, 4     `_3 // 10: Restore SP to input value
764.	`MUL   R2, R1    `_3 // 11: Calculate: N! = N * (N-1)! into R2
765.	`BX    R3        `_2 // 12: Return N! in register R2
766.	

Listing 15.5: Recursive factorial function

Listing 15.5 implements a recursive function example. The main program that calls function *FactN* is exactly the same as that in Listing 15.4 that called a non-recursive function.

- Line 739: Initialize the "stack pointer" which points to RAM memory for saving working contents of variables in recursive function.
- Line 740: The argument N is loaded into register R2.
- Line 741: Function call to FactN
- Line 742: Infinite loop to do nothing. If this was a real application, we would do more coding here or return control to the operating system.
- Line 752: Function entry point (i.e., target of the BL call). Test for argument N == 1.
- Line 753: Special argument value: FactN (1) results in an immediate value of "1" returned in register R2.
- Lines 754, 755: Save argument N value on stack.
- Lines 756, 757: Save return address LR value on stack.
- Line 759: Function FactN calls itself with N = N - 1.
- Lines 760, 761: Reload return address LR value from stack into R3.
- Lines 762, 763: Reload argument N value from stack into R1.
- Line 764: Calculate the factorial: R2 contains (N-1)! and R1 contains N (originally in R2, but now in R1).
- Line 765: "Return" to the calling program using the address in register R3.

Interesting breakpoint addresses:

- 5 (entry point to FactN): SW[9:0] is set to 0001010010
- 11 (multiply instruction): SW[9:0] is set to 0010010010

The Stack

A stack is a general data storage technique used by computer programs for the storage of temporary data. In the recursive FactN function, the stack is used for the storage of the return address and the argument value for each call.

Programs typically divide data storage memory into two general categories: heap and stack. The "heap" generally consists of memory for variables that do not change in size, while the stack is for data that grows and shrinks while the program is running. Arrays and tables are generally stored in the heap, and subroutine and function variables are often stored on the stack. PUSH and POP "instructions" and stack implementation will be discussed in much more detail in chapters 16 and 18.

The amount of memory used by stack operations is usually larger than expected, and sometimes the stack can be "blown" and will overwrite memory used for other purposes leading to intermittent and difficult software problems to recognize. Even in the simple example in Listing 15.5, the stack must have at least 80 bytes of memory (8 bytes for each function call times 10 calls to FactN).

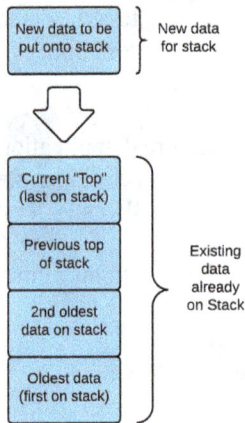

Figure 15.6: Stack "concept"

A stack is referred to as a last-in first-out (LIFO) data structure, but we still need some type of pointer to where the last element of data is stored in memory. In the example in Listing 15.5, register SP is used but any of the other available general purpose registers would work just as well.

The concept of a stack has been used in computer programs from the beginning, and an actual "stack register" and PUSH and POP instructions appeared on mini-computers in the 1970s. The ARM instruction set does not have instructions named PUSH and POP, but uses other instructions to implement the same thing.

Highlights and Comparisons

Branch instructions are an essential part of a computer program, but they can easily be overused leading to poor performance and programs that are a nightmare to maintain. The Cond field, which is a part of every ARM instruction, is a great feature that minimizes the number of branch instructions needed within a program.

There is a common programming technique known as a "jump tape" or a "computed goto" that can be nicely implemented by modifying the PC directly. However, a better approach is to use other instructions, such as BX, that can fulfill that need as well.

Review Questions

1. * Compare Verilog functions to functions in software? Likewise, how is a Verilog task different from a subroutine?
2. What are two reasons why ARM branch instructions use relative memory addresses rather than absolute addresses?
3. Too much branching leads to spaghetti code which is difficult to maintain, but how does having a lot of branch instructions in a program actually degrade its performance?
4. *Since recursive subroutines are less efficient, what is their advantage?

Computer Architecture Tutorial Using an FPGA

— 16 —
Indexed Memory Addresses

Although RAM is Random Access Memory, it is very often accessed sequentially. Many software applications work with data structures such as character strings, arrays, and tables that are processed in the order they appear in memory. To improve performance, most CPUs can automatically increment and decrement memory addresses for their load and store instructions.

Chapter 14 introduced the LDR and STR instructions using the "indirect" addressing mode where any of the 16 general purpose registers can point to a byte address in memory. These instructions have two other modes that combine the indirect memory address in the register with an offset value in the instruction. As you might suspect, the offset is located in the second operand field of the instruction, register shifting is possible, and the base register can even be updated.

Chapter 16 implements the enhancements to the ARM load and store instructions in the following three stages:

1. **Assembler:** The asmem1 function from Chapter 14 will be upgraded to support offsets, base register updates, register shifts, and PUSH/POP mnemonics.
2. **Execute:** The DataMOD module from Chapter 14 will now support LDR and STR instructions in the pre-index and post-index modes.
3. **Application Demonstrations:** The factorial program will be changed in two ways:
   a. A "look up" answer table will be built using a pre-indexed STR and accessed with a post-indexed LDR.
   b. The recursive function *FactN* will be modified to use PUSH and POP "instructions."

Introductions

No new instructions are introduced in this chapter, but significant enhancements to the operations of the LDR and STR instructions will be added. These advancements simplify accessing data in arrays:

1. Sequentially stepping through the data elements in an array
2. Indexing into an array based on element size (power of two)
3. PUSH and POP macros will use STR and LDR instructions to perform stack operations.

Indexed Addressing

There are three addressing modes for the LDR and STR instructions which move data between a single register and a memory location:

- Indirect: The base register contains the complete memory address. Example: LDR R0,[R4].
- Pre-Indexed: The memory address is calculated by adding an offset to the contents of a base register. Example: LDR R0,[R4,#56].
- Post-Indexed: The base register contains the complete memory address, but then the base register contents will be updated after the address is used in the current instruction. Example: LDR R0,[R4],#56.

It's easy to remember which assembler syntax distinguishes "pre" from "post" because it is similar to common algebraic expressions. The brackets enclose the part which is to be done first. In A×[B+56], we add B and 56 before (pre) using it to multiply by A. While in [A×B]+56, we first multiply A and B, and finally we add 56 (post).

Pre-Indexed Example: STRB R2,[R4,#1]!

The [R4,#1] expression indicates pre-indexed format, where the address of the byte to be stored is determined by the sum of the contents of base register R4 and immediate the value 1. The exclamation point (!) indicates that the address generated by the sum of the base register and immediate offset will also update the base register.

Figure 16.1 shows the machine code instruction that is generated by an assembler. How large of an immediate offset can be added to or subtracted from a base register in pre-indexed format?

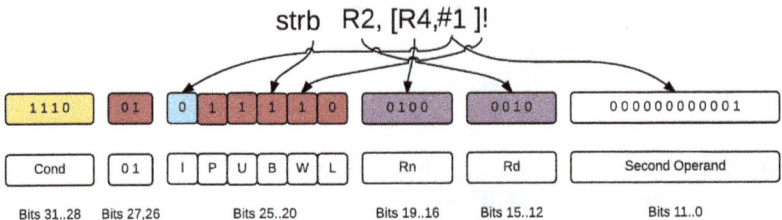

Figure 16.1: Machine code generated for pre-indexed mode store byte instruction

Twelve bits are available in the second operand which provide a maximum immediate offset of 4095 bytes. The field definitions of the LDRB/STRB machine code instructions depicted in figures 16.1 and 16.2 are the following:

Computer Architecture Tutorial Using an FPGA

- Cond (bits 31..28): Status indicating whether the instruction should be executed (1110 indicates always execute).
- 0 1 (bits 27,26): Opcode indicating that this is an LDR or STR.
- IPUBWL (bits 25..20): Bits indicating specifically what is to be done by the LDR/STR instruction.
  - I (bit 25): "1" indicates scaled Index (register with a shift), "0" indicates offset is immediate 12-bit constant.
  - P (bit 24): "1" indicates Pre-indexed, "0" indicates post-indexed.
  - U (bit 23): "1" indicates Up (add the offset), "0" indicates subtract the offset.
  - B (bit 22): "1" indicates Byte (LDRB), "0" indicates word (LDR).
  - W (bit 21): "1" indicates Write back. The exclamation point (!) to the assembler sets this bit.
  - L (bit 20): "1" indicates Load (LDR), "0" indicates store (STR).

- Rn (bits 19..16): Base register, R0 through R15
- Rd (bits 15..12): Register to contain result, R0 through R15
- Offset (bits 11..0): Immediate range of +4095 to -4095. If the I-bit is set (1), this is a scaled index (register with a shift). See Figure 16.4.

Post-Indexed Example: LDRB R5,[R4],#-1

Figure 16.2 illustrates the machine code instruction generated for "LDRB R5, [R4],#-1" which is a post-indexed instruction.

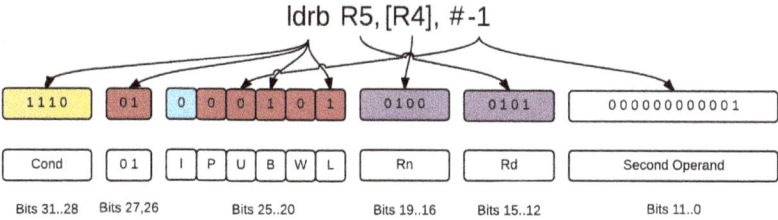

Figure 16.2: Machine code generated for post-indexed mode load byte instruction.

In post-indexed mode, the write-back "option" (!) to update the base register is not an option. The exclamation symbol is not allowed by assemblers since it is always implied. If an offset is present, it will always be used to update the base register. It really would not make much sense to calculate a new index offset and not ever use it.

Load Initial Data: LDR R1,[PC, #56]

Although a Von Neumann architecture supports instructions and data being interspersed within the memory, most programs allocate their program and data areas to separate address ranges. This is many times done so that data doesn't accidentally get executed as if it was an instruction. In embedded systems, the program's code is often in Read Only Memory (ROM) while the working data is in RAM (different types of memory, but on the same bus).

An exception to this practice is data initialized to particular values at assembly time, such as text messages to be displayed. In this case, the "data" will be located within the ROM and can be loaded from an offset to the current PC address. As pointed out in Chapter 15, the offset is relative to the address of the current LDR instruction plus 8 (due to pipelining).

Arrays, Tables, Vectors, Matrices

An array is an ordered list of adjacent storage locations in memory. It could be a list of bytes, words, or even a more complicated combination of bytes and words. Tables, vectors, and matrices are other names commonly associated with arrays, and many times only differ by the number of dimensions (number of rows, number of columns, etc.). A register is typically used to "index" into an array as well as sequentially stepping through each value from the beginning to the end.

As you recall from the format of the data processing instructions in Chapter 12, the second operand isn't simply an immediate constant, but instead can be a register with a shift. For example, the instruction "LDR R2,[R4,R3 LSL #2]" loads R2 from a word in an array whose beginning is pointed to by R4, and its "word position" is in R3. Since the ARM memory is byte-addressable, the word position must be multiplied by four (i.e., a scale factor) to get the byte address (same as shifting left two bit positions, LSL #2).

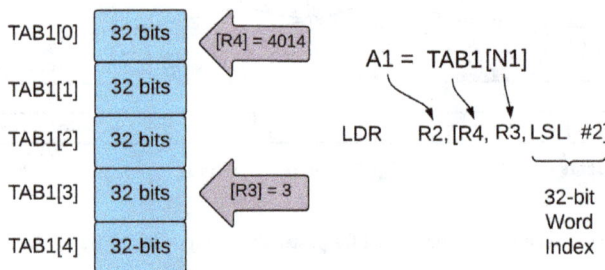

Figure 16.3: Indexing into an array of integers

Figure 16.3 illustrates the conversion of the C statement *A1 = TAB1[N1]* into assembly language. The address of the beginning of the TAB1 array is in register R4. Register R3 represent the index variable N1 (value is 3 in the illustration). If the array contained bytes, then the scale factor would be 1 (a shift of 0). If the array contained half-words (pairs of bytes), then the scale factor would be 2, (a shift of 1).

An exclamation point can be placed after the load instruction in Figure 16-3 to update the base address after the word is loaded. However, this would update register R4 which is supposed to be pointing to the beginning of the TAB1 array.

Assembly of LDR and STR Instructions with Indices

The preceding examples show the typical ARM assembly language notation using brackets and the exclamation point. The Verilog tasks and functions do not permit these characters, so I'll build upon the alternate format from Chapter 14. The following macros will be defined to indicate how and when the base register should be updated. Note: There is no `IAW value because `IA implies the base register will always be updated.

- `IB: Increment Before
- `IBW: Increment Before with Writeback
- `IA: Increment After (Writeback is assumed)

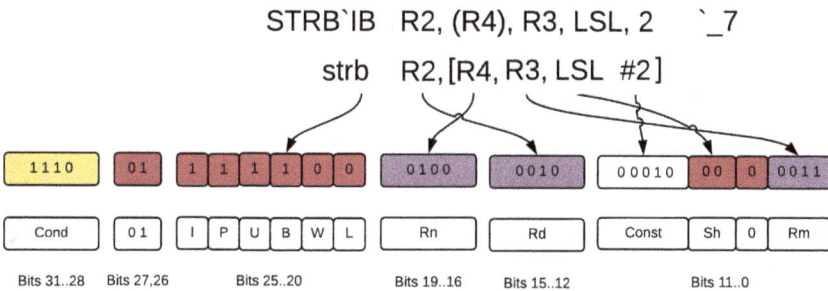

STRB`IB  R2, (R4), R3, LSL, 2    `_7

strb R2, [R4, R3, LSL #2]

| 1110 | 01 | 1 | 1 | 1 | 1 | 0 | 0 | 0100 | 0010 | 0 0010 | 00 | 0 | 0011 |

| Cond | 01 | I | P | U | B | W | L | Rn | Rd | Const | Sh | 0 | Rm |

Bits 31..28   Bits 27,26   Bits 25..20   Bits 19..16   Bits 15..12   Bits 11..0

Figure 16.4: Verilog macro and typical assembler format of STRB pre-index mode instruction

Listing 16.1 contains the changes to the assembly of the LDR/STR instructions. It adds new options for pre-indexed and post-indexed modes which include updating a base register.

```
1. // Listing 16.1 has assembly of load/store instructions with scaled offsets
2. // 1) Assembles ARM load/store instructions with shifts in offsets
3. // 2) Dumps 32-bit words from program memory in hexadecimal
4. //
5. // Modules and macros contained in this file:
6. // 1) CPU_UI: User Interface that dumps 16 bits of word at a time
7. // 2) Macros for assembling LDR/STR/LDRB/STRB instructions
8. // 3) ProgMod: Memory containing "ARM" program
```

Listing 16.1 A: File to test the assembly of load and store instructions

```
50.  //
51.  //--------------- Macro definitions for assembly language ----------------
52.  //
61.  `define  STR    asmem (3'b000,     // Store full 32-bit R register
62.  `define  STRB   asmem (3'b100,     // Store low-order byte in register
63.  `define  LDR    asmem (3'b001,     // Load full 32-bit R register
64.  `define  LDRB   asmem (3'b101,     // Load lower 8 bits and zero fill
65.  `define  IB     6'b110000,         // Increment Before
66.  `define  IBW    6'b110010,         // Increment Before with write back
67.  `define  IA     6'b100000,         // Increment After (Implies write back)
```

Listing 16.1 B: Pre and Post-index options for LDR and STR macros for assembly language

The macros and asmem task are identical to those in Chapter 14. The asmem1 function will be updated and new function asmem2 added to process the features of the pre and post indexed modes. The code is very similar to that used for processing the second operand of the data processing instructions, except the shift can only be a constant, not a register. Also, the range of immediate offset is ±4095 instead of a shifted 8-bit value.

```
121.  // Task asmem is called by load/store opcode macros `LDR, `STR, ...
122.  // The number of parameters will vary between three and eight.
123.     // `LDR`EQ`IA R1,(R2),R3,LSL,4 `_8 // General format with 8 parameters
124.     // `LDR R1,(R2) `_3 // Many parameters are optional
125.
126.     task asmem ();
127.        input [15:0] P0,P1,P2,P3,P4,P5,P6,P7;
128.        {progMem[IP+3],progMem[IP+2],progMem[IP+1],progMem[IP]}
129.           <= asmem1(P0,P1,P2,P3,P4,P5,P6,P7);
130.        IP = IP + 4;
131.     endtask
```

Listing 16.1 C: The asmem task is the same as in Chapter 14

The IPUBWL bits determine what the load and store instructions do. Bits I, P, B, W, and L are set in the argument with the LDR, LDRB, STR, STRB, IB, IBW, and IA macro calls (Listing 16.1 B). The I and U bits are set in the asmem2 function appearing in Listing 16.1 D:

- Lines 139 - 151: Handle the Cond status bit field (`EQ, `GT, ...) and increment conditions (`IA, `IB, `IBW):
- Line 160: Offset is scaled. The 'H68 sets the 01 opcode, I (scaled index), and U (up, i.e., add) flags.
- Line 163: Offset is a positive constant. The 'H48 sets the 01 opcode and U (add) flag.
- Line 165: Offset is a negative constant. The 'H40 sets the 01 opcode.

```
132.
133.  // Function asmem1 constructs the opcode, condition, and status fields.
134.    // `LDR Format 1: Always load register, but don't update base register
135.    // `LDR`IA Format 2: Always load register, and also update base register
136.    // `LDR`EQ Format 3: Load register only if Z-flag, don't update base register
137.    // `LDR`EQ`IA Format 4: Load register and update base register if Z-flag set
138.
139.    function [31:0] asmem1 ();
140.      input [15:0] P0,P1,P2,P3,P4,P5,P6,P7;
141.      if (P1 >= R0)                           // Format 1
142.        asmem1 = 'hE<<28 | asmem2(P0|8'h10,P1,P2,P3,P4,P5);
143.      else
144.        if (P1 > 'hF)                         // `IA,`B,`IBW; Format 2
145.          asmem1 = 'hE<<28 | asmem2(P0|P1,P2,P3,P4,P5,P6);
146.        else
147.          if (P2 >= R0)                       // Format 3
148.            asmem1 = P1<<28 | asmem2(P0|8'h10,P2,P3,P4,P5,P6);
149.          else                                // Format 4
150.            asmem1 = P1<<28 | asmem2(P0|P2,P3,P4,P5,P6,P7);
151.    endfunction
152.
153.  // Function asmem2 constructs the operand fields for function asmem1.
154.    // `LDR R1,(R2),R3,LSL,4 Format 1: Offset is scaled register
155.    // `LDR R1,(R2),-4 Format 2: Offset is constant
156.
157.    function [31:0] asmem2 ();
158.      input [15:0] PUBWL, Q1, Q2, Q3, Q4, Q5;
159.      if (Q3>=R0&&Q3<=R15)                    // Offset is scaled register
160.        asmem2 =
              {8'H68|PUBWL[4:0],Q2[3:0],Q1[3:0],Q5[4:0],Q4[1:0],Q3[4:0]};
161.      else
162.        if (Q3[12]==0)                        // Positive constant offset
163.          asmem2 = {8'H48|PUBWL[4:0],Q2[3:0],Q1[3:0],Q3[11:0]};
164.        else                                  // Negative constant offset
165.          asmem2 = {8'H40|PUBWL[4:0],Q2[3:0],Q1[3:0],-Q3[11:0]};
166.    endfunction
```

Listing 16.1 D: Functions asmem1 and asmem2 for pre and post indexed LDR/STR

Compile, download, and examine the instructions created in Listing 16.1. The test setup for viewing the 7-segment display is the same as in previous chapters:

1. Push KEY[0] to initialize.
2. Set the switches to desired memory address (0 up to 24 hex). Note: These are byte addresses, four bytes per instruction.
3. Push KEY[1] to update and show machine code instruction.
4. Repeat steps 2 and 3 to see the contents of other addresses.

179.	`STRB`IBW	R2, (R4), 1	`_5 // 00: E5E42001
180.	`LDRB`IA	R5, (R4), -1	`_5 // 04: E4545001
181.	`STRB`IB	R2, (R4), R3, LSL, 2	`_7 // 08: E7C42103
182.	`STR`EQ`IBW	R1, (R2), R3, LSL, 10	`_8 // 0C: 07A21503
183.	`STRB`IB	R2, (R4), R3, ROR, 2	`_7 // 10: E7C42163
184.	`LDR	R4, (R5)	`_3 // 14: E5954000
185.	`LDR`GT	R4, (R5)	`_4 // 18: C5954000
186.	`LDRB	R4, (R5)	`_3 // 1C: E5D54000
187.	`STR	R1, (R2)	`_3 // 20: E5821000
188.	`STRB	R1, (R2)	`_3 // 24: E5C21000

Listing 16.1 E: Assembly of assortment of load and store instructions

Sample assembly of indexed LDR/STR instructions:

- Line 179: Figure 16.1, strb R2,[R4, #1]!
- Line 180: Figure 16.2, ldrb R5,[R4], #-1
- Line 181: Figure 16.4, strb R2,[R4,R3,LSL #2]
- Line 182: Condition and base increment, streq R1,[R2,R3,LSL #10]!
- Line 183: Circular shift, strb R2,[R4,R3,ROR #2]

Push and Pop

Most CPUs today have Push and Pop instructions that place and remove register contents on and from the "run-time stack," an area of user memory devoted to temporary data. The set of 32-bit ARM instructions does not have these instructions explicitly, but implements them using the Load Multiple (LDM) and Store Multiple (STM) instructions described in Chapter 18. Most CPUs have instructions that push or pop a single register at a time, and some architectures have an instruction that pushes or pops all of the user general purpose registers. However, by using the LDM and STM instructions, the ARM can select any combination of the 16 registers to store and reload.

In Chapter 16, I will temporarily use the indexed memory LDR and STR instructions to implement POP and PUSH, respectively. These will be functionally compatible with those in Chapter 18, except only one register will be saved or restored at a time. Assembling PUSH and POP in Listing 16.3 is rather simple and straightforward. The asSP task will have either one or two arguments, depending on whether there are any conditions (EQ, GT, ...). No functions will need to be called. Most of the fields within the instruction format will be predetermined because the base register is the SP and the offset is a constant of 4 bytes.

- Lines 61, 62: The IPUBWL bits are determined by whether it is a push or pop.
- Lines 117, 119: ALways execute the push and pop.
- Lines 120, 122: A condition exists, but otherwise its the same.

Computer Architecture Tutorial Using an FPGA

```
  1.    // Listing 16.3 has assembly of PUSH and POP for moving one register
  2.    // 1) Assembles PUSH/POP pseudo instructions using STR/LDR
  3.    // 2) Dumps 32-bit words from program memory in hexadecimal
  4.    //

 61.    `define PUSH asSP (7'b1010010,        // Append 1 register to stack
 62.    `define POP asSP (7'b1001001,         // Remove 1 register from stack

112.    // Task asSP is called to do PUSH or POP for only one register
113.
114.    task asSP ();
115.      input [31:0] P0,P1,P2,P3,P4,P5,P6,P7;
116.      begin
117.        if (P1>=R0)                       // No conditions
118.          {progMem[IP+3],progMem[IP+2],progMem[IP+1],progMem[IP]} <=
119.            'hE<<28 | {P0,SP[3:0],P1[3:0],12'H4};
120.        else                              // EQ, GT, ...
121.          {progMem[IP+3],progMem[IP+2],progMem[IP+1],progMem[IP]} <=
122.            P1[3:0]<<28 | {P0,SP[3:0],P2[3:0],12'H4};
123.        IP = IP + 4;
124.      end
125.    endtask
126.
```

Listing 16.3 A: Assembly of PUSH and POP using STR and LDR

Let's again check that the assembly process creates the intended ARM machine code. Compile and download the code in Listing 16.3. Verify that the PUSH and POP commands generate the matching load and store instructions on lines 139 through 144 of Listing 16.3 B.

- Line 139: `PUSH R2 is really `STR`IBW   R2, (SP), -4   which for most assemblers is STR   R2, [SP, #-4]!
- Line 142: `POP R1 is really `LDR`IA   R1, (SP), 4   which for most assemblers is LDR   R1, [SP], #4

```
138.
139.    `PUSH        R2              `_2 // 0: E52D2004
140.    `PUSH        LR              `_2 // 4: E52DE004
141.    `POP         LR              `_2 // 8: E49DE004
142.    `POP         R1              `_2 // C: E49D1004
143.    `PUSH`NE     R2              `_3 // 10: 152D2004
144.    `POP`GT      LR              `_3 // 14: C49DE004
145.
```

Listing 16.3 B: Machine code of PUSH and POP using STR and LDR

Execution of Indexed Load and Store Instructions

Now that we can build the ARM load and store machine codes in all three index modes, how do we execute them in a running program? What modification must be made to the existing Verilog coding of the DataMod module?

1. The offset must be included in the calculation of the "effective address" of the memory to be read or written:
   a. Define a Memory Address Register (MAR) which contains the sum of the base register Rm and the offset.
   b. The offset may be as simple as a straight 12-bit field providing a difference of ± 4095 bytes.
   c. The offset may be a scaled index which is almost as complicated as the second operand in the data processing instructions. Another instance of the shiftMod module will be created to calculate this value.
   d. The parameter list of the DataMod module must be increased to include the Rm index register value and the op2Raw second operand.
2. The base register, Rn, may need to be updated.
   a. A MARup register will be created to contain the sum of the base register and offset
   b. The output signal will be expanded to 64 bits to include both the data loaded from memory for Rd as well as the new Rn base register value.

```
 1.  // Listing 16.4 Demonstrates ARM program with LDR/STR used with arrays
 2.  // 1) Assembles most ARM instructions including PUSH/POP using LDR/STR
 3.  // 2) Executes demonstration of table generation and indexing
 4.  // 3) Dumps any "one" 32-bit register at a time.
 5.  // 4) Breakpoint address can be set to "stall" assembler program
 6.  //
 7.  // Modules and macros contained in this file:
 8.  // 1) CPU_UI: User Interface that dumps 32-bit words, 16 bits at a time
 9.  // 2) CPU: 32-bit CPU with data processing and multiplication instructions
10.  // 3) DataProcIns: 16 "ARM like" DP instructions using 32-bit registers
11.  // 4) Op2Mod: Fill in second operand including all shift possibilities
12.  // 5) ShiftMod: Calculate all shift possibilities for Op2Mod
13.  // 6) MultiplyIns: Module to calculate multiplication
14.  // 7) DataMod: RAM memory and load/store instructions
15.  // 8) Macros for assembling ARM data processing and multiplication instructions
16.  // 9) ProgMod: Program to demonstrate table of factorials
```

Listing 16.4 A: Directory header for Listing_16-4.txt file

140.	assign SelRAM = ok2exe && IR[27:26]==2'b01;
141.	
142.	ProgMod (R[PC], CPU_state==fetch, reset, IR);
143.	Op2Mod (RmVal, RsVal, op2raw, iFlag, op2Val);
144.	DataProcIns (opCode, RnVal, op2Val, CPU_state==execute && SelDP, CPSR, RdVal_DP, RdVal_DPU, CPSR_DP);
145.	MultiplyIns (opCode, RdVal, RmVal, RnVal, RsVal, CPU_state==execute && SelMul, CPSR, RdV64_M, CPSR_M);
146.	DataMod (IR[25:20], R[RdID], RnVal, RmVal, op2raw, CPU_state==execute && SelRAM, RdVal_RAM);

Listing 16.4 B: Instantiate ALU modules for instruction execution, including load/store.

The modified DataMod module is instantiated on line 146 shown in Listing 16.4 B. It has the following seven arguments:

1. IR[25:20]: IPUBWL; Immediate, Pre-index, Up (add), Byte, Write back, Load
2. R[RdID]: Data value needed for STR and STRB
3. RnVal: Memory byte-address from Rn register
4. RmVal: Register value needed for offset calculation
5. op2raw: Lower 12 bits of instruction needed for immediate offset
6. CPU_state==execute && SelRAM: Clock "tick" that says when the memory transfer will take place.
7. RdVal_RAM: Data loaded from memory by LDR and LDRB

The writeBack state of the instruction cycle has two registers to update: Rn and Rd. The Verilog conditional operator selects the updated values based upon the opcode value of the current instruction (flags RdVal_DPU, SelDP, SelMul, SelM64, SelRAM). The load and store instructions do not update the CPSR.

202.	writeBack: // Update specific register
203.	begin
204.	{R[RnID],R[RdID]} <= (RdVal_DPU & SelDP) ? {R[RnID],RdVal_DP} :
205.	(SelM64) ? RdV64_M :
206.	(SelMul) ? {RdV64_M[31:0],R[RdID]} :
207.	(SelRAM) ? {RdVal_RAM} :
208.	{R[RnID],R[RdID]};
209.	if (SU) CPSR <=
210.	SelDP ? CPSR_DP :
211.	SelMul ? CPSR_M :
212.	CPSR;
213.	{R[LR],R

Listing 16.4 C: Update general purpose R registers and CPSR

```
343.  //
344.  //--------------- ALU for load/store instructions and RAM memory ---------------
345.  //
346.  module DataMod (IPUBWL, Rd, Rn, Rm, op2raw, clk, Rd64);
347.     input [5:0] IPUBWL;
348.     input [31:0] Rd,Rn,Rm,op2raw;
349.     input clk;
350.     output reg [63:0] Rd64;              //
351.     wire [31:0] op2shifted;              // Value if 2nd op. is in Rm
352.     wire [31:0] offset;
353.     wire [31:0] MAR, MARup;
354.     reg [7:0] RAM[0:149];                // Read/Write Random Access Memory
355.
356.     ShiftMod LDRSTR (Rm, op2raw[6:5], op2raw[11:7], op2shifted);
357.     assign offset = (IPUBWL[5]) ? op2shifted : op2raw;
358.     assign MAR = (~IPUBWL[4]) ? Rn :     // ~P
359.        (IPUBWL[3]) ? Rn + offset : Rn - offset;
360.     assign MARup = (~IPUBWL[1] && IPUBWL[4]) ? Rn :   // ~W && P
361.        (IPUBWL[3]) ? Rn + offset : Rn - offset;
362.
363.     always @ (posedge(clk))
364.        if (IPUBWL[0])                    // Load
365.           if (IPUBWL[2])                 // byte
366.              Rd64 <= {MARup,24'b0,RAM[MAR]};
367.           else                           // Load word
368.              Rd64 <=
                     {MARup,RAM[MAR+3],RAM[MAR+2],RAM[MAR+1],RAM[MAR]};
369.        else                              // Store
370.           begin
371.              if (IPUBWL[2])              // byte
372.                 RAM[MAR] <= Rd[7:0];
373.              else                        // Store word
374.                 begin
375.                    RAM[MAR+3] <= Rd[31:24];
376.                    RAM[MAR+2] <= Rd[23:16];
377.                    RAM[MAR+1] <= Rd[15:8];
378.                    RAM[MAR] <= Rd[7:0];
379.                 end
380.              Rd64 <= {MARup,Rd};
381.           end
382.  endmodule
```

Listing 16.4 D: ALU module for RAM memory and load/store instructions

The DataMod module was introduced in Listing 14.3 D where it contained the read/write RAM memory and the "indirect" load/store instruction execution. It is now upgraded in Listing 16.4 D to include the pre and post indexing modes. In

Chapter 17, it will be updated again to accommodate memory mapped I/O operations. The following enhancements should be noted:

- Line 346: Parameters Rm and op2raw have been included so full pre and post indexing can be calculated. Rd64 returns 64 bits: the 32-bit Rd register with the data and an updated Rn base register.
- Line 351: The offset in the second operand may contain a scaled value (register Rm shifted)
- Line 353: MAR is the Memory Address Register to use to load/store the data; MARup is the updated value to be returned to the Rn register
- Lines 356 - 361: Calculate the pre and post index values.
- Line 356: The same module is used to perform the register shift as for the second operand of the data processing instructions. Note: This uses a different instance of ShiftMod, but did not have to.
- Line 357: Immediate data will use the straight 12 bits of the second operand, which is different from that calculated for data processing instructions.
- Line 358: The P bit selects whether pre or post indexed mode
- Line 359: The Up bit selects whether add or subtract the offset
- Line 360, 361: Determine whether the Rn register will be updated
- Line 363 - 381: Basically the same as in Chapter 14 except now the calculated MAR address is used instead of only the Rn register value.

Indexed Load and Store Application Examples

In the first application, the factorial program will be changed to "look up" the answer in a table rather than calculate it during the function call. Figure 16.5 provides a Java or C like example, and Listing 16.4 E provides the equivalent assembly language.

```
int R1, R2, R3, R4, R9, R12, Fact[0:12];
R2 = 1;
//   Build the table of factorials: Fact[1]=1, Fact[2]=2, Fact[3]=6, ...
for (R1=1; R1<=12; R1=R1+1)
   begin
      R2 = R2 * R1;
      Fact[R1] = R2;
   end
//   Use the table: Look up values for Fact[4], Fact[9], and Fact[12]
R4 = Fact[4];
R9 = Fact[9];
R3 = 12;
R12 = Fact[R3];
```

Figure 16.5: Generate and use table of factorials

The variable names are certainly not creative, but they do match the register ID numbers used in the assembly language program in Listing 16.4 E.

```
785.
786. //      Build table of factorials: X, 1, 2, 6, 24, ...
787.
788.      `MOV      R2, 1      `_3 // 1: Factorial will be created in R2
789.      `MOV      R1, 0      `_3 // 2: R1 will be multiplier for factorials
790.      `MOV      R8, 20     `_3 // 3: Pointer to beginning of table
791. FLoop =         IP;              // Loop: Calculate the factorial
792.      `ADD      R1, 1      `_3 // 4: R1 will increment from 1 to 12
793.      `MUL      R2, R1     `_3 // 5: On each pass, multiply by one higher
794.      `STR`IBW  R2, (R8), 4   `_5 // 6: Store calculated value
795.      `CMP      R1,12      `_3 // 7: Test for highest in table
796.      `B`MI     FLoop      `_3 // 8: Go back until multiplier = 12.
797.
798.      `MOV      R8, 20     `_3 // 9: Pointer to factorial table
799.      `LDR      R4, (R8), 4*4   `_4 // A: Look up factorial of 4 = 'h18
800.      `LDR      R9, (R8), 9*4   `_4 // B: Look up factorial of 9 = 'h58980
801.
802.      `MOV      R3,12      `_3 // C: 12 ! will be 32'h1C8CFC00
803.      `LDR      R12, (R8), R3, LSL, 2   `_6 // D: Factorial of value in R3
804.
805.      `B        IP         `_2 // E: Infinite loop
806.
```

Listing 16.4 E: Application example using factorial table built in RAM

- Lines 790, 798: In this example, register R8 will be set pointing to Fact[0], which itself will not be given a value.
- Lines 791 - 796: Loop to build table
- Line 794: Use pre-indexed mode with write back to fill next position in Fact[ ] table
- Lines 799, 800: Load Fact[4] and Fact[9] using immediate offsets.
- Line 802: R3 will be pointing to the 12th element of array Fact
- Line 803: Use a scaled index to load Fact[12].

The second example is in Listing 16.5 where the recursive factorial function uses PUSH and POP "instructions" to save register contents on the stack. This code is easier to write and understand than the same code in Listing 15.5 which explicitly used LDR and STR instructions for stack operations.

- Lines 804, 805: The PUSH instructions can be used without even knowing how the stack is built.
- Lines 808, 809: Register contents can, of course, be restored into different registers than when saved.

Computer Architecture Tutorial Using an FPGA

```
784.
785. //      Program to calculate single precision factorial in R2
786. //      12! = 479,001,600 = 1C8C,FC00 H
787.
788.        `MOV    R0, 0     `_3 // 0: Reset "boot" address
789.        `MOV    SP, 120   `_3 // 1: Initialize SP Stack Pointer
790.        `MOV    R2, 12    `_3 // 2: 12 ! = 12*11*...2*1 = 32'h1C8CFC00
791.        `BL     FactN     `_2 // 3: Call factorial function
792.        `B      IP        `_2 // 4: Infinite loop ends program
793.
794. //      FactN: Function that calculates factorials, N! = FactN (N).
795. //      Input:    R2: Value of N. Range = [1, 12]
796. //                LR: Return address
797. //                SP: Stack must have at least 120 bytes available
798. //      Output:   R2: Calculated value of N. Range = [1, 479001600]
799. //                R1,R3: Used as scratch, i.e., not saved
800.
801. FactN  =       IP;           // Function FactN entry address
802.        `CMP    R2, 1     `_3 // 5: Test for special, 1! = 1
803.        `BX`EQ  LR        `_3 // 6: Return with factorial in R2
804.        `PUSH   R2        `_2 // 7: Save the input value of N
805.        `PUSH   LR        `_2 // 8: Save return address
806.        `SUB    R2, 1     `_3 // 9: Prepare argument of (N-1)
807.        `BL     FactN     `_2 // A: Get value for (N-1)! in R2
808.        `POP    R3        `_2 // B: Reload return address into R3
809.        `POP    R1        `_2 // C: Restore input value of N into R1
810.        `MUL    R2, R1    `_3 // D: Calculate: N! = N * (N-1)! into R2
811.        `BX     R3        `_2 // E: Return N! in register R2
812.
```

Listing 16.5: Recursive factorial function with PUSH and POP

Highlights and Comparisons

Most CPU architectures have special features for accessing data in arrays in memory. Not only are there multiple index registers present, but also auto increment and decrement capability.

- Arrays can store lists and tables of data as either words or bytes. Many tables have more complicated structures.
- Load and store instructions can efficiently step through arrays using an auto-increment or auto-decrement option.
- The "stack" is a place for temporary storage. Many computers have instructions named PUSH and POP, but the ARM implements these functions using other instructions. In Chapter 18, multiple registers will be pushed and popped using one "block transfer" instruction each.

Review Questions

1. *The second operand in the LDR and STR instructions is very similar to that of the data processing instructions. Both can be an immediate value or a register with a shift. What are two differences between the formats?
2. *The LDR and STR instructions allow a negative offset. Is this negative direction set using "two's complement" or "sign and magnitude" format?

Exercises

1. The stack I implement for PUSH and POP is consistent with that common to most operating systems, including Windows® and Linux®. Details about stack operations will be covered in more detail in Chapter 18, but for now, let's simply change the stack direction. Instead of decreasing the memory address with each new entry, let's increase it. Modify PUSH and POP, set the initial SP value to a low value, and let the stack grow "upward."
2. Implement a circular buffer queue. It will need two pointers, IN (where to store the next data) and OUT (where to load the queued data). You will also need the size of the buffer in order to know when to wrap around to the beginning of the buffer.

― 17 ―
Input / Output

Data is moved between a computer and external devices through Input/Output (I/O) channels. In many computer architectures, there are special I/O instructions that read data from or write data to I/O channels. In the ARM, the I/O channels are "memory mapped" where they are accessed using the same instructions that access RAM memory. This approach has been fairly common starting with mini-computers in the 1970s.

Chapter 17 will extend the LDR and STR instructions to read from the switches and write to the 7-segment displays. This example is only an introduction as there exist many types of I/O devices and even a variety of I/O processors that can assist in data acquisition and distribution. FPGAs are often used as "front ends" that pre-process data before it enters a computer system.

There will be no need to modify the assembler since the very same load and store instructions from previous chapters will be used for the I/O operations. Chapter 17 implements the enhancements and demonstrates the I/O in the following two steps:

1. **Execute:** The DataMod module will be enhanced to also read the switches and write to the 7-segment displays.
2. **Application Demonstration:** The I/O will be demonstrated by calculating any factorial from 2 through 12 set on the input switches and directly displaying the results on the 7-segment displays.

Introductions

No new instructions are introduced in this chapter because I/O in the ARM is memory mapped and uses the same instructions that read and write memory.

Execution of I/O Instructions

Since we can already build the ARM machine code for the I/O instructions, how do we execute it in a running program? How can we integrate I/O with the existing memory access instructions? Basically, we will divide the "memory" address space between memory and I/O devices. The following coding changes are needed:

1. Modify the parameter list of the DataMod module to include the switches and the hexadecimal display.

2. Modify the DataMod module to move data between a register and either memory or an I/O device depending on the effective "memory" address.

```
1.  // Listing 17.1 Demonstrate ARM program with memory mapped I/O
2.  // 1) Assembles ARM instructions including memory mapped LDR/STR
3.  // 2) Executes demonstration of factorial program
4.  // 3) Input data from switches, output to 7-segment displays
5.  // 4) Breakpoint address can be set to "stall" assembler program
6.  //

61.  //---------------- "ARM" CPU imitation ----------------
62.  //
63.  module CPU (KEY, reset, clk, PCbp, dmpID, hexDisp, LEDR);
64.    input [1:0] KEY;
65.    input reset, clk;
66.    input [4:0] PCbp;               // Breakpoint instruction address
67.    input [3:0] dmpID;              // Register to dump
68.    output [9:0] LEDR;
69.    output [31:0] hexDisp;          // Register contents for display
70.  // assign hexDisp = R[dmpID];

142.   ProgMod (R[PC], CPU_state==fetch, reset, IR);
143.   Op2Mod (RmVal, RsVal, op2raw, iFlag, op2Val);
144.   DataProcIns (opCode, RnVal, op2Val, CPU_state==execute && SelDP, CPSR,
       RdVal_DP, RdVal_DPU, CPSR_DP);
145.   MultiplyIns (opCode, RdVal, RmVal, RnVal, RsVal, CPU_state==execute &&
       SelMul, CPSR, RdVal64_M, CPSR_M);
146.   DataMod (IR[25:20], R[RdID], RnVal, RmVal, op2raw, CPU_state==execute
       && SelRAM, dmpID, hexDisp, RdVal_RAM);

343.  //
344.  //---------------- ALU for load/store instructions and RAM memory ----------------
345.  //
346.  module DataMod (IPUBWL, Rd, Rn, Rm, op2raw, clk, iPort, oPort, Rd64);
347.    input [5:0] IPUBWL;
348.    input [31:0] Rd,Rn,Rm,op2raw;
349.    input clk;
350.    input [3:0] iPort;
351.    output reg [31:0] oPort;
```

Listing 17.1 A: Instantiate ALU modules with change for memory mapped I/O

- Line 346: Two new parameters on the DataMod definition: iPort is parameter 7 and oPort is parameter 8
- Line 146: DataMod instantiation assigns dmpID (switches [3:0]) to iPort and hexDisp to oPort.
- Line 70: The hexDisp parameter will no longer be used to dump register contents (assignment commented out)

The DataMod module now has the following nine parameters:

1. IR[25:20]: IPUBWL bits: Immediate, Pre-index, Up (add), Byte, Writeback, Load
2. R[RdID]: Data value needed for STR and STRB
3. RnVal: Memory byte-address from Rn register
4. RmVal: Register value needed for offset calculation
5. op2raw: Lower 12 bits of instruction needed for immediate offset
6. CPU_state==execute && SelRAM: Clock "tick" that says when the memory or I/O transfer will take place.
7. dmpID: Lower four switches, SW[3:0], formerly used to select register to display.
8. hexDisp: 32-bit value that will be transferred to eight hex digits in user interface
9. RdVal_RAM: Data loaded from memory or I/O by LDR and LDRB

```
352.   output reg [63:0] Rd64;              //
353.   wire [31:0] op2shifted;              // Value if 2nd op. is in Rm
354.   wire [31:0] offset;
355.   wire [31:0] MAR, MARup;
356.   reg [7:0] RAM[0:149];                // Read/Write Memory
357.
358.   ShiftMod LDRSTR (Rm, op2raw[6:5], op2raw[11:7], op2shifted);
359.   assign offset = (IPUBWL[5]) ? op2shifted : op2raw;
360.   assign MAR = (~IPUBWL[4]) ? Rn :    // ~P
361.     (IPUBWL[3]) ? Rn + offset : Rn - offset;
362.   assign MARup = (~IPUBWL[1] && IPUBWL[4]) ? Rn :   // ~W && P
363.     (IPUBWL[3]) ? Rn + offset : Rn - offset;
364.
365.   always @ (posedge(clk))
366.     if (IPUBWL[0])                     // Load
367.       if (MAR>=1000)                   // Range to read input devices
368.         Rd64 <= {MARup,28'b0,iPort};
369.       else                             // Range to read memory
370.         if (IPUBWL[2])                 // byte
371.           Rd64 <= {MARup,24'b0,RAM[MAR]};
372.         else                           // Load word
373.           Rd64 <=
                 {MARup,RAM[MAR+3],RAM[MAR+2],RAM[MAR+1],RAM[MAR]};
```

Listing 17.1 B: Read from either memory or I/O port depending on memory address range

- Line 367: Assign memory to addresses below 1000 and I/O devices to addresses above 1000
- Line 368: The data on the iPort parameter will be read.
- Lines 369 - 373: For memory addresses below 1000, use the RAM array as before.

```
374.    else                              // Store
375.        begin
376.            if (MAR>=1000)            // Range of I/O devices
377.                oPort <= Rd;
378.            else                       // Range for writing to memory
379.                if (IPUBWL[2])        // byte
380.                    RAM[MAR] <= Rd[7:0];
381.                else                   // Store word
382.                    begin
383.                        RAM[MAR+3] <= Rd[31:24];
384.                        RAM[MAR+2] <= Rd[23:16];
385.                        RAM[MAR+1] <= Rd[15:8];
386.                        RAM[MAR] <= Rd[7:0];
387.                    end
388.            Rd64 <= {MARup,Rd};
389.        end
390. endmodule
```

Listing 17.1 C: Write to either memory or I/O port depending on memory address range

- Line 376: Assign memory to addresses below 1000 and I/O devices to addresses above 1000
- Line 377: "Store" register data to output port (assigned to 7-segment displays in user interface.
- Lines 378 - 387: For memory addresses below 1000, use the RAM array as before.
- Line 388: New value for Rn is sent back to CPU module. Rd does not change for store/output instructions.

In this example, I arbitrarily set the address ranges to 0 through 999 for "real" memory and addresses above 999 to I/O devices. For a real computer, we would need address ranges for Read Only Memory (ROM), read/write memory (typically referred to as RAM), and each I/O device would be assigned a few addresses. In addition to an actual data channel, most I/O devices have status registers that can be read and written.

I/O Application Demonstration

The I/O instructions will be demonstrated in a single program that calculates the factorial of a number entered on the switches and displayed on the seven-segment displays.

- Line 799: Address of beginning of program to calculate factorials
- Line 800: Base register R6 is loaded with "memory" address in the I/O range (that is any address >= 1000 in this example)
- Line 801: Register R2 is loaded from switches because (R6) is in I/O range

- Line 808: Since address in R6 is in I/O range, output goes to display.
- Line 809: Branch back to get another factorial to calculate.

793.			
794. //		Program to calculate factorial of numbers between 2 and 12 in SW[3:0]	
795. //		2! = 2 = 2 H	
796. //		6! = 720 = 2D0 H	
797. //		12! = 479,001,600 = 1C8C,FC00 H	
798.			
799. Fact	=	IP;	// Program to calculate factorials
800.	`MOV	R6,1,2	`_4 // 1: Load 'h40000000 into base register
801.	`LDR	R2,(R6)	`_3 // 2: Load SW[3:0] contents
802.	`SUB	R1, R2, 1	`_4 // 3: Initialize R1 as multiplier (loop counter).
803. FLoop	=	IP;	// Loop: Calculate the factorial
804.	`MUL	R2, R1	`_3 // 4: On each pass, multiply by one lower
805.	`SUB	R1, 1	`_3 // 5: Decrement multiplier for next pass.
806.	`CMP	R1, 2	`_3 // 6: Compare to loop ending condition.
807.	`B`GE	FLoop	`_3 // 7: Go back until multiplier = 2.
808.	`STR	R2, (R6)	`_3 // 8: Copy result to output device for display
809.	`B	Fact	`_2 // 9: Go back for new value for factorial
810.			

Listing 17.1D: Calculate factorial of number input in SW[3:0]

Go ahead and compile, download, and run the example program in Listing 17.1. Just set the breakpoint address in SW[8:4] to binary 01001 and try multiple factorial values by using the following sequence:

1. Set the value for the factorial on SW[3:0].
2. Push KEY[0] to run the program.
3. Push Key[1] to stop it.
4. Try another value by repeating the above three steps.

Status and Timing

Some I/O devices are so fast that a computer cannot keep up with them. That's one application for FPGAs where they act like a "front end" to preprocess the data before it even reaches the computer I/O channels. We take advantage of the parallelism within the FPGA to gain performance.

Other devices are so slow or intermittent, that the data doesn't change for weeks at a time. Most I/O devices fit somewhere in between where the CPU has time to do computations while it waits for each I/O byte to be transferred.

In addition to the data ports, I/O devices typically have status registers also memory mapped to particular addresses. These registers have bits that can be checked by a computer program to see the availability of new data on the input data port and when output data has been accepted by the device.

Polling and Interrupts

A simple type of I/O programming is where a computer program focuses on a particular I/O port or set of ports. It constantly checks the status bits for new data to be processed. This "polling" approach of continuously checking the port leaves the CPU little time to do anything else.

A more efficient technique is to allow the device to "interrupt" the computer program whenever new data appears. This enables the CPU to be doing meaningful computations rather than simply watching the I/O device's status register. Obviously, interrupt programming is more complicated than polling.

1. A hardware interrupt is the action of the CPU branching to a predefined address in a manner very similar to the SWI instruction (or even somewhat like a subroutine being called using a BL instruction).
2. Because a running program is being interrupted in the middle of making some calculations, the interrupt handler must be careful not to destroy any register contents or data in memory used by the running program. When complete, the interrupt handler must return control, as well as all register contents, to the program that was running when the interrupt started.
3. There must also be ways to prohibit interrupts, such as one interrupt interrupting another before it had time to save the register contents.

Review Questions

1. Name two advantages of using memory mapped I/O instead of having specific instructions dedicated only to I/O operations.
2. Name two disadvantages of using memory mapped I/O. Why do some computers have specific instructions dedicated only to I/O operations?

Exercise

1. Another way to implement memory mapped I/O is to create another module dedicated only to I/O and make it active only when a particular address range is reached.
   a. Generate a module named IOMOD and give it the same parameters that the DataMod module was given in this chapter.
   b. Instantiate it along with the other modules.
   c. Make a new select named SelIO and make it active only for LDR/STR instructions AND the address range you choose for I/O devices.
   d. Restrict the SelRAM to be active only for the non-I/O address range.

— 18 —
Block Transfers

The contents of as many as sixteen registers can be moved to or from memory with one "block transfer" instruction. The STM (Store Multiple) and LDM (Load Multiple) instructions move data between memory and registers. These instructions are very similar to the STR and LDR instructions described in previous chapters. An assembly language example is STMIA R8,{R0-R7} where R8 is the base register providing the indirect memory location, and the set of registers is identified within a pair of braces like {R0-R7}.

I will not be providing the implementation of these instructions, but leaving that as "an exercise for the student." Chapter 18 provides sufficient background on how these instructions work as well as suggestions for implementation of both the assembly and the execution of their machine codes.

Introductions

The term "block transfer" in the ARM architecture is different than its meaning in most CPUs, where it refers to copying multiple bytes of data from one location in memory to another. In the ARM, it refers to the group of eight "load and store multiple" instructions. Each instruction's name is built from three parts shown in Figure 18.1.

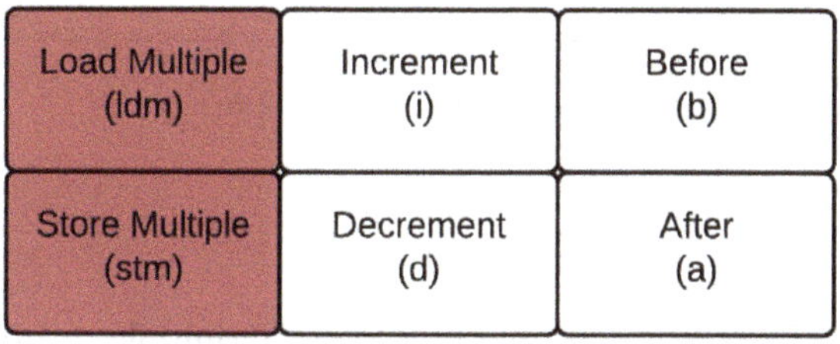

Load Multiple (ldm)	Increment (i)	Before (b)
Store Multiple (stm)	Decrement (d)	After (a)

Figure 18.1: Three parts of the block transfer instructions

- ldmib: Load Multiple Increment Before
- ldmia: Load Multiple Increment After
- ldmdb: Load Multiple Decrement Before
- ldmda: Load Multiple Decrement After
- stmib: Store Multiple Increment Before
- stmia: Store Multiple Increment After

- stmdb: Store Multiple Decrement Before
- stmda: Store Multiple Decrement After

Let's use the sample data in Figure 18.2 to describe how these instructions work. We have five words of data starting at byte address 40100. Register R8 will be used as the base register currently pointing to memory address 40108 which contains data value C0000.

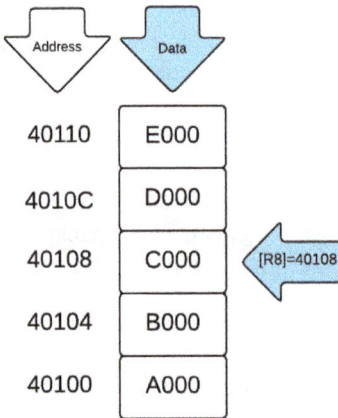

Figure 18.2: Block transfer sample

The four load multiple combinations are shown in Table 18.1. In the first line in the example, the address in base register R8 will be incremented after (IA) it is used to load (LDM) the data value C000 into register R0. Then the same instruction loads data value D000 into register R1, and the address is incremented after that also.

Notice that the address is being incremented, but not directly in R8, but in a CPU internal memory address register. In other words, the contents of register R8 are unchanged by these examples.

	R0	R1
ldmia R8,{R0,R1}	C000	D000
ldmib R8,{R0,R1}	D000	E000
ldmda R8,{R0,R1}	B000	C000
ldmdb R8,{R0,R1}	A000	B000

Table 18.1: Examples of loading two words from memory

Block Transfer Instruction Format

The use of the STM (Store Multiple) and LDM (Load Multiple) instructions have the following general characteristics.

- All 32 bits of each register are always moved (i.e., no single bytes).
- A register must be specified as a "base register" which points to the memory address
- Obviously, not all registers will be moved within one memory clock cycle, but only one instruction is executed.
- Any combination of from one to sixteen registers can be specified.
- In assembly language, the order of registers doesn't matter as each of

Computer Architecture Tutorial Using an FPGA

the following examples selects the same set of registers: R1, R2, R3, R5, R6, and R7.

  - o {R1-R3,R5-R7}
  - o {R6,R2,R7,R3,R5,R1}
  - o {R6-R7,R5,R1-R3}

- The register contents are stored in memory in the order of the register ID numbers (i.e., R0 is always stored at a lower address than R1, regardless of whether the base register is being incremented or decremented).

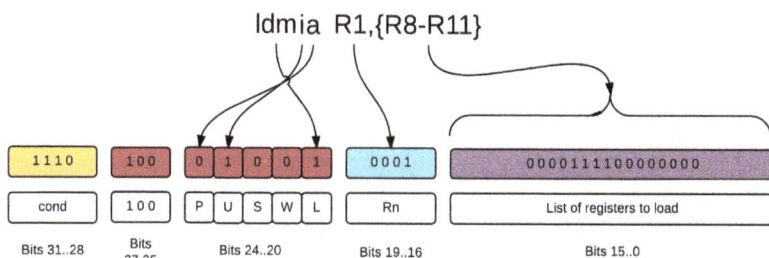

Figure 18-3: Assembly of the LDMIA R8,{R0,R1} instruction machine code

The LDM and STM instructions are similar to the LDR and STR instructions previously used, but a register list is now included:

- Cond (bits 31..28): Status indicating whether the instruction should be executed (1110 indicates always).
- 1 0 0 (bits 27..25): Indicates this is a LDM/STM instruction.
- PUSWL (bits 24..20): Bits indicating specifically what is to be done by LDM/STM instruction.
  - o P (bit 24): "1" indicates pre-indexed (before), "0" indicates post-indexed (after).
  - o U (bit 23): "1" indicates increment address, "0" indicates decrement address.
  - o S (bit 22): special use only by operating system .
  - o W (bit 21): "1" indicates write-back, i.e., update the base register, "0" indicates no update.
  - o L (bit 20): "1" for load (LDM), "0" for store (STM).

- Rn (bits 19..16): Indicates which register is the base register.
- List (bits 15..0): Which registers (R0 through R15) are loaded/stored.

As you might expect, an exclamation point is typically used in ARM assembly language to indicate the base register is to be updated, and that is shown in Figure 18.4.

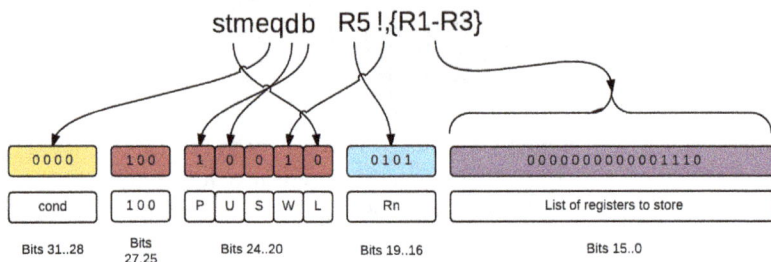

Figure 18-4: Assembly of STMEQDB R5!,{R1-R3}

Stacks and Queues

The block transfer instructions are typically used to save register contents onto a stack and later restore the data back into the same registers. They are often used in a subroutine to open up a workspace of registers without destroying the values being used by the calling routine. Characteristics of a stack:

- A common metaphor of stack operation is the placing and getting of cafeteria trays and plates. You place new trays on top and also remove trays from the top. Who would try to take the tray on the very bottom or from the middle of a stack?
- Data is stored onto and retrieved from the stack in a LIFO (Last In, First Out) manner. Stack usage is very easy: You "push" new data onto the stack and "pop" the most recent data from the top of the stack.
- A running program can have several stacks active at one time, where each is pointed to by a different base register. However, most operating systems provide a "run-time stack" that uses the SP register (R13 in the ARM). The memory is allocated by the operating system, and the pushing and popping user does not have to know the details of where in memory the stack is actually located and exactly how it works.
- Pushing the contents of an ARM register onto the run-time stack commonly results in the SP register being decreased by 4. Most operating systems fill their stacks from high memory addresses to lower addresses.
- The amount of memory allocated to the run-time stack and the SP pointer contents are set up by the operating system when it starts each program. It is possible for a user program to "blow" the stack by pushing more data onto it than it was allocated.

Are the block transfer instructions limited to just LIFO stacks? No, of course not.

First-in first-out (FIFO) queues can also be easily implemented, but you'll need two base registers: one pointing to "in" and one pointing to "out."

Ascending or Descending, Full or Empty

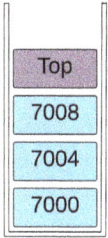

Figure 18.5:
Ascending
stack

When implementing a stack, there are two factors to consider regarding the stack pointer contents:

- Full or Empty: Does the stack pointer point to the last data pushed onto the stack (i.e., the location is full) or to the next location to receive data (i.e., the location is empty)?
- Ascending or Descending: When new data is pushed onto the stack, is the stack pointer incremented (ascending) or decremented (descending)?

Figure 18.5 illustrates an ascending stack which corresponds to the analogy of stacks of plates and trays in a cafeteria. This example shows the stack starting at address 7000 and incrementing for each word that is pushed onto the stack

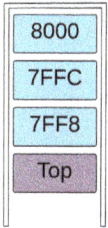

Figure 18.6:
Descending
stack

Figure 18.6 illustrates a descending stack, where the stack pointer is decremented when data is pushed onto the stack. This illustration shows the stack starting at address 8000 and descending for each word pushed onto the stack. An analogy for this approach is a stack of helium balloons put into a small elevator shaft.

So far in this discussion, I've used the term "stack pointer" to refer to the SP register, R13, but it could also refer to any of the other registers that could be pointing to other stacks separate from the run-time stack.

The block transfer instructions with names like LDMIA are very descriptive from a hardware viewpoint. They describe exactly what the instruction is to perform: Increment After, for example. On the other hand, names like PUSH and POP are more relevant to the software data structure. How can the STM and LDM machine codes be used to implement pushing and popping multiple registers at a time? There is also a middle ground of names for block transfer instructions that refers to a stack's full/empty ascending/descending structure.

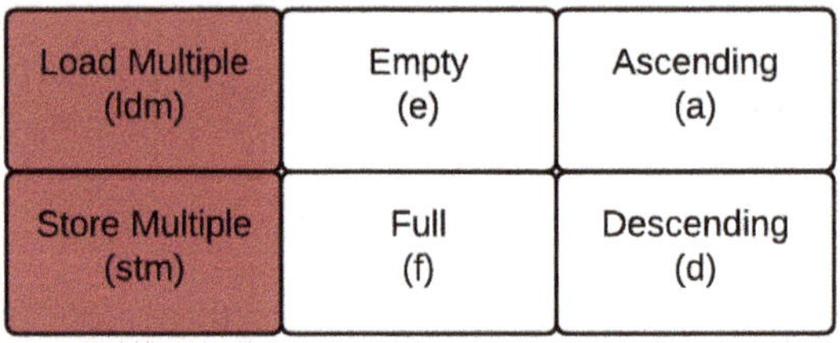

| Load Multiple (ldm) | Empty (e) | Ascending (a) |
| Store Multiple (stm) | Full (f) | Descending (d) |

Figure 18.7: Block transfer instructions alternate names

- ldmea: Load Multiple Empty Ascending
- ldmed: Load Multiple Empty Descending
- ldmfa: Load Multiple Full Ascending
- ldmfd: Load Multiple Full Descending
- stmea: Store Multiple Empty Ascending
- stmed: Store Multiple Empty Descending
- stmfa: Store Multiple Full Ascending
- stmfd: Store Multiple Full Descending

Table 18.2 equates the hardware-descriptive instruction names to the data-structure-descriptive names for each of the eight block transfer instructions. It may seem like a small point, but which of the following would be less confusing for programming an Empty/Ascending stack? Push data onto the stack with a STMEA, and pop data with a LDMEA? Or would you prefer pushing data with STMIA and popping data with LDMDB? The assembler doesn't care, and I think fewer programming errors would be made using STMEA/LDMEA than STMIA/LDMDB.

Stack type	Save on stack	Reload from stack
Empty Ascending	stmea (stmia)	ldmea (ldmdb)
Empty Descending	stmed (stmda)	ldmed (ldmib)
Full Ascending	stmfa (stmib)	ldmfa (ldmda)
Full Descending	stmfd (stmdb)	ldmfd (ldmia)

Table 18.2: "Stack relevant names" for the block transfer instructions

Push and Pop "Instructions"

The SP register points to the "top" of the stack, an area of memory where temporary data may be stored. In the mid-1970s, computers, such as the DEC PDP 11, implemented the concept of a stack using PUSH, POP, and CALL machine code instructions.

Computer Architecture Tutorial Using an FPGA

The 32-bit ARM architecture does not have machine code instructions that are specifically named PUSH and POP, but assemblers do generate code for PUSH and POP "instructions." So what kind of stack are we really using for the run-time stack provided by Linux and Windows, and what machine code should the assembler really generate for PUSH and POP?

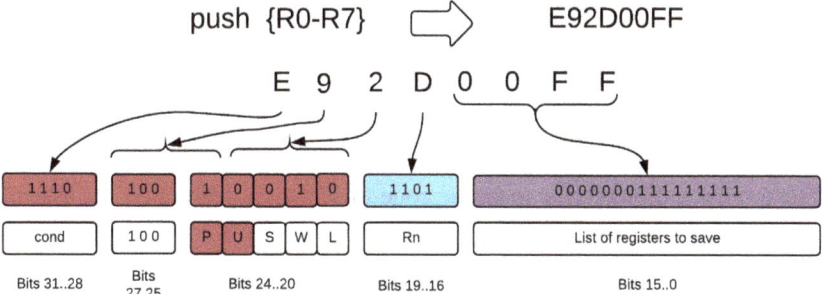

Figure 18.8: Machine code generated by assembler for PUSH {R0-R7}

Let's examine the bits in the machine code to see what instruction the assembler really generated for the PUSH.

- Cond (bits 31..28): Status indicating instruction should be executed (1110 indicates always).
- 1 0 0 (bits 27..25): Opcode indicating that this is a LDM/STM instruction.
- PUSWL (bits 24..20): Bits indicating specifically what is to be done by LDM/STM instruction.
   - P (bit 24): "1" indicates pre-indexed (before)
   - U (bit 23): "0" indicates decrement address.
   - S (bit 22): "0" indicates don't update the CPSR status bits.
   - W (bit 21): "1" indicates write-back, i.e., update the base register
   - L (bit 20): "0" indicates store instruction(STM).
- Rn (bits 19..16): Indicates register R13 (i.e., the SP) is used as the base register.
- List (bits 15..0): Indicates which registers (R0 through R7 in this example) are to be stored.

Therefore, we find that the assembler should generate a STMDB SP!,{R0-R8} instruction, which from table 18.2 is the same instruction as STMFD SP!,{R0-R8}. If we do the same thing for the POP instruction, we get the following in Figure 8.9.

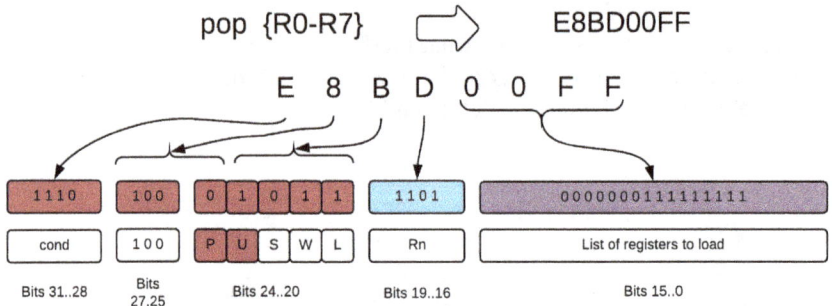

Figure 18.9: Machine code generated by assembler for POP {R0-R7}

Let's examine the bits in the machine code to see what instruction the assembler really generated for the POP.

- Cond (bits 31..28): Status indicating instruction should be executed (1110 indicates always).
- 1 0 0 (bits 27..25): Opcode indicating that this is a LDM/STM instruction.
- PUSWL (bits 24..20): Bits indicating specifically what is to be done by LDM/STM instruction.
  - P (bit 24): "0" indicates post-indexed (after)
  - U (bit 23): "1" indicates increment address.
  - S (bit 22): "0" indicates don't update the CPSR status bits.
  - W (bit 21): "1" indicates write-back, i.e., update the base register
  - L (bit 20): "1" indicates load instruction (LDM).
- Rn (bits 19..16): Indicates register R13 (i.e., the SP) is used as base register.
- List (bits 15..0): Indicates which registers (R0 through R7 in this example) are loaded.

We find that the assembler generated a LDMIA SP!,{R0-R8} instruction for POP {R0-R8} which from Table 18.2 has the same machine code as LDMFD SP!,{R0-R8}.

Since PUSH {R0-R8} is STMFD SP!,{R0-R8} and POP {R0-R8} is LDMFD SP!,{R0-R8}, the type of run-time stack is usually descending (D), and the top of the stack is full (F). This convention has been the most popular implementation used by most computer operating systems. It allows program code and local variables to grow upward from low memory addresses, and stack data to grow downward from high memory, and hopefully the two pointers never meet and run out of memory. Could it have been implemented differently? Sure. Programs can even have multiple stacks within them. PUSH and POP use the SP register, but other stacks could use other registers.

Assembler for Block Transfer Instructions

Using Verilog macros, tasks, and functions, the block transfer instructions can be assembled into ARM machine code. Of course, we cannot support the exact format used by typical ARM assemblers. It's your option, of course, but I recommend the following approach:

1. Make a task named "asbt" having eight parameters, This way the same endings, such as `_4, can be used with the new block transfer instructions as with the other instructions.
2. Have two macros named STM and LDM that will call the asbt task. The task macro will call a function to be named asbt1 as was done for the other instructions.
3. Define macros IB, IA, DB, and DA to modify the STM and LDM macros to provide STMIB, STMIA, etc.
4. Also define the four macros, EA (Empty Ascending), FD (Full Descending), etc. that parallel the above macros IB, IA, etc.
5. Define macros named PUSH and POP that call a task named astack, which will call the previously defined function asbt1. The following three parameters will be "hard coded:" base register SP (same as R13), writeback option bit, and FD (Full Descending)
6. The writeback (i.e., update) will either have to be implemented with macros such as W or U or by combinations such as IBW and IAW.
7. The parentheses around the base register are not needed (or can even be detected) by the asbt task, but they do provide consistency between LDM syntax and that previously used for the LDR instruction.
8. The bit identifying each register in the set can be generated with the Verilog << operator and ORed together to form the lower 16 bits of the instruction.
9. The bit pattern for a range of registers can be implemented with a subtraction. For example: The bit pattern for R3 TO R8 is 16'b0111111000 which is produced by 2<<R8[3:0] - 1<<R3[3:0]. The separator "TO" can be equated to some value such as 1 so as to distinguish it from the registers R0 through R15.

"Verilog" Assembler	Typical Assembler
`LDM`IA (R8), R0, R12, R5 `_6	ldmia R8, {R0, R12, R5};
`STR`DB`EQ   (R14), R5  `_5	strdbeq R14, {R5};
`LDM`FA (R8), R0, TO, R5, R10 `_7	ldmfa R8, {R0-R5, R10};
`PUSH   LR, R0, TO, R5, R10  `_6	push {LR, R0-R5, R10};

Table 18.3: Compare assembler syntax for multiple load/store instructions

Block Transfer Instruction Imitation

The block transfer instructions are similar to the LDR and STR instructions that were started in Chapter 14 and finished in Chapter 16. The main difference is the execution of the block transfers will require a modification to the instruction cycle to support staying in the writeback state for as many as 16 clock cycles. It's your option, of course, but I recommend the following approach:

1. The DataMod module contains the RAM. It will either have to be modified to include the block transfer instructions, or a new module named BlockDataMod be created that works in unison with it.
2. The bit list in the LDM/STM instructions indicates which of the registers wil be loaded or stored. The number of bytes read or written will be to a block of memory having between 4 and 64 bytes. The lower register numbers will be at the lower memory addresses (i.e., if registers R1, R3, and R7 are stored in memory addresses 1000 through 100C, then R1, R3, and R7 will be stored in addresses 1000, 1004, and 1008, respectively, irregardless of which instruction is used.
3. The writeBack state will need to be modified to hold off moving to the fetch state until all of the selected registers have been stored or reloaded. A technique similar to that used in the decode state for performing the breakpoint can be used.

Highlights and Comparisons

The machine code generated and the operation of the ARM imitation should, of course, be compared to that of a real ARM processor. There are several ARM assemblers available on the Internet and various ARM based computers. An inexpensive, but very powerful combination, is using a Raspberry Pi computer running the Linux operating system along with the GNU set of utilities.

Review Question

1. The 'PUSH   R5 assembly language instruction can be implemented with a STMFD   SP!,{R5} instruction. What STR instruction can produce the same effects?

Exercise

1. Modify the recursive factorial assembly language program in Listing 16.5 to use the new PUSH and POP with multiple registers. Don't forget to leave plenty of room for the expansion of the stack.

— 19 —
Thumb

Thumb is an alternate "instruction set" for the ARM CPU. It uses the same registers, same memory, and same CPU, but has a 16-bit instruction format. It's like having a different "face" available for the same body. In some applications and configurations, Thumb instructions provide a performance improvement over the equivalent ARM instructions.

Chapter 19 does not implement the assembler or the machine code instruction execution for the Thumb format. Instead, it provides some background and suggestions for you to practice your Verilog skills and build upon your computer architecture knowledge gained in the preceding chapters.

For a first impression of the 16-bit Thumb format, take a look at the subtraction instruction in Figure 19.1. Compare it to what you already know about the subtraction instruction within the 32-bit ARM data processing instruction format.

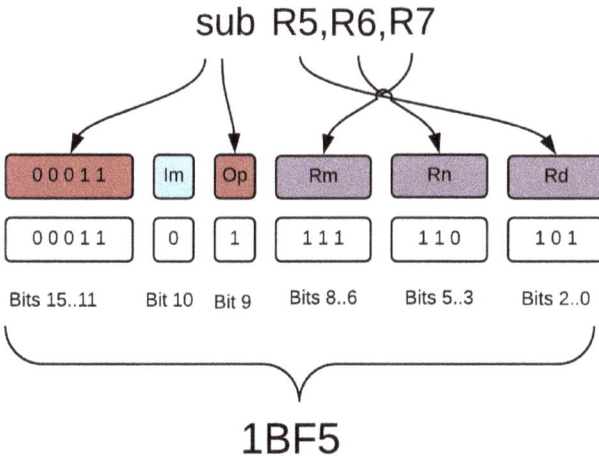

sub R5,R6,R7

00011	Im	Op	Rm	Rn	Rd
00011	0	1	111	110	101
Bits 15..11	Bit 10	Bit 9	Bits 8..6	Bits 5..3	Bits 2..0

1BF5

Figure 19.1: Thumb add/subtract machine code example

- Bits 11-15 indicate this is an add/subtract operation.
- Bit 9 indicates whether it is add (0) or subtract (1).
- Three registers can be used in this add/subtract format. In the example, [R6] is added to [R7] and the result is stored into R5.
- Bit 10 is the "immediate" flag. Figure 19.2 shows it being set, but the range of immediate values is only 0 through 7 (3 bits).

sub R5, R6, #7

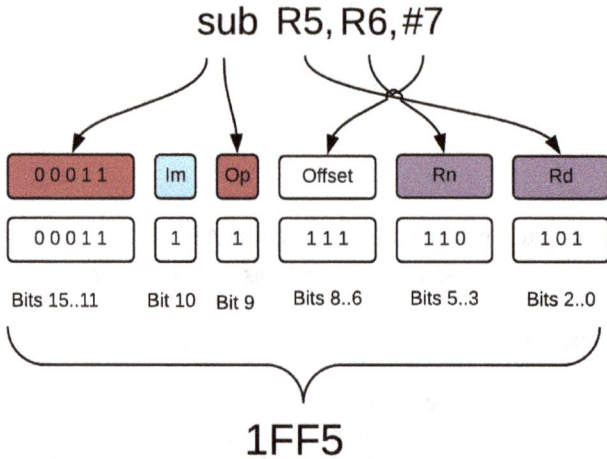

00011	Im	Op	Offset	Rn	Rd
00011	1	1	111	110	101
Bits 15..11	Bit 10	Bit 9	Bits 8..6	Bits 5..3	Bits 2..0

1FF5

Figure 19.2: Thumb add/subtract with immediate constant

Thumb instructions are really repackaged ARM instructions, but how can a 32-bit ARM instruction be stuffed into a 16-bit package? Obviously, some things have to be eliminated and other things assumed.

1. No conditions for execution are present (i.e., no Cond field).
2. There is no S status update flag. Note: Thumb arithmetic instructions always update the CPSR.
3. The same registers are used as in 32-bit mode, but these fields are only three bits wide, thereby restricting them to R0 through R7.
4. There is no shift built into the data processing instructions.

Thumb Background

There are three variations of Thumb available:

1. Thumb: The original only had instructions that were 16-bits in size.
2. Thumb-2: This enhancement added some 32-bit instructions to the original 16-bit instructions, providing almost all the capability as code written using the regular ARM instruction set.
3. ThumbEE: This Execution Environment modification of Thumb-2 is optimized for programming languages like Java that use an intermediate level language.

What is the advantage of using the Thumb set of instructions? Performance. Because Thumb instructions are only 16-bits wide, they take up less room in memory and will load faster in hardware configurations having a 16-bit data bus or smaller. Today's configurations, having 32-bit data buses and large memories, could actually run slower in Thumb-coded programs due to the Thumb's lack of features present in the 32-bit ARM format as well as the overhead due to switching between ARM and Thumb modes.

Are the Thumb instructions a subset of the set of ARM instructions? This is close to being true, but not exactly true. Most of the Thumb instructions are a limited version of corresponding 32-bit ARM instructions.

Since Thumb instructions are converted to 32-bit ARM instructions before being executed, isn't performance degraded by this extra step of conversion? No, there isn't an extra step. When in Thumb mode, the ARM CPU substitutes the Thumb instruction decoding for the regular 32-bit ARM instruction decoding.

Thumb Imitation

If you decide to practice your Verilog skills by implementing the Thumb instruction format, I recommend only starting with a partial Thumb implementation. Do enough to demonstrate switching between ARM and Thumb modes along with a few instructions like ADD and SUB.

Just like in the ARM imitation, there should be two parts to implementing the Thumb format: the assembly language and the machine code execution. Each of these two parts will need a way to switch between the ARM and Thumb modes.

1. Switch assembler between Thumb and ARM modes: Write two new macros, `THUMB and `ARM.
2. Switch machine code execution between Thumb and ARM modes: Implement the BX instruction.

For the assembly language part, I recommend creating a new *integer* variable named ThumbMode. Write two new macros named `THUMB and `ARM that will simply set ThumbMode to 1 or clear it to zero, respectively. Then, when any of the macros ADD, SUB, or BX call the asdp task, it can examine the ThumbMode variable to determine the type of instruction to build. If ThumbMode is 0, it calls asdp1 to build a 32-bit instruction and if ThumbMode is 1, it calls a new function named asdpthumb to build a 16-bit Thumb instruction.

For the instruction execution part, I recommend using bit 5 of the CPSR to determine whether instructions are to be interpreted as 16-bit or 32-bit. However, do not set and clear this bit directly. Switching should be done using the BX instruction, just like it is done in a real ARM processor.

You may want to start by implementing only the ADD, SUB, and BX instructions. I've provided the details of their formats in the first three figures in this chapter. Later, you may choose to include more Thumb instructions. Those formats are readily available from several Internet sites.

Branch Exchange

The Branch Exchange (BX) instruction appears in both the ARM and Thumb instruction sets. Its main purpose is to change between the two modes, but it also loads a new PC address. The instruction shown in Figure 19.3 becomes a BLX (Branch, Link, and Exchange) if bit 7 in the instruction is set. In that case, R14 (the LR link register) will be loaded with a return address.

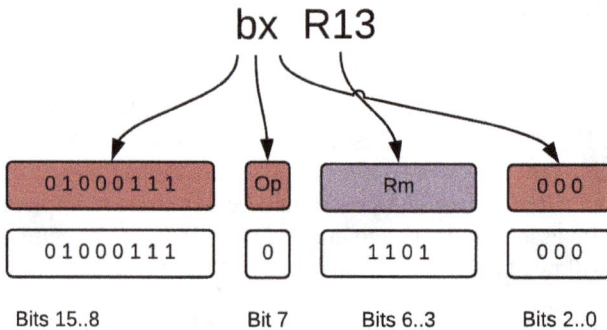

Figure 19.3: Branch exchange (BX) is used to enter and exit Thumb mode.

Thumb instructions are intermingled with the 32-bit ARM instructions. How does the CPU know which format to use when decoding each instruction? If the T-bit (CPSR bit 5) is set to 1, the CPU assumes the PC register (Program Counter) is pointing to a Thumb format instruction, while if the T-bit is zero, the instruction is assumed to be a 32-bit ARM instruction. Note: Do not switch between the Thumb state and full ARM instruction format by directly setting this bit. The BX (Branch Exchange) instruction is used to change states.

The 32-bit ARM instructions are word-aligned on addresses that are multiples of 4 bytes, while the 16-bit Thumb instructions are half-word-aligned on addresses that are multiples of 2 bytes. In other words, the value in the Rm register (R13 in Figure 19.3) is always an even number, implying bit 0 must always be zero. The ARM processor thereby repurposes bit 0 as a flag to set or clear Thumb mode.

When you code the execution of the BX instruction, copy bit 0 in the Rm register to bit 5 of the CPSR. Also do not use that bit to calculate the branch-to address. Example: If you want to jump to the address of label "Test" and switch into Thumb mode, load "Test + 1" into the Rm register before executing the BX instruction.

Computer Architecture Tutorial Using an FPGA

```
1.
2.  // Program code that switches to Thumb mode and back
3.
4.      `ADR   R9,T_Mode+1   `_3        // Load relative address of T_mode into R9
5.      `BX   R9   `_2                  // Bit 0 of R4 = 1. Switch to Thumb mode
6.      `THUMB   `_1                    // Switch "assembler" to create Thumb code
7.  T_Mode = IP;                        // Beginning of section of Thumb code
8.      `ADD   R5,R6,R7   `_4           // Typical 16-bit Thumb instruction
9.      `ADR   R4,A_Mode   `_3          // Load relative address of A_mode into R9
10.     `BX   R4   `_2                  // Bit 0 of R4 = 0. Switch to ARM mode
11.     `ARM   `_1                      // Switch "assembler" to create 32-bit ARM
                                        code
12. A_Mode = IP;                        // Beginning of section of ARM code
13.     `ADD   R10,R11,R2   `_4         // Typical 32-bit ARM instruction
14.
```

Figure 19.4: Sample code to switch between ARM and Thumb modes

Repackaging 32-bit instructions into a 16-bit "bag" results in a "shoe horn" format. The Thumb instruction format does not have the clean, somewhat orthogonal, design of the 32-bit ARM instruction format. Implementing the assembler as well as the execution imitation will be a fair amount of work.

Highlights and Comparisons

The ARM format has unique features such as conditional execution for every instruction and register shifts included within every data processing instruction. The Thumb format does not have room for these features in its 16-bit format, and is therefore much more like other CPU architectures than the ARM 32-bit format.

1. All arithmetic operations (ADD, SUB, ..) set the status flags with Thumb (i.e., no S bit).
2. Only the branch instructions examine the CPSR register to see if an instruction should be executed (i.e., no Cond field in the instruction format).
3. Shift instructions are individual instructions and not built into the data processing instructions (i.e., the second operand is not very complicated).

If the Thumb format is so much simpler, why didn't I begin the book with that format? One detailed look at how the instructions are "shoe horned" into 16 bits, with operations like ADD having more than one opcode, and some opcodes are broken up into multiple fields, will give you the answer.

Review Questions

1. If the ARM architecture was developed with today's technology available (memory speeds, widths, costs, etc.), do you think Thumb mode would be included? Why or why not?
2. What hardware configurations (processor speed, memory bus width, etc.) would offer Thumb the greatest advantage over regular ARM instructions?
3. * We switch between Thumb mode and ARM mode with the BX instruction. Although we could set the T-bit in the CPSR directly, why would this lead to problems if we did?

— 20 —
NEON ALU

One way to improve performance is to perform multiple operations in parallel. The NEON coprocessor provides SIMD (Single Instruction Multiple Data) capability where the same instruction, such as multiplication, is performed on multiple pairs of numbers "at the same time."

The NEON coprocessor is an independent processor having its own set of registers and instructions. It does not contain its own memory for program and data storage, but "watches" all the instructions that the ARM processor is fetching. Basically, a NEON SIMD operation is performed by the following programming sequence:

1. One or more coprocessor instructions move source data into the desired NEON registers.
2. Perform the desired SIMD operation using a NEON coprocessor "vector" instruction (i.e., VADD, VORR, VEOR, ...).
3. Use another NEON instruction to move the result into either one of the ARM's registers or ARM memory.

Chapters 20, 21, and 22 present three more aspects of computer architecture that are very important: parallel processing of data, interleaved memory transfers, and floating point. I'm just providing enough information to see if you are interested in going deeper. Doing so will provide practice for your Verilog skills and build upon your computer architecture knowledge gained in the preceding chapters.

The NEON architecture is an excellent candidate to practice your FPGA development skills:

1. The parallelism of the FPGA lends itself well to implementing the parallelism of the NEON coprocessor.
2. Creating a module and then instantiating it many times also lends itself to the NEON "lanes" architecture.

NEON Overview

The NEON coprocessor contains a 2048-bit register data "file" along with logical, integer, and floating point operations. How are these 2048 bits grouped? They are grouped as multiple registers, either 16 128-bit Quad registers, 32 64-bit Double registers, or 32 32-bit Single registers.

All three of these register types share the same physical bits. Each NEON instruction specifies the grouping, so it is even possible to change the register size from one instruction to the next. Figure 20.1 shows the correspondence of the first two Q registers with the D and S registers.

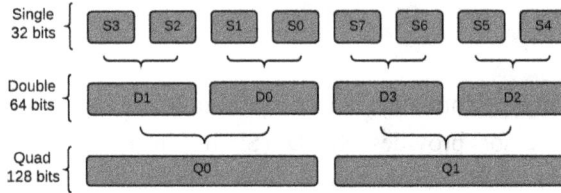

Figure 20.1: Overlap of NEON registers: Bits 0 of S3, 0 of D1, and 0 of Q0 are the same physical bit.

In Chapter 6, bit-by-bit logical operations were shown using the AND which worked on multiple bits simultaneously. NEON also supports bit-by-bit logical operations on as many as 128 bits at a time using the Q registers. Figure 20.2 has an assembly language program using the NEON coprocessor to perform 128 inclusive OR operations at a time. It's a simple example that changes the case of sixteen ASCII characters with one instruction.

1.	LDR	R6, =cvt	@ Load pointer to 0x20202020
2.	LDR	R1, =msg	@ Load pointer toASCII data
3.	VLDM	R6, {Q1}	@ Load 128 bit conversion pattern.
4.	VLDM	R1, {Q0}	@ Load first 16 characters
5.			
6. @	Convert to lower case ASCII by setting bit 5 to 1 in each byte		
7.			
8.	VORR	Q0, Q0, Q1	@ Convert 16 letters to lower case.
9.	VSTM	R1, {Q0}	@ Restore with lower case letters.
20. cvt:	.ASCII	" "	@ String of 16 hex 20s (blanks)
21. msg:	.ASCII	"Test Message"	@ Memory buffer of character data

Figure 20.2: Changing case of 16 letters using NEON instructions.

The following lines of code in Figure 20.2 are of particular interest:

- 3: Vector Load Multiple (VLDM) copies 16 byte pattern into 128-bit NEON Q1 register. This pattern is in ARM memory at location pointed to by ARM register R6 and contains 16 bytes of x020 (bit 5 is set in each byte).
- 4: Another VLDM loads NEON register Q0 with the first 16 bytes from the message buffer. This is similar to the ARM LDMIA instruction which can load 16 bytes into four ARM registers.

Computer Architecture Tutorial Using an FPGA

- 8: The Vector inclusive OR (VORR) instruction operates on 128 bits simultaneously, thereby replacing the equivalent of four ARM ORR instructions.
- 9: The Vector Store Multiple VSTM instruction is similar to and thereby replaces the ARM STMIA instruction, but uses a list of NEON data registers. ARM register R1 points to the ARM memory where the NEON data is to be stored.

Scalars and Vectors

A scalar value consists of a single number, and a vector value consists of a group of numbers. Examples of vectors are the position and velocity of an object in a three-dimensional coordinate system which would have X, Y, and Z components, for example. An example of a scalar is the mass of an object (i.e., one number).

Why all the bother? Are vectors used that extensively that it's worth adding confusion to push for a performance gain? Consider the following:

1. In physics and engineering, quantities like position, velocity, and acceleration are all vectors that are measurable quantities.
2. Many scientific and engineering problems are solved using matrix transformations and inversions which require many vector-type multiplications and additions.
3. Graphics applications which display the 3-D world mapped onto a 2-D screen require many matrix multiplications which are efficiently processed by vector instructions.
4. Digital signal processing and many analog to digital conversions work efficiently with extensive vector processing.

Lanes

NEON also provides integer arithmetic operations. How do the results from an arithmetic operation such as addition or multiplication compare to one of the logical operations? Logical operations are bit-by-bit, and the results stay in each bit "column," but arithmetic operations must expand to use more bits. Even an example such as $1_2 + 1_2 = 10_2$ shows addition can have a carry that requires another bit column. In order to provide multiple simultaneous parallel arithmetic operations, the NEON coprocessor forces arithmetic operations to stay in fixed-sized "lanes" and does not allow the results from one lane to carry (or overflow) into the next.

NEON arithmetic assembly language operations are of the form "Operation.TypeLanesize," such as VADD.U16 and VSUB.S8. NEON arithmetic lane type and size combinations are specified by the type and size suffix:

- S8, S16, S32, S64: Signed integer: High order bit is the sign bit.
- U8, U16, U32, U64: Unsigned integer: High order bit is a data bit.
- I8, I16, I32, I64 : Integer, either signed or unsigned
- F32: Floating point, single precision

In a NEON arithmetic operation, the total number of bits involved is specified by whether a D (64 bits) or a Q (128 bits) register is used. The number of parallel lanes is therefore either 128 or 64 divided by one of the above lane sizes. For example, "VADD.U16   D0,D1" has four lanes (64/16). Although there can be 128 simultaneous 1-bit logical operations performed, the number of simultaneous arithmetic operations is reduced to the number of lanes.

The maximum number of lanes is sixteen (128-bit Q register divided by an 8-bit lane width). Of course, the minimum number of lanes is one, where a 64-bit D register is divided by a 64-bit lane width.

Consider the example in Figure 20.3, where a "VADD.S8   D0, D1. D1" doubles each of the eight values in the 8-bit lanes. The first four lanes, which double 1, 2, 10, and 50 produce correct results. The fifth lane, $100 + 100 = 200$, would be fine if an unsigned U8 was used, but the binary for 200_{10} is 11001000_2 which is a negative number. The S8 implies these are signed values. The remaining three lanes are wrong, whether we consider them to be signed or not.

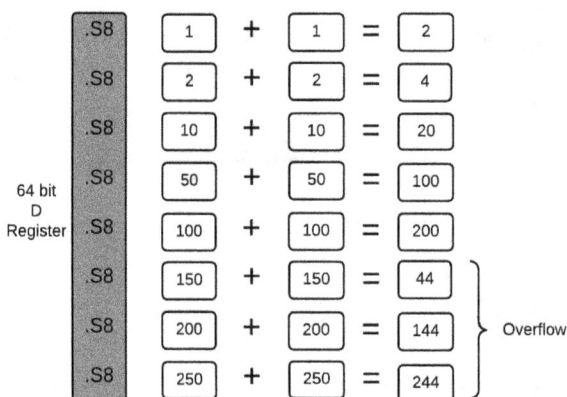

Figure 20.3: Overflow of addition of two 8-bit numbers

In the ARM CPSR, the carry and overflow status bits could be set for ARM arithmetic, but that is not the case for NEON. There is no status bit to help us. Is there anything that can be done about overflow conditions giving ridiculous looking results? NEON provides two features to alleviate or at least soften the overflow problem: Saturation (Q) and Wide (W) lanes.

VADD.U8

VQADD.U8   VADDW.U8

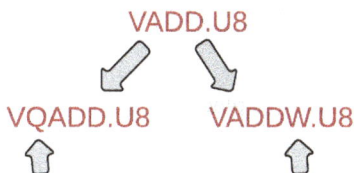

Figure 20.4: NEON Saturation and Wide lanes opcodes

Saturation Integer Arithmetic

By using VQADD instead of VADD, the NEON coprocessor still can't provide the correct answer in the case of an overflow, but it will keep the answer as close as possible. For signed S8 operations, the extreme limits for an 8-bit signed integer are -128 and 127. For unsigned U8 operations, the limits are 0 and 255. What are the limits for the generalized integer code of I8? Since the limits aren't specified in I8, NEON cannot know the limits. Therefore, the I8, I16, I32, and I64 suffixes are not allowed to be used with the Q saturation modification.

Figure 20.5 compares unsigned saturated integer addition to unsaturated addition. If 16-bit lanes are used with VQADD or any of the other saturated arithmetic operations, the range for unsigned numbers would be 0 through 65,635.

150	+	150	=	44	
200	+	200	=	144	Overflow (not saturated)
250	+	250	=	244	
150	+	150	=	255	
200	+	200	=	255	Unsigned .U8 saturated
250	+	250	=	255	

Figure 20.5: Effect of using unsigned saturation addition

Promotions to Wide Lanes

Although addition of two large binary numbers can overflow by a single bit and be lost, it is not unusual for the product of two very large numbers to require double the number of bits of the factors. Most CPUs have a special "long" multiply instruction to accommodate this situation. In the ARM, the UMULL instruction multiplies two 32-bit integers that produces a 64-bit product that is then stored in two registers. NEON, however, takes this a step further and provides "long" modifications for most of its instructions. As an example, the VADD instruction has the following modifications:

- VADDL: Long: The result is placed in a lane that is twice the size of the operands.
- VADDN: Narrow: The result is placed in a lane that is half the size of the operands.
- VADDW: Wide: The result is the same size as that of the first operand, but the second operand is half the size of the first.

Highlights and Comparisons

Computer application developers and users always want better performance, and electronics designers have generally been able to fulfill those expectations for decades. "Moore's Law" implies that computer performance will double every 18 months, mostly due to improvements in the packing density of transistors on integrated circuits. Oddly enough, this has proven to be the case for over three decades, far longer than many of us thought possible.

Not all of the performance improvements over the years have come from improvements in the structure of integrated circuits. Some have come from improvements in CPU design as well as running operations in parallel.

Obviously, I did not provide you with sufficient information to build either the assembler or the execution code for NEON. If you are interested, I recommend you do the following:

1. Get in a little practice using the NEON vector instructions using assembly language. Even a Raspberry Pi would be a great platform. Use the debugger to set up initial register contents as well as examine the results.
2. You can find the machine code format of the NEON instructions on the Internet and by examining a hex dump of the assembler code you generated above.
3. Have fun. Instantiate multiple copies of the same module to get multiple lanes. Use *generate* with parameters to produce various sizes of the same operation.

Review Questions

1. * If the lane size for integer addition within NEON could be one bit wide, it would be exactly the same as which logical operation?
2. * When does a VADD.U8 produce a different result than a VADD.S8? How about the same question for S16 verses U16 or S32 verses U32 for any of the VADD, VSUB, etc. NEON instructions?

— 21 —
NEON Memory Transfers

$\mathbf{U}$sing parallel computation, the NEON processor can substantially improve total throughput in processing data stored in tables and arrays.

However, setting up the proper combinations of operands and loading them into the right sets of registers may be awkward and inefficient in itself. NEON has interleaved load and store instructions to meet this challenge of array data which is stored in various patterns in memory. They can transfer up to four quad words (512 bits) at a time between the NEON's registers and the ARM's memory.

The purpose of Chapter 21 is to introduce features of interleaved load and store instructions and give ideas for their assembly and execution. Practicing your Verilog skills by implementing these instructions is valuable experience whether or not you implement the NEON ALU instructions described in Chapter 20.

NEON Uses Little Endian

Chapter 14 described the difference between big and little endian for moving bytes between the ARM registers and memory. By default, the ARM load and store instructions use little endian, but the ARM also has a status bit that can be set to enable big endian format.

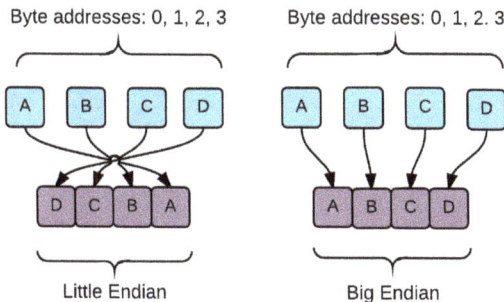

Figure 21.1: Movement of bytes from memory to a 32-bit ARM register

The NEON vector load instructions, such as "VLDM   R1, {D0}," presented in Chapter 20 only uses little endian format. As seen in Figure 21.2, the first byte from the memory buffer is loaded first into the "little end" (i.e., bit 0) of the register, and the following bytes keep moving in until the "big end" is reached.

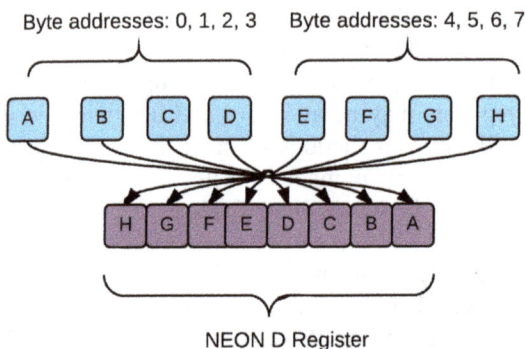

Figure 21.2: NEON loads data into registers in little endian format

As many as three NEON registers can be loaded with one VLDM instruction. If register R1 points to a 24-byte memory array containing the sequence of ASCII characters A through W, and a "VLDM   R1, {D0, D1, D2}" is executed, then the NEON register contents are loaded in the pattern in Figure 21.3.

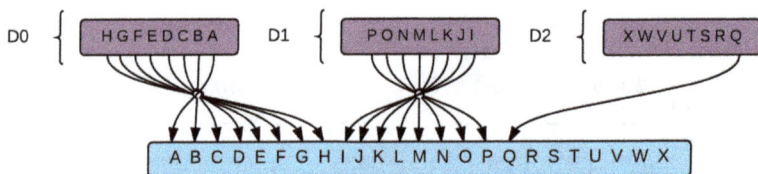

Figure 21.3: VLDM   R1,{D0-D2}   or   VSTM   R1,{D0-D2}

Interleaved Load and Store

What if we are working with an array of 32-bit integers or 16-bit half words? What if the data in arrays do not match the format of NEON register sizes and patterns? Basically, we would have to do a fair amount of ARM programing to rearrange the data before loading it into NEON's registers, but there's another option. NEON offers "interleaved" load and store instructions that move sublists of array elements:

- As few as one 64-bit D register and as many as 4 adjacent 128-bit Q registers are processed with one instruction (minimum of 64 bits and maximum of 512 bits).
- The memory array is composed of multiple elements of the same size which can be 8, 16, 32, or 64 bits in length.
- Elements are moved sequentially between ARM memory and one or more NEON registers.
- An ARM general purpose register acts as an index and contains the byte address of the next element to load/store in the ARM's memory.
- The ARM index register may be updated if the write-back bit is set ([R4]!, for example).

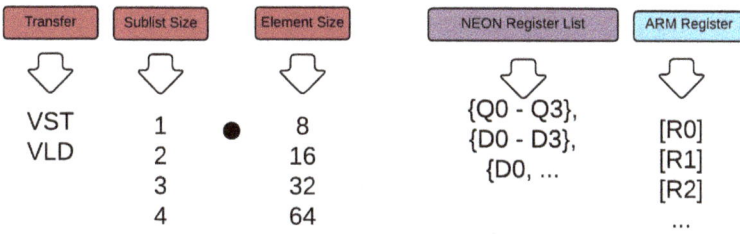

Transfer	Sublist Size	Element Size	NEON Register List	ARM Register
VST	1	8	{Q0 - Q3},	[R0]
VLD	2	16	{D0 - D3},	[R1]
	3	32	{D0, ...	[R2]
	4	64		...

Figure 21.4: Interleaved store and load instructions

The operation of interleaved storage instructions are analogous to eating dinner. I will eat all the carrots, beans, and fish on my plate, but in what order should I eat them? Do I eat all the carrots before moving on to eat all the beans, and then all the fish, or do I mix them together by taking a bite of carrots, then a bite of beans, then a bite of fish, and back to the carrots until everything is gone?

A few examples will help explain how the interleaved store instructions work. Assume registers D0 through D3 have been loaded with the A to W pattern as previously described for the VLDM instruction. They have the following contents:

- D0: "HGFEDCBA"
- D1: "PONMLKJI"
- D2: "XWVUTSRQ"

The VST3.8 instruction says I will still be moving data in groups of 8-bits (the element size), but I will be taking them from a subset of three adjacent NEON registers. This is analogous to eating a bite of carrots, then a bite of beans, and finally a bite of fish, before going back to get the next round of carrots, beans, and fish, etc.

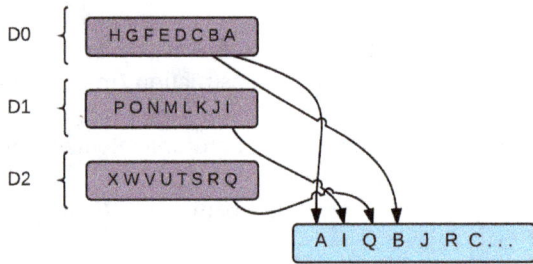

Figure 21.5: VST3.8 {D0-D2},[R1]

Figure 21.6 is the same as 21.5 except I'm taking bigger "mouthfuls": 16 bits instead of 8. The VST3.16 says I will be moving data as 16-bit elements, and I will be taking them from a subset of three adjacent registers.

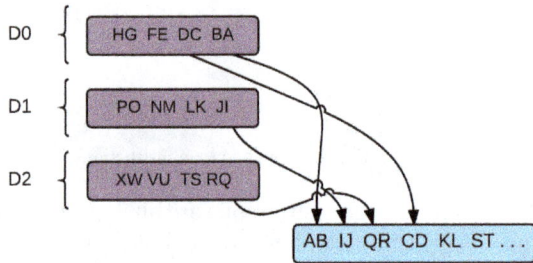

Figure 21.6: VST3.16 {D0-D2},[R1]

Figure 21.7 again uses 3 registers, but now 32 bits are moved at a time.

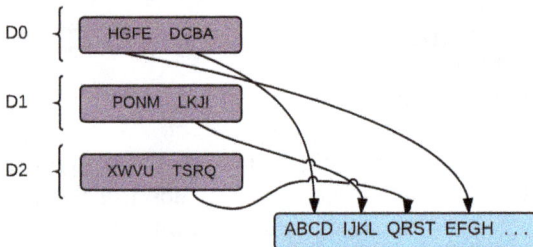

Figure 21.7: VST3.32 {D0-D2},[R1]

Table 21.1 summarizes the above examples. The first line is the regular store multiple. In the second line, VST1.8, the "1" in VST1 indicates I want all the bits transferred from one register before moving onto the next. The "8" is the element size which is analogous to the size of a mouthful, and here indicates that VST1 will be transferring 8 bits at a time. It is like eating all my carrots before starting on my beans. As you might suspect, the VST1.8 storage pattern will look identical to that from VSTM.

Array in Memory	Instruction
ABCDEFGHIJKLMNOPQRSTUVWX	VSTM R1,{D0-D2}
ABCDEFGHIJKLMNOPQRSTUVWX	VST1.8 {D0-D2},[R1]
AIQBJRCKSDLTEMUFNVGOWHPX	VST3.8 {D0-D2},[R1]
ABIJQRCDKLSTEFMNUVGHOPWX	VST3.16 {D0-D2},[R1]
ABCDIJKLQRSTEFGHMNOPUVWX	VST3.32 {D0-D2},[R1]

Table 21.1: Memory contents after various "store" instructions

Can I use a VST2 instruction? Yes, but not with three registers in the list. My list would have to be either {D0,D1} or {D0-D3}. "Left overs" are not allowed. The number of registers in the list can be between one and four, but must be a multiple of the sublist size specified adjacent to the "VST."

As in Chapter 20, I did not provide you with sufficient information to build either the assembler or the execution code for these NEON instructions. If you are interested, I recommend you do the following:

1. Get in a little practice using the NEON interleaved instructions using assembly language. Do the examples I presented in this chapter, then try a few new combinations of your own. Use the debugger to set up initial register contents as well as examine the results.
2. You can find the machine code format of the NEON instructions on the Internet as well as examining a hex dump of the assembler code you generated above.
3. These interleaved instructions are another example to practice you Verilog skills to instantiate multiple copies of the same module and the *generate* command with parameters to produce various sizes of the same operation.

Highlights and Comparisons

The efficiencies of the parallelism provided by the NEON arithmetic instructions can be lost if too much time is wasted getting all of the data into the correct pattern in memory. The interleaved vector load and store instructions help alleviate this problem of data rearrangement. This capability is not new with

NEON, but similar capability existed on the mainframe computers of the 1960s with their scatter/gather hardware.

Review Questions

1. * If R and V are two 16-bit integer vectors dimensioned as R[1000] and V [1000,2]. If you needed to calculate R[i] = V[i,1] × V[i,2], which interleaved load and store instructions would you most likely use?
2. * If R and V are two 32-bit floating point vectors dimensioned as R[1000] and V [1000,3]. If you needed to calculate R[i] = V[i,1] × V[i,2] × V[i,3], which interleaved load and store instructions would you most likely use?

— 22 —
Floating Point Format

The main theme associated with the NEON coprocessor is the simultaneous execution of multiple pairs of numbers associated with a single instruction. Chapter 20 only described a few of the instructions that are available, but I do want to whet your appetite for further study and possible implementation. For instance, the NEON coprocessor provides excellent support for floating point numbers used in engineering and scientific applications.

The purpose of Chapter 22 is to provide background information on the IEEE standard 754 floating point format used within the NEON coprocessor.

Fixed Point

A decimal point separates a decimal whole number from a decimal fraction. Likewise, a binary point separates a binary whole number from a binary fraction. How can we represent fractions in binary?

A decimal number is really a short notation for a polynomial of powers of 10. For example: 137_{10} is $1 \times 10^2 + 3 \times 10^1 + 7 \times 10^0$ which is $100 + 30 + 7$. Fractions continue that pattern where .036 is $0 \times 10^{-1} + 3 \times 10^{-2} + 6 \times 10^{-3}$. Likewise, a binary number is really a short notation for a polynomial of powers of 2. For example: 101.01_2 is $1 \times 2^2 + 0 \times 2^1 + 1 \times 2^0 + 0 \times 2^{-1} + 1 \times 2^{-2}$.

In the example below, it seems reasonable to make the first bit to the right of the binary point equal to 2^{-1} which is ½. The second bit to the right of the binary point is 2^{-2} which is $½^2$ which is ¼.

- $1111._2 = 1 \times 2^3 + 1 \times 2^2 + 1 \times 2^1 + 1 \times 2^0 = 8 + 4 + 2 + 1$
- $111.1_2 = 1 \times 2^2 + 1 \times 2^1 + 1 \times 2^0 + 1 \times 2^{-1} = 4 + 2 + 1 + ½$
- $11.11_2 = 1 \times 2^1 + 1 \times 2^0 + 1 \times 2^{-1} + 1 \times 2^{-2} = 2 + 1 + ½ + ¼$

The binary point can be "fixed" at any location. Normally, application software places the binary point to the right of bit 0, which is for whole numbers and integers. Just like the arithmetic hardware circuit doesn't know if the software is using signed or unsigned numbers, it doesn't know where the software places the binary point. It's all the same in a one's or two's complement binary format. If we "fix" the binary point to the left of bit 31 in a 32-bit format, we get the following:

- ½ => $10000000000000000000000000000000_2 = 80000000_{16}$
- ¼ => $01000000000000000000000000000000_2 = 40000000_{16}$

- $\frac{3}{4} \Rightarrow 1100000000000000000000000000000_2 = C0000000_{16}$
- $7/8 \Rightarrow 1110000000000000000000000000000_2 = E0000000_{16}$
- $9/16 \Rightarrow 1001000000000000000000000000000_2 = 90000000_{16}$
- $1/32 \Rightarrow 0000100000000000000000000000000_2 = 08000000_{16}$
- $63/64 \Rightarrow 1111110000000000000000000000000_2 = FC000000_{16}$

If we assume the binary point is "fixed" between bits 2 and 3, Figure 22.1 shows the value associated with each bit.

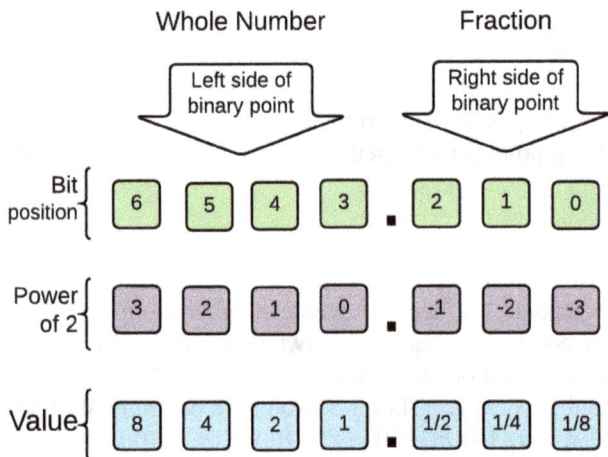

Figure 22.1: Value of each bit with three bits right of binary point.

Figure 22.2 shows another example of "fixed" binary point where the whole number 5 (fraction value of 0) is divided by 2 a couple of times. Dividing by 2 (i.e., shifting right one bit position) gives us a value of 2.5, and dividing by 2 again gives us 1.25.

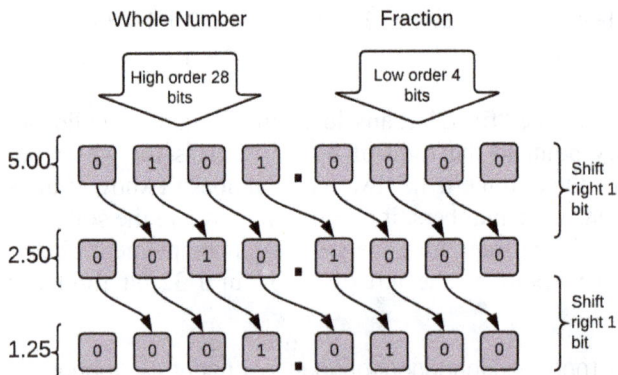

Figure 22.2: Divide by two by shifting right one bit position

Computer Architecture Tutorial Using an FPGA

One of the things we learn about fractions is that we can't represent all of them exactly. We can come close, depending on how many bits we have available, but some numbers will not be exact, and every time we use them in a calculation, the error will grow in the result. This is something we've become accustomed to in base ten where $1/3 = 0.33333...$, $1/7 = 0.142857142857...$, and $1/11 = 0.090909...$ are among an infinite number of fractions that lose precision when representing them in certain bases. Note that this is really due to the way we use bases. In base three: $1/3_{10} = 0.1_3$ exactly while $1/10_{10} = 0.00220022..._3$.

Floating Point

There are two problems with fixed point fractions: the point doesn't stay fixed for multiplication and the range is too limited for most scientific and engineering applications.

Almost all of the mainframe computers of the 1960s were available with floating point hardware. Many of the minicomputers of the 1970s did as well depending on the manufacturer and model. When the microcomputers emerged in the 1970s and 1980s, they did not initially support floating point hardware. The early ones were eight bit processors like the Motorola 6800 and the Intel 8080. These microprocessors did support 16-bit (double byte) arithmetic and addressing but no floating point hardware. Floating point could be supported in software, of course, but it was extremely slow. Floating point coprocessors like the Intel 8087 and 80287 were developed as options to accompany the Intel 8086 and 80286 CPUs, respectively. Many of the complex instruction set microcomputers (CISC) that followed actually contained floating point arithmetic on the same chip.

Moving real number data (i.e., floating point) from one computer system to another was not impossible, but certainly more difficult than necessary, and it was even prone to error. The problem with the floating point formats present in the mainframes and minicomputers was that although they were almost identical in concept, their implementations were incompatible. In the 1960s, even the size of a floating point number varied: 32 bits, 36 bits, 48 bits, and 60 bits were common, and double precision added another four sizes. Some computers used one's complement; some used two's complement. Most had the exponent in base 2, while one used base 16 and another base 8.

In the 1960s, ASCII was defined to address incompatibility among character sets in different computers. Likewise in 1985, the Institute of Electrical and Electronics Engineers (IEEE) standard 754 was defined to address the incompatibility among floating point formats used by various computer manufacturers. This standard was later refined in 2008, as well as becoming standard ISO/IEC/IEEE 60559:2011.

Floating Point Implements Scientific Notation

When we look at real numbers expressed in scientific notation such as 6.0221409×10^{23}, 9.10938356×10^{-31}, and $-1.60217662\times10^{-19}$ used in science, we observe the following:

1. The number is positive or negative
2. The significant (left of the $\times10$)
    - Is in base 10
    - Contains a decimal point
    - Has a precision related to the number of digits

3. The exponent (right of the $\times10$)
    - Can be negative or positive
    - Is in base 10
    - Is a whole number (i.e., although exponents like 5.23 are certainly allowed in mathematics, we only use integers in scientific notation)

So how are these base 10 real numbers with a wide range of values implemented in floating point? Figure 22.3 illustrates the floating point components and their locations within IEEE standard 754's single precision format.

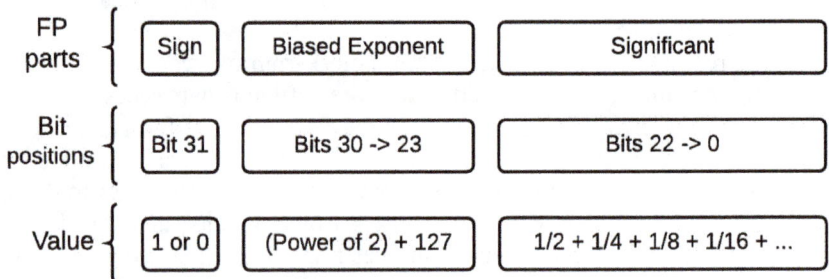

FP parts	Sign	Biased Exponent	Significant
Bit positions	Bit 31	Bits 30 -> 23	Bits 22 -> 0
Value	1 or 0	(Power of 2) + 127	1/2 + 1/4 + 1/8 + 1/16 + ...

Figure 22.3: Single precision floating point fields in IEEE 754 format

Normalization

Does 220 equal 2.2×10^2 and equal 2200×10^{-1}? Of course. What about binary? Is 110_2 equal to $1.10_2\times2^2$ and equal to $1100_2\times2^{-1}$? That is also true. In scientific notation, a number is expressed in "normalized" form when it has exactly one non-zero digit left of the decimal point. When a floating point value is "normalized," it has exactly one non-zero digit to the left of the binary point. This restriction leads to the following three advantages:

Computer Architecture Tutorial Using an FPGA

- Each real number is represented by a unique floating point value. Of the above three decimal choices, only 2.2×10^2 is in scientific notation. Of the above three binary numbers, only $1.10_2 \times 2^2$ is eligible for floating point format.
- Since there are only two binary symbols, and the digit left of the binary point cannot be "0," it must therefore be a "1." For this reason, the IEEE 754 format doesn't include this bit in the 32-bit format, and thereby "gains" an extra bit for more precision.
- In "normalized" floating point format, the number of significant digits will be consistent and maximized. Note: I didn't say that the precision of all floating point numbers is equal.

Conversion to IEEE 754 Floating Point

Let's look at a couple of examples to see how a floating point number in IEEE 754 format is constructed. The first example will be easier to convert since it is only a whole number and requires only the steps listed below:

1. Convert the number to base 2.
2. Normalize it.
3. Bias the exponent by adding 127; store it into bits 23 through 30.
4. Store the fractional part of the normalized binary number into bits 0 through 22. Nothing is done with the "1" left of the binary point.

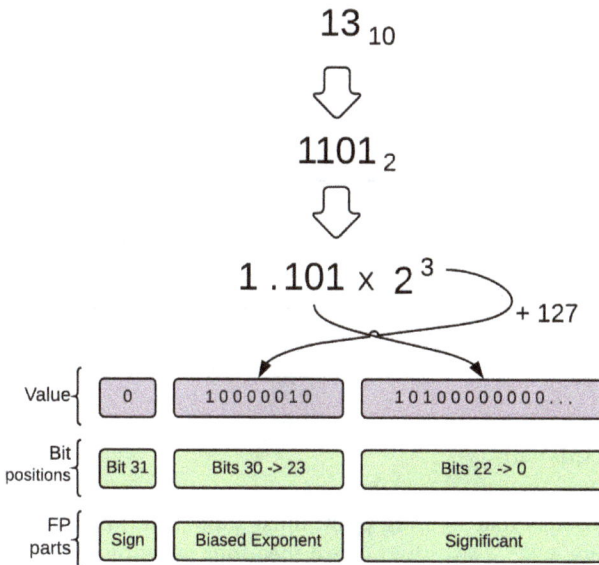

$$13_{10}$$

$$1101_2$$

$$1.101 \times 2^3$$

$+ 127$

Value	0	10000010	10100000000...
Bit positions	Bit 31	Bits 30 -> 23	Bits 22 -> 0
FP parts	Sign	Biased Exponent	Significant

Figure 22.4: Pack 13.0 into single precision floating point fields in IEEE 754 format

Let's take a more thorough examination of the construction of floating point representation using the more complicated example shown in Figure 22.5. Although several different programing approaches can be taken, the following the steps are pretty common:

1. Set the sign bit: 1 if negative, 0 if positive.
2. Convert the base 10 exponent into a base 2 exponent
3. Convert the fraction to base 2 as the significant
4. Normalize the significant
5. Bias the exponent (adding 127); store it into bits 23 through 30.
6. Store the significant of the normalized binary number into bits 0 through 22. Note: Nothing is done with the 1-bit that is to the left of the decimal point.

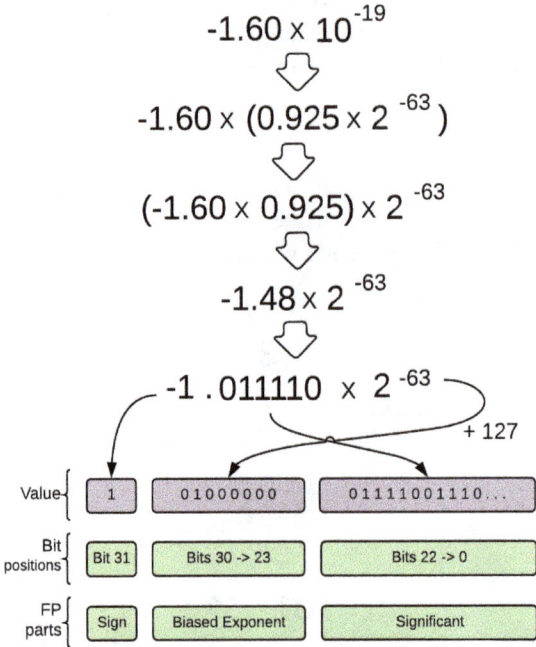

$$-1.60 \times 10^{-19}$$

⇩

$$-1.60 \times (0.925 \times 2^{-63})$$

⇩

$$(-1.60 \times 0.925) \times 2^{-63}$$

⇩

$$-1.48 \times 2^{-63}$$

⇩

$$-1.011110 \times 2^{-63}$$

+ 127

Value	1	01000000	01111001110...
Bit positions	Bit 31	Bits 30 -> 23	Bits 22 -> 0
FP parts	Sign	Biased Exponent	Significant

Figure 22.5: Convert scientific notation into floating point format

Why Bias the Exponent?

The obvious answer is that floating point must support a range of both positive and negative exponents. Appendix C describes four ways of indicating negative numbers: sign/magnitude, bias, one's complement, and two's complement. They all work, but why not just pick one and use it consistently? One would think that

Computer Architecture Tutorial Using an FPGA

even though differences arise among different computer manufactures, at least there would be consistency within a single machine.

Both one's and two's complement use the same arithmetic unit for signed and unsigned numbers. They can also extend to virtually any size "word" by combining multiple bytes using the carry flag. The bias format, on the other hand, used in the exponent, enables the integer compare instruction (CMP) to work on floating point numbers. One way to look at it is that floating point is a package, and integer format is homogeneous.

Where Did the Most Significant "1" Bit Go?

We put the sign bit into bit position 31. We biased and put the base 2 exponent into bits 23 through 30. We put the fractional part of the normalized binary number into bits 0 through 22, but we discarded what seems to be the most significant bit of all. Every additional bit included in a binary number doubles its range, so if a bit is "always" going to be the same in the floating point format, why not allow that bit position in the 32-bit word to either extend the range of the exponent or the precision of the significant? Secondly, if the bit is not there in the format, it is nearly impossible to make a non-normalized floating point number.

A Note on Normalization

I have to admit that I took a bit of liberty in the above floating point description. I did so because that description has been the one that my students have found the easiest and quickest to accept. You may find other descriptions where the bias is 128 (hex 80) because that is one half of the 256 range provided in the 8-bit exponent field. The exponent is then decremented because normalization from a hardware viewpoint has the significant being less than one.

Traditionally, a normalized floating point number is defined as one where the significant is shifted until its high order bit is a 1 bit (i.e., the significant is greater than or equal to ½, but less than 1). Unlike IEEE 754, most floating point formats did not remove the high order bit even though it "always" had to be a 1. It was even possible to generate non-normalized floating point numbers, but their use in arithmetic usually produced undesirable results.

Not a Number (NaN)

The IEEE 754 format includes a "value" known as NaN (Not a Number) which results from operations like square root of a negative number or division by zero. There are other special cases where the exponent is either all one bits or all zero bits for special "numbers" like positive and negative infinity.

Significant or Mantissa?

The terms "significant" and "mantissa" refer to the fractional part in the floating point format and are used somewhat interchangeably in the literature. The term "mantissa" has been used to describe floating point format for several decades beginning in the mid 1940s. The term "significant" is preferred in the IEEE 754 documentation apparently because "mantissa" has been associated with logarithms for centuries, and the "fraction" in the floating point format really isn't a logarithm.

Review Questions

1. * What are the advantages of fixed point over floating point?
2. * "By hand, without a computer," convert the following real numbers into single precision IEEE 754 floating point and provide the answers in hexadecimal.
    a. 128.0
    b. 9.25
    c. -9.25
    d. 0.03125
    e. 128.03125
    f. 0.0
    g. -0.0
3. * "By hand, without a computer," convert the following IEEE 754 floating point numbers from hexadecimal back into real numbers in base 10.
    a. 42a80000
    b. C1A80000
    c. 424C8000
    d. BF100000
    e. 3DCCCCCD
4. Which of the four ways to represent negative numbers described in Appendix C allows for both a positive and a negative zero?
5. In IEEE 754 floating point format, zero is represented by both the significant and the exponent being zero. If this was not the case, what value, expressed as a power of 2, would a word of all zero bits represent?
6. If the exponent is not zero, why is it impossible to have a non-normalized IEEE 754 format value?

Conclusion

Now, "The Times They Are a-Changin'" in computer architecture. Neural net applications associated with deep learning are gaining strongholds, and quantum computing is on the horizon. Hybrid configurations consisting of a specialized neural net or a quantum "computer" combined with a traditional CPU architecture are appearing. Today's computer engineers should be familiar with at least two architectures: CPUs for traditional applications and at least one other such as FPGAs for implementing embedded systems and neural net applications.

During times of change, hybrids appear. The gasoline/electric hybrids in the automotive industry provide flexibility and range for electric vehicles until a vast infrastructure of electric changing stations is available. Hybrids of analog computers alongside digital computers were somewhat popular until processor speeds increased in the 1980s. Hybrids combining a CPU with specialized I/O processors have been popular for decades.

The objective of this book is to provide a "hands on" experience to learn computer architecture and the Verilog language, not build a fully operational ARM emulator. The Terasic DE10 <u>Standard</u> development board includes an Intel Cyclone® V which contains both an FPGA and dual-core ARM processors. For those still wanting a "soft" processor, CPU emulators can be found on the Internet. Intel also provides the Nios® II embedded soft processor for FPGA downloads.

I took some liberties in my Verilog coding, and I don't want to lead you astray with some bad habits. My internal documentation (i.e., the comments with the Verilog coding) was a bit too brief. I did have to keep it somewhat short in order to get the code on the print pages, and it was closely linked to the descriptions in the book, but if it was by itself, it is too brief. More comments would let those reading the code to know what is supposed to be happening.

Secondly, some of my timings are a bit rough. Let this be a good incentive for you to look into the simulation aspect of Verilog. Production code has to be tight.

Finally, I was changing the parameter list on a couple of the modules. I did that because I built up the capability of the modules as we got closer to meeting the instruction format of the ARM processor. While "playing" with code to get new ideas, this is fine, but don't do it in any production code that others might incorporate. I recall the Late Rear Admiral Grace Murray Hopper telling us that there were two important considerations when dividing a program into components:

1. Make the interfaces as simple and clean as possible and
2. Once defined, never change the interface. If you need to enhance the interface, make a new one with a different name, but don't change what is already in production.

Blank page

Appendix A
Intel® Quartus Prime®

Intel® Quartus Prime® design software is used in the examples presented in this book. The "Lite" edition is a free download from Intel.

Quartus II was the predecessor of this software and was available from Altera Corporation, which was bought by Intel in 2015. Altera also had a free download that was known as its "Web" edition.

- Intel has three editions: Pro, Standard, and Lite. The Lite edition will work with all examples in this book and includes support for most of Intel's lower cost FPGAs. This includes the MAX 10 and Cyclone IV E FPGAs which are in the Terasic DE10-Lite and DE2-115 boards, respectively.

- Various versions are available for download. For this book, I used version 18.1 which supports both the MAX 10 and Cyclone IV E. Intel's latest version does not support the DE2-115 board containing the Intel Cycle IV E FPGA. For the older Altera DE2 FPGA boards having the Cyclone II, version 13.0 or earlier is required.

Figure A.1: Version 18.1 is highest version supporting the Cyclone IV E.

I recommend the following steps for setting up your computer for doing the examples in this book. If you already have either Quartus Prime® "Lite" or one of the paid editions already working, then the second step is not needed.

1. Download the source code and pin-assignments files from GitHub for the examples used in this book. It basically involves entering "https:// github.com/ robertdunne/ FPGA-ARM" from an Internet browser, and then unzipping the downloaded directory file. See Appendix F for details.

2. Download and install the free Quartus Prime Lite 18.1 and the USB Blaster device driver from Intel. Many of my students easily download and install this software, but for those who don't regularly set up software, I have included a detailed step-by-step description in file **InstallQP.PDF** located in the source code file directory downloaded from GitHub in the previous step. This document describes the following steps:
   a. Selecting the right version of Quartus Prime and downloading it to your computer
   b. Installing Quartus Prime from the downloaded files
   c. Installing the USB Blaster device driver

Of course, other FPGA boards and software can be used with this book. Basically, any FPGA or CPLD board with 10 slide switches, 10 LEDs, 6 seven-segment displays, 2 push-button switches, and a high-speed internal clock will be adequate, but slight modifications will be needed to run the examples in this book

Appendix B
DE2 & DE10-Lite Boards

The examples in this book use the Terasic DE2-115 and DE10-Lite FPGA boards. Either one is fine. Both are part of Intel's FPGA Academic Program (formerly known as FPGA University Program). All examples in this book use the Intel® Quartus® Prime Lite Edition software, as described in Appendix A.

I have also used the original DE2 board in the classroom for over ten years, and it works with all examples in this book; however, an earlier version of the Intel® Quartus® Web Edition software is needed (Appendix A). Other FPGA and CPLD boards with similar switches and displays will also work, but slight modifications to the examples may be necessary.

Copy of Figure 11.4: DE10-Lite I/O devices

The DE10-Lite is a capable board that can be purchased from many retailers on the Internet for a very attractive price. Almost every example in this book targets its complement of 10 slide switches, 10 LEDs, 6 seven-segment displays, and 2 push-button switches.

The DE2-115 has a more comfortable complement of 18 slide switches, 24 LEDs, 8 seven-segment displays, and 4 push-button switches. I have used it in the classroom for many years.

Copy of Figure 7.3: Hexadecimal output on 7-segment displays

The Quartus Prime software is used to both develop the applications in this book and download (a.k.a, program) the applications into the FPGA boards using a

USB port. Pay particular attention to the following details:

1. There are hundreds of possible FPGA devices, and exactly the right one must be selected within the Quartus software.
2. The "USB Blaster" device driver is used to download the application from Quartus Prime to the DE2 and DE10-Lite boards.
3. A particular pin-assignments file (CSV format) identifies which pins on the FPGA are assigned to which variable names in the application. Each board model has its own unique pin-assignment file.

The Intel Cyclone IV E, resident on the Terasic DE2-115 development board, and the Intel MAX 10 DA, resident on the Terasic DE10-Lite development board are the target FPGAs used in this book. Chapter 1 walks through the details of setting up a project with the correct configuration. Appendix A helps with the details of obtaining the appropriate components of Quartus Prime software that are needed for each board: FPGA part number "EP4CE115F29C7" for the DE2-115 and "10M50DAF484C7G" for the DE10-Lite.

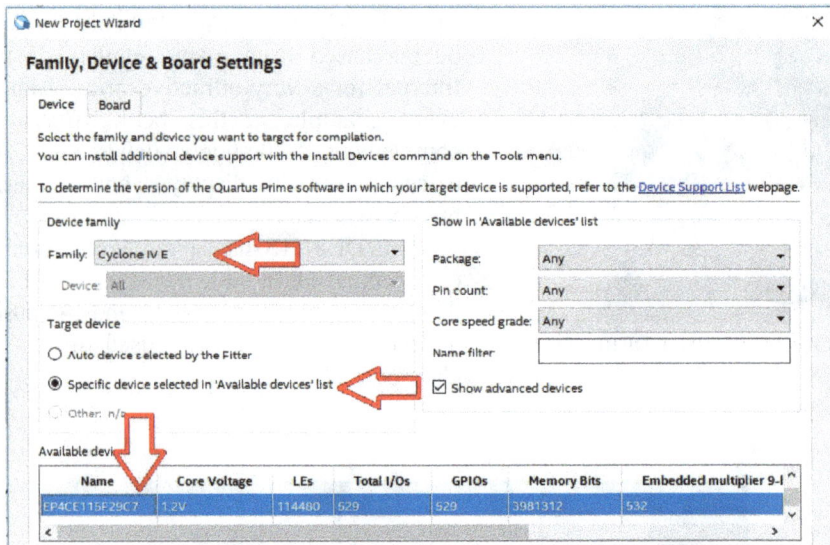

Copy of Figure 1.8: Select target device

Terasic Board	FPGA Family	Intel Part Number
DE2-115	Cyclone IV E	EP4CE115F29C7
DE10-Lite	MAX 10 DA	10M50DAF484C7G

Copy of Table 1.1: FPGA families and part numbers on Terasic boards

The **"USB Blaster" device driver** is used to download the application from Quartus Prime to the FPGA development boards. Chapter 1 walks through a procedure of downloading, also known as programming. The DE10 boards have three USB ports, and it's important to be connected to the one closest to the power cable. If the USB-Blaster device driver cannot be found, do *not* select a different driver, such as the "Ethernet-Blaster."

Copy of Figure 1.26: Upper left corner of DE2-115 with USB cable attached

A particular **pin-assignments file** (CSV format) identifies which pins on the FPGA are assigned to which variable names in the application. For those who cannot locate the pin assignment CSV file, I have included three copies on GitHub: one for the DE10-Lite, one for the DE2-115, and one for the original Altera DE2. See Appendix F for details.

Signal Name	Direction	Location on MAX 10 DA
LEDR[0]	Output	PIN_A8
LEDR[1]	Output	PIN_A9
LEDR[2]	Output	PIN_A10
SW[0]	Input	PIN_C10
SW[1]	Input	PIN_C11
SW[2]	Input	PIN_D13

Copy of Table 3.1: Sample of data in "DE10_Lite_pin_assignments.csv" file

It's not required, but if you want to verify that the pins have been assigned, select "Assignments" from the top menu bar, and then click on "Pin Planner." Figure 3.6 shows the pin configuration of the FPGA and the pin location associated with each Node Name (i.e., pin name). Pin locations could have been entered manually using this page, but for a large project, the CSV file is much more convenient.

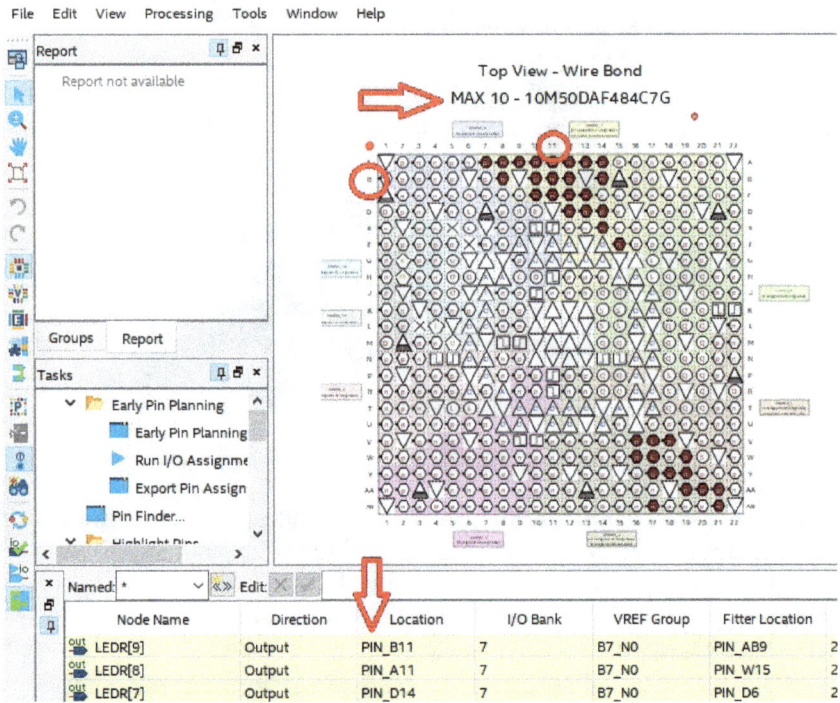

Copy of Figure 3.6: Pin assignments for DE10-Lite board

Computer Architecture Tutorial Using an FPGA

Appendix C
Binary Numbers

To be precise, it's not the numbers that are binary, but the written representation of numbers. For example, we currently count eight planets in the solar system. This has been "written down" as 8, VIII, 10_8, 1000_2, as well as a variety of other representations throughout history.

What's Binary?

Binary means *two* like a binary star system consisting of a pair of stars. In the case of binary "numbers," the *two* refers to the base, also known as the radix, which indicates how many different symbols (or digits) can be used. In our every day decimal (base 10) system, there are ten symbols available {0, 1, 2, 3, 4, 5, 6, 7, 8, 9} so we can represent a number in a form like 3274, 1620, and 36. While in binary (base 2), we have only two symbols available {0, 1} so we are restricted to representing numbers in a form like 1100, 10101, 1, and 111. Other popular bases that have been used in the computer industry are octal (base 8) having eight symbols {0, 1, 2, 3, 4, 5, 6, 7} and hexadecimal (base 16) having sixteen symbols {0, 1, 2, 3, 4, 5, 6, 7, 8, 9, A, B, C, D, E, F}.

Why Binary?

The simple answer is that the logical building blocks (i.e., electronics in today's systems) are simpler and more efficient in binary than they are in our everyday decimal. The electronic logic circuits have two states: High and Low (voltage levels) which can model a variety of binary states like True and False, Yes and No, and of course One and Zero. This system follows the logic attributed to Aristotle thousands of years ago.

In the following table, we compare the written representations of counting from 0 to 12 in five different bases. Notice how the rightmost column (one's place) is incremented through all of the possible symbols available in the base before the next column to its left is incremented.

base 10 10 symbols {0123456789}	base 2 2 symbols {01}	base 3 3 symbols {012}	base 4 4 symbols {0123}	base 5 5 symbols {01234}
0	0	0	0	0
1	1	1	1	1
2	10	2	2	2
3	11	10	3	3
4	100	11	10	4
5	101	12	11	10
6	110	20	12	11
7	111	21	13	12
8	1000	22	20	13
9	1001	100	21	14
10	1010	101	22	20
11	1011	102	23	21
12	1100	110	30	22

Table C.1: Counting from 0 to 12 in bases 10, 2, 3, 4, and 5

Column	3	2	1	0
Base 10	$10^3=1000$	$10^2=100$	$10^1=10$	$10^0=1$
Base 2	$2^3=8$	$2^2=4$	$2^1=2$	$2^0=1$
Base 3	$3^3=27$	$3^2=9$	$3^1=3$	$3^0=1$
Base 4	$4^3=64$	$4^2=16$	$4^1=4$	$4^0=1$
Base 5	$5^3=125$	$5^2=25$	$5^1=5$	$5^0=1$

Table C.2: Value of each column in bases 10, 2, 3, 4, and 5

The Problems with Binary

The problems with binary are not with computers, but with us humans:

1. We are comfortable with base ten and have used it daily for most of our lives.
2. Binary numbers are awkward for us due to the large number of columns required. Who would prefer replacing the decimal representation of 7094, 1620, 1108, 6600, 3033, and 7800 with their binary equivalents 1101110110110, 11001010100, 10001010100, 1100111001000, 101111011001, and 1111001111000?
3. Conversion between binary and decimal is difficult to do "in our heads." The difficulty stems from the fact that 10 is not an integer power of two.

Superscripts and Subscripts

In math books, the base (or radix) used to represent a number is given as a subscript. For example: a number written in decimal would be like 257_{10} and in binary it would be like 10000001_2. If no subscript is provided, we assume it is decimal unless it is stated in the text that the numbers are expressed in a different base such as binary. When working with computer programs, whether assembler or higher level, subscripts are not commonly available so binary is generally entered as 'b10101 in Verilog and 0b10101, 10101b, or %10101 in other computer applications.

Superscripts indicate a number raised to a power. For example, 4^3 means $4{\times}4{\times}4$ equaling 64 and 2^8 is $2{\times}2{\times}2{\times}2{\times}2{\times}2{\times}2{\times}2$ equaling 256. Also recall that 2^0, 10^0, 16^0, and any non-zero number raised to the zeroth power equals one.

A decimal number is really a short notation for a polynomial of powers of 10. For example: 137_{10} is $1{\times}10^2 + 3{\times}10^1 + 7{\times}10^0$ which is $100 + 30 + 7$. Likewise, a binary number is really a short notation for a polynomial of powers of 2. For example: 110101_2 is $1{\times}2^5 + 1{\times}2^4 + 0{\times}2^3 + 1{\times}2^2 + 0{\times}2^1 + 1{\times}2^0$. By the way, this polynomial structure is the main reason we label and count bits within a byte or word from right to left starting with zero.

Bit Position	3	2	1	0
Power of 2	$2^3{=}8$	$2^2{=}4$	$2^1{=}2$	$2^0{=}1$
Binary example	1	0	1	1
$1011_2 = 1 \times 2^3 + 0 \times 2^2 + 1 \times 2^1 + 1 \times 2^0 = 8 + 0 + 2 + 1 = 11_{10}$				

Table C.3: Bit position example: $1011_2 = 2^3 + 0 + 2^1 + 2^0 = 8 + 0 + 2 + 1 = 11_{10}$

Conversion to Any Base

A popular way to convert a number to a particular base is successive division. The remainders from each division will provide the digits (i.e., symbols) beginning with the rightmost digit. For example, converting the number 3274 to decimal follows:

1. $3274 / 10 = 327$ Remainder 4
2. $327 / 10 = 32$ Remainder 7
3. $32 / 10 = 3$ Remainder 2
4. $3 / 10 = 0$ Remainder 3

So, the "number" 3274 is represented in decimal as the sequence of remainders "3" "2" "7" and "4." By the way: This technique of successively dividing a number by the desired base works regardless of how the "computer" internally stores numbers. It could be binary, decimal, or any conceivable internal structure that would permit division.

Converting the same number 3274 to binary follows:

1. $3274 / 2 = 1637$ Remainder 0
2. $1637 / 2 = 818$ Remainder 1
3. $818 / 2 = 409$ Remainder 0
4. $409 / 2 = 204$ Remainder 1
5. $204 / 2 = 102$ Remainder 0
6. $102 / 2 = 51$ Remainder 0
7. $51 / 2 = 25$ Remainder 1
8. $25 / 2 = 12$ Remainder 1
9. $12 / 2 = 6$ Remainder 0
10. $6 / 2 = 3$ Remainder 0
11. $3 / 2 = 1$ Remainder 1
12. $1 / 2 = 0$ Remainder 1

So, the "number" 3274 is represented in binary as the sequence of remainders "1" "1" "0" "0" "1" "1" "0" "0" "1" "0" "1" and "0." As an exercise, try converting 3274 to base five by successively dividing by five until the quotient is zero ($3274/5 = 654$ remainder 4, ...). The answer will be 101044_5.

Multiplying and Dividing by Shifting

If we want to multiply by ten "in our heads" in our everyday decimal system, we just append a zero. For example to multiply 709 by 10, we append "0" to "709" and get "7090." Likewise, when we multiply by 100 (i.e., 10^2), we append two zeroes, and for 1000, we append 3 zeroes, etc. For dividing by powers of ten, we do the reverse: we remove zeroes on the right. What if there are not enough zeros present on the right? Then we move the decimal point. For example to divide 1108 by 100, we move the decimal point to the left two places giving us 11.08.

When we shift a number to the left in base two, we are multiplying by a power of two, and when we shift to the right, we are dividing by a power of two. This means that conversion into and from binary format is done very efficiently using shifting rather than division. Converting the same number 3274 (110011001010_2) to binary by shifting is below. Note: The notation ">> 1" means shift 1 bit position to the right, and the "Carry out" refers to the rightmost bit that is lost when the value is shifted.

1. $110011001010 >> 1 = 11001100101$ with Carry out 0
2. $11001100101 >> 1 = 1100110010$ Carry out 1
3. $1100110010 >> 1 = 110011001$ Carry out 0
4. $110011001 >> 1 = 11001100$ Carry out 1
5. $11001100 >> 1 = 1100110$ Carry out 0
6. $1100110 >> 1 = 110011$ Carry out 0

Computer Architecture Tutorial Using an FPGA

7. $110011 >> 1 = 11001$ Carry out 1
8. $11001 >> 1 = 1100$ Carry out 1
9. $1100 >> 1 = 110$ Carry out 0
10. $110 >> 1 = 11$ Carry out 0
11. $11 >> 1 = 1$ Carry out 1
12. $1 >> 1 = 0$ Carry out 1

Converting Digits Into a Number

To convert "written digits" into a number, run the above process in reverse: Do successive multiplications. For example in base 10: the sequence of digits "1" "6" "2" "2" could be used to "build" the number 1622 as follows:

1. Start with 0
2. $0 \times 10 + 1 = 1$
3. $1 \times 10 + 6 = 16$
4. $16 \times 10 + 2 = 162$
5. $162 \times 10 + 2 = 1622$

In binary, it is simply a matter of shifting to the left by one bit position to "multiply" by two. In the following example, the number expressed as a sequence of digits "110011001010" is built by a series of logical left shifts notated by "$<< 1$" combined with a logical OR notated by "+":

1. Start with 0
2. $0 << 1 + 1 = 1$
3. $1 << 1 + 1 = 11$
4. $11 << 1 + 0 = 110$
5. $110 << 1 + 0 = 1100$
6. $1100 << 1 + 1 = 11001$
7. $11001 << 1 + 1 = 110011$
8. $110011 << 1 + 0 = 1100110$
9. $1100110 << 1 + 0 = 11001100$
10. $11001100 << 1 + 1 = 110011001$
11. $110011001 << 1 + 0 = 1100110010$
12. $1100110010 << 1 + 1 = 11001100101$
13. $11001100101 << 1 + 0 = 110011001010$

Negative Binary Numbers

When we include negative numbers, we effectively double how many numbers we have to be able to represent in binary. For every positive number, we have a corresponding negative number. This requires an additional bit, a "sign" bit, that has to be associated with every binary number in registers and storage.

Rather than append an additional bit to each numeric storage type, computer manufacturers have chosen to steal a bit from the positive range. Instead of an 8-bit byte supporting numbers in the range of 0 through 255, it supports -128 through +127 for "signed" bytes. Likewise, signed half-words have a range of -32,768 to +32,767 rather than 0 through 65,535 for the unsigned format. The range is actually the same, but it has been shifted by 50%.

There have been four popular formats for representing signed numbers in binary computers:

- **Bias:** Add ½ the total range to all numbers
- **Sign and magnitude:** High order (leftmost) bit is the sign: 1 for negative
- **One's complement:** Complement (i.e., toggle) all bits for negative.
- **Two's complement:** Add 1 to one's complement value

The question is, which one is popular in today's computers? Being even more specific, which are present in the ARM and NEON coprocessor? Three are used: two's complement represents signed integers in the ARM CPU while both sign/magnitude and bias are used in the floating point format. Table C.4 gives 8-bit binary examples where positive and negative 26_{10} are represented four ways. I've also included zero, including the rather unexpected negative zero case.

Decimal	+ 26	– 26	+ 0	– 0
Sign & Magnitude	00011010	10011010	00000000	10000000
One's Complement	00011010	11100101	00000000	11111111
Two's Complement	00011010	11100110	00000000	00000000
Biased	10011010	01100110	10000000	10000000

Table C.4: Comparison of +26, –26, +0, and –0 in four signed byte formats

Nine's complement

How can we subtract using an "adding machine"? This question was not new with electronic computers, but goes back to the days when accountants and human "computers" used mechanical adding machines. It involves converting the algebraic expression "A − B" to "A + (−B)" which transforms the question into how should we represent −B?

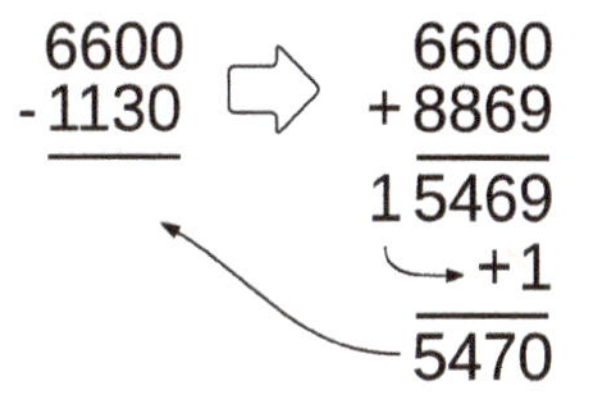

```
 6600           6600
-1130    ⇨    +8869
 ────          ─────
              1 5469
               ↳+1
              ─────
               5470
```

Figure C.1: Nine's complement example of subtraction by addition

Accountants, working in base ten, could represent a negative number by subtracting each of its digits from nine (one less than the base). Figure C.1 is an example where the negative of 1130 is 8869 in nine's complement (each 8 comes from 9 − 1, the 6 comes from 9 − 3, and the 9 comes from 9 − 0).

Obviously, since we're adding, rather than subtracting, the result is larger than we want, but if you do the algebra, you'll notice that the correct answer can be achieved. Notice how the first sum in Figure C.1 had a "carry out" that did not fit in the number of columns we were using. If you add this carry in a second step as shown, the correct answer appears. If there is no carry, do not add it, and there will be a large number, but it is really a negative number.

One's complement

One's complement is the same as nine's complement except the base is now two: every digit is subtracted from 1, instead of 9. Actually, this technique works in any base. Do the algebra if you like to prove it. Because there's only two symbols in base 2, one's complement is achieved by simply inverting each bit as is shown in Figure C.2.

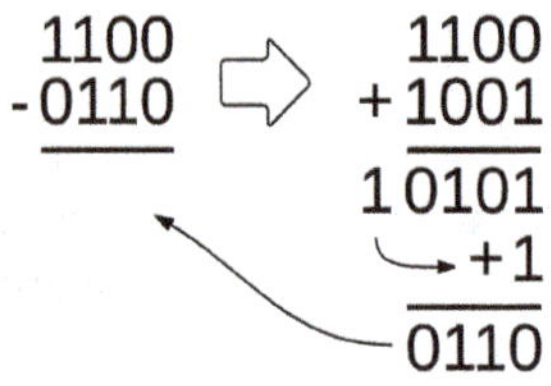

$$
\begin{array}{r}
1100 \\
-0110 \\
\hline
\end{array}
\Rightarrow
\begin{array}{r}
1100 \\
+1001 \\
\hline
1\,0101 \\
\;\hookrightarrow +1 \\
\hline
0110 \\
\end{array}
$$

Figure C.2: One's complement
example of subtraction by addition

Just like in nine's complement a subtraction is converted into an addition. Here the negative of "0110" is calculated to be "1001" where the value in each column is calculated by subtracting it from one less than the base. Notice that it's still a two-step process where the carry out is added back to obtain the correct answer.

If you follow the above naming convention, you would think that two's complement involves numbers expressed in base 3. Actually, the expression "two's complement" refers to the technique that eliminates the second step during a subtraction.

Two's complement

In two's complement, the negative of a number is generated by adding one to the one's complement and ignoring any caries. For example, the negative of 0110 is $1001 + 1 = 1010$.

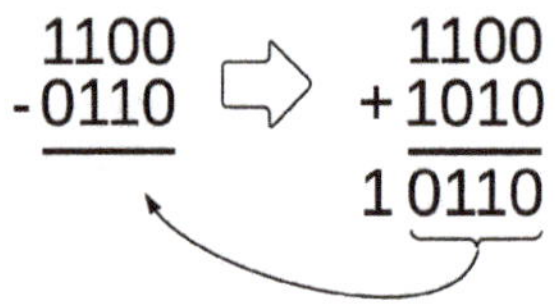

$$
\begin{array}{r}
1100 \\
-0110 \\
\hline
\end{array}
\Rightarrow
\begin{array}{r}
1100 \\
+1010 \\
\hline
1\,0110 \\
\end{array}
$$

Figure C.3: Two's complement
example of subtraction by addition

Rather than adding the "carry out" as a second step, a 1-bit is added preemptively when the negative is generated. Then during the subtraction, the carry is just ignored, making two's complement subtractions twice as fast as one's complement subtractions.

Binary Fractions

Binary supports binary fractions in the same manner that decimal supports decimal fractions. A decimal point separates decimal whole numbers from decimal fractions. Likewise, a binary point separates binary whole numbers from binary fractions. In the following example, it seems reasonable to make the first bit to the right of the binary point equal to 2^{-1} which is ½. The second bit to the right of the binary point is 2^{-2} which is $\frac{1}{2}^2$ which is ¼. For further details, please see Chapter 13 on multiplication and Chapter 22 on floating point formats.

Computer Architecture Tutorial Using an FPGA

- $1111.0_2 = 1 \times 2^3 + 1 \times 2^2 + 1 \times 2^1 + 1 \times 2^0 = 8 + 4 + 2 + 1$
- $111.1_2 = 1 \times 2^2 + 1 \times 2^1 + 1 \times 2^0 + 1 \times 2^{-1} = 4 + 2 + 1 + \frac{1}{2}$
- $11.11_2 = 1 \times 2^1 + 1 \times 2^0 + 1 \times 2^{-1} + 1 \times 2^{-2} = 2 + 1 + \frac{1}{2} + \frac{1}{4}$

Hexadecimal

Hexadecimal is a compact form of binary representation where we have sixteen symbols {0,1,2,3,4,5,6,7,8,9,A,B,C,D,E,F} to represent numbers. If it wasn't for binary, there would be negligible need for hexadecimal in the computer industry. A hexadecimal number is really a short notation for a polynomial of powers of 16. For example: $5A732C_{16}$ is $5 \times 16^5 + 10 \times 16^4 + 7 \times 16^3 + 3 \times 16^2 + 2 \times 16^1 + 12 \times 16^0$ where A and C are digits representing values of 10 and 12, respectively.

1. Inputting and displaying numbers in the computer's natural binary notation is very efficient for the computer, but clumsy and inefficient for us humans. Who is comfortable reading and entering numbers like 100001101010010 or 1101101101101, and even much longer ones up to 64 bits in length?
2. Decimal is a rather compact form of representing numbers, and we are very comfortable with it because we use it in our daily lives. We can convert between decimal and binary by using successive divisions by ten. However, that is slow and cumbersome to do "in our heads." A division by sixteen is simply a four bit shift, but a division by ten cannot be achieved by shifting bits.
3. Do we humans actually need to use binary? As people working with computers at a detailed architectural level, we have to see the actual bits. We have to look at status words, PC addresses, instruction formats, and memory dumps.

Figure C.4 shows a binary number being "mapped" to hexadecimal digits, four bits at a time. starting from the right side.

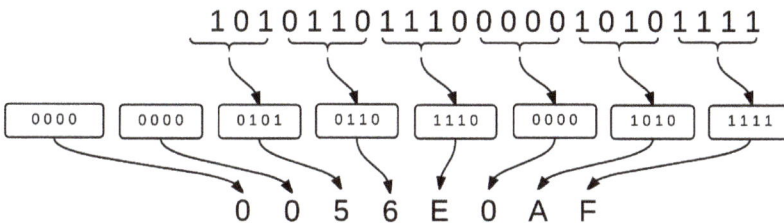

Figure C.4: Binary reduced to hexadecimal

Table C.5 shows counting from 0 to 17 in decimal, binary, hexadecimal, and octal. Notice how one hexadecimal digit fits exactly in four bits.

base 10	base 2	base 16	base 8
10 symbols	2 symbols	16 symbols	8 symbols
{0123456789}	{01}	{0123456789ABCDEF}	{01234567}
0	0	0	0
1	1	1	1
2	10	2	2
3	11	3	3
4	100	4	4
5	101	5	5
6	110	6	6
7	111	7	7
8	1000	8	10
9	1001	9	11
10	1010	A	12
11	1011	B	13
12	1100	C	14
13	1101	D	15
14	1110	E	16
15	1111	F	17
16	10000	10	20
17	10001	11	21

Table C.5: Counting from 0 to 17 in four different bases

You probably also noticed that octal fits nicely in three bits. Octal has the advantage of being much more compact than binary, is easy to convert to binary, and it doesn't have any letters mixed in with the digits. Who really likes numbers looking like 3AF6?

Octal was very popular for decades in the computer industry. When character codes (see Appendix D) contained 64 characters, the "byte size" was six bits, which was displayed by two 3-bit octal digits. However, with the 8-bit bytes which are popular today, the 4-bit hexadecimal digits are a much better fit.

Appendix D
ASCII

Why ASCII? Why not Baudot, BCD, Display Code, Fieldata, Unicode, XS3, or any other character code?

What is a Character Code

Binary computers store and manipulate bits (binary digits). Numbers are represented by "groups of bits" as either integers or real numbers. That's fine for science and engineering applications, but what's stored in "groups of bits" for business applications, such as correspondence, reports, and mailing lists? How is this text data consisting of letters, digits, and punctuation represented by "groups of bits"? A character code is a set that assigns each text character to a unique number.

This was not so much of a problem 3000 years ago. Several of the ancient languages including Assyrian, Hebrew, and Greek were "computer ready," but our modern written languages, such as English, are not. In these ancient languages, every symbol used to compose words was also used to compose numbers. The symbols alpha and beta in Greek were assigned both sounds to form words as well as numeric values to write numbers. In English, letters and digits are separate (i.e., the letter "R" does not have a numeric value). This means there was no "standard" for storing text data as a series of numbers.

In the 1960s, several companies were manufacturing mainframe computer systems. They were competing for sales and were interested in locking customers into their unique designs rather than making computer data files and applications portable from one system to another. There were basically two problems with character codes in the 1960s:

- Each character was stored in a byte, but the number of bits composing a byte varied from system to system.
- Each character was assigned a unique numeric code, but each computer system had a different set of character code assignments.

Several mainframe computer systems had 6-bit bytes, which supported a set of 64 different characters. BCD, Display Code, Fieldata, and XS3 are examples of 6-bit codes. Each of these sets contained 26 upper case letters, 10 digits, and a few punctuation marks and control characters. In order to include lower case letters, IBM switched from a 6-bit code to an 8-bit EBCDIC (Extended BCD Interchange Code) in the mid 1960s. The size of the byte determines how many different characters can be represented as listed below:

- 6 bits: 64 characters
- 7 bits: 128 characters
- 8 bits: 256 characters
- 16 bits: 65,536 characters

The second compatibility problem was that the unique assignments were inconsistent among the different character code sets and computer systems. It took a presidential decree to alleviate some of the inconsistencies. On March 11, 1968, President Johnson signed ASCII (American Standard Code for Information Interchange) into existence.

Character Code	Letter A	Digit 5	blank
IBM BCD	11	05	30
CDC Display Code	01	20	2D
Univac Fieldata	06	25	05
XS3	14	08	33
EBCDIC	C1	F5	40
ASCII	41	33	20
Unicode	41	33	20

Table D.1: Example of three characters expressed in various character codes (in hexadecimal)

The 7-bit ASCII code from 1968 was fine for the English language, but it could not even support all the characters used in French, Spanish, and other Latin languages. In 1985, character set ISO 8859 was defined as an 8-bit code with 256 character codes defined, where the first 128 are identical to 7-bit ASCII. The remaining 128 character codes were assigned to accent characters for the Latin languages and a variety of special symbols like copyright and trademark.

Table D.2 contains the 96 "printable" characters of the ASCII definition. The first 32 character codes (hex 00 through 1B) are assigned to control characters that were popular in the early days of data communications. They included codes such as SOH (Start Of Header, 01), STX (Start of TeXt, 02), and ACK (ACKnowledge, 06). Many of these codes still exist in text files today, such as CR (Carriage Return, 0D) and LF (Line Feed, 0A) which together mark the end of text lines in Windows® systems. Line feed, by itself, also known as New Line (NL, 0A) marks the end of text lines in Unix and Linux systems.

Other popular control characters still in use today are BS (Back Space, 08), HT (Horizontal Tab, 09), FF (Form Feed / page eject, 0C), and ESC (ESCape, 2B). Table D.2 begins with hexadecimal code 20 representing the "printable" blank character.

Hex Code	Symbol	Hex Code	Symbol	Hex Code	Symbol
20		30	0	40	@
21	!	31	1	41	A
22	"	32	2	42	B
23	#	33	3	43	C
24	$	34	4	44	D
25	%	35	5	45	E
26	&	36	6	46	F
27	'	37	7	47	G
28	(	38	8	48	H
29	)	39	9	49	I
2A	*	3A	:	4A	J
2B	+	3B	;	4B	K
2C	,	3C	<	4C	L
2D	-	3D	=	4D	M
2E	.	3E	>	4E	N
2F	/	3F	?	4F	O

Table D.2 A: ASCII and ISO codes in hexadecimal

Hex Code	Symbol	Hex Code	Symbol	Hex Code	Symbol
50	P	60	`	70	p
51	Q	61	a	71	q
52	R	62	b	72	r
53	S	63	c	73	s
54	T	64	d	74	t
55	U	65	e	75	u
56	V	66	f	76	v
57	W	67	g	77	w
58	X	68	h	78	x
59	Y	69	i	79	y
5A	Z	6A	j	7A	z
5B	[	6B	k	7B	{
5C	\	6C	l	7C		
5D	]	6D	m	7D	}
5E	^	6E	n	7E	~
5F	_	6F	o	7F	

Table D.2 B: ASCII and ISO codes in hexadecimal

What about those written languages like Hebrew and Greek that were "computer ready" thousands of years ago? Were they still computer ready in 1968 when ASCII was defined? They were by themselves, but to include them alongside ASCII and ISO 8895, a new character set has been defined: Unicode. Casually speaking, Unicode is considered to be a 16-bit code supporting 65,536 different character code symbols, enough to encompass all the written symbols composing thousands of different languages. The first 128 characters of Unicode are the same as the ASCII character set.

Appendix E
Assembly Language

Assembly language is the computer programming language closest to a computer's machine code language. Programming in assembly language is an excellent "hands-on" approach to "getting comfortable" with computer architecture.

This book uses Verilog macros, functions, and tasks to mimic assembly language programming as a substitute to entering machine code instructions in binary or hexadecimal. However, this approach can only be pushed so far, and regular software development tools will eventually be needed.

There's more to assembly language programming than simply translating text lines to machine code instructions. The purpose of this appendix is to acquaint the reader with various aspects and considerations related to developing an application in assembly language.

1. **Target device:** Basically this is which CPU is to be used, but also includes the circuit board, peripheral devices, and associated "on board" software.
2. **Software Development Environment:** Most assemblers in use today are part of a larger application development suite that includes the C++ compiler.
3. **Directives:** These commands, sometimes referred to as pseudo-ops, are assembly time commands to the assembler, such as those to organize memory and set up external linkages to other software.
4. **Documentation:** Good documentation is important while a system is being developed and crucial as a system is maintained over the years.
5. **Higher Level Language Interface:** Today, assembly language is seldomly used alone, but is a part of a larger application written in a higher level language like C++.
6. **Operating System Interface:** The operating system provides a variety of services such as handling I/O devices and writing to files.
7. **Hardware Interface:** From a practical perspective, most assembly language used today is to control specific hardware devices.
8. **Debugger:** Assembly language programming and testing can be tricky. Being able to stop execution, single step through the code, and examine register contents is extremely important during program development and maintenance.
9. **Software Design Patterns:** Design patterns provide guidance and a general method for developing common software applications. They are important in the long-term maintenance of large programs.

Software Development

From a programmer's perspective, software development is a vicious cycle of modify the program, test the program, modify the program, test the program, modify the program, test the program until we are satisfied with the test results.

An assembly language program consists of lines of text which can be created and modified using a simple text editor. The program is then tested by translating it into ARM machine code.

There are many assemblers compatible with the ARM instruction set, and one of the most popular is the GNU assembler available from the Free Software Foundation, Inc., Boston, Massachusetts. The GNU assembler (a.k.a., *gas* or just *as*) is part of the C++ software appearing on many Linux distributions. There are four steps in creating and testing an assembly language program.

1. **Edit (make the source code):** The source code must be entered into a text file.
2. **Compile (make the object code):** Each "machine language" instruction executed by the ARM CPU is composed of several fields ("groups of bits"), which are expressed in mnemonic names and decimal, hexadecimal, or binary numbers.
3. **Link (make the executable program):** The "ld" linker program combines multiple object files into a single executable file.
4. **Execute (run the program):** This step is the objective of the previous three steps, but how do you know if your program is doing what you wanted it to do or is even doing anything at all? You'll need some type of I/O (Input/Output).

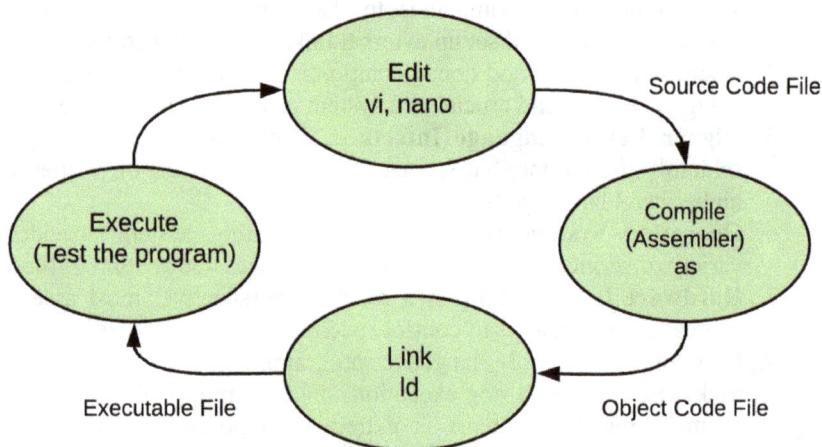

Figure E.1: Work flow of developing an assembly language program

Computer Architecture Tutorial Using an FPGA

Large Programs

In programs that are more than about 100 lines of code, you'll want to divide them into multiple source files as illustrated in Figure E.2. The linker is then really linking more than one object file in order to produce the complete working program. The following illustration shows three source files, but in reality hundreds of files could be involved to produce a large program.

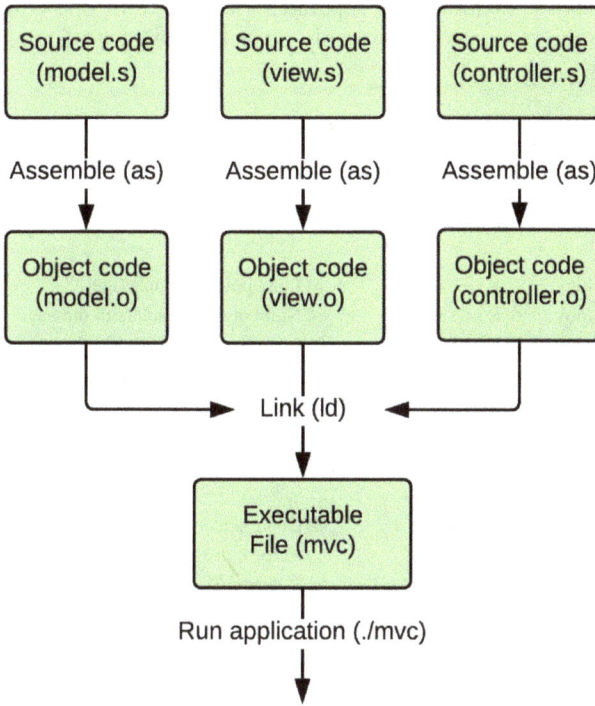

Figure E.2: Dividing a program into modules

In figure E.2 we see how a large program can be divided up into multiple source files and individually compiled into individual object files. These object files then have to be linked together to form a working program that can be run.

Source Code

A machine language instruction is composed of multiple integer fields indicating which operation is to be performed and on what data. Assembly language substitutes names for these integers, and usually generates one machine code instruction for each line of assembly language coding. Almost every assembly language program for all CPUs consists of four columns:

1. Label: Name associated with instruction's memory address
2. Op-code: The operation being performed (add, sub, shift, ...)
3. Operands: Location of the data (usually a register combined with a constant, another register, or memory address)
4. Comment: Describes why the instruction is being used

Each column is separated by one or more blanks or tab characters. How wide is a column? Typical columns are 8 to 10 characters wide with the exception of the rightmost column which contains comments. The assembler doesn't care if it's one blank, two blanks, or more that separates the data from one column to the next. We line up the columns of assembly code for the ease of reading by the programmers.

```
1.          .global   _start           @ Program starting address for linker
2.
3. @        Program to display Hello World.
4.
5. _start:  ldr       R1,=msg          @ Load pointer to message.
6.          mov       R2,#12           @ Number of characters in message
7.          mov       R0,#1            @ Code for stdout (usually the monitor)
8.          mov       R7,#4            @ Linux service command code to write
9.          svc       0                @ Call Linux command.
10.
11. @       Exit and return full control back to Linux.
12.
13.         mov       R7,#1            @ Command 1 terminates programs.
14.         mov       R0,#0            @ Zero "exit status" implies success.
15.         svc       0                @ Return full control to Linux.
16.
17.         .data                      @ Begin "data" section of memory
18. msg:    .ascii    "Hello World\n"  @ ASCII string to display
19.         .end
```

Figure E.3: Program to display "Hello World" on the monitor

Figure E.3 shows a small assembly language program that writes "Hello World" on a monitor screen.

- Line 1: Assembler directive to pass an address to the linker
- Line 3: Global comment
- Line 5: The label defines a "branch to" location in the program's memory.
- Lines 5 - 8: Arguments are assigned values in preparation for a call to the operating system
- Line 9: Request service of the operating system

- Line 17: Assembler directive for organizing memory
- Line 18: Assembler directive to build string of ASCII text characters

Comments

There is more to a well-written program than the machine code itself. Program design and maintenance requires documentation. "Internal documentation" is the description of the program appearing in the program itself and consists of two types of comments:

1. **Global**: These comments describe what a section of code is doing. They normally consist of more than one line of text and are not on the same physical text lines as the actual machine code instructions. Global comments are used in both assembly language as well as higher level languages.
2. **Local**: These comments share the text line with the actual machine code instructions. They are rarely needed in higher level languages, but are very important in assembly language to explain not what the code is doing but *why* the line of code is doing it.

Importance of comments: They're not necessary for a program to successfully run, and many programmers use very few comments. They're necessary for program maintenance, whether it be by a new programmer next week or by the original programmer a month or even several years later.

Language Interface

The *Procedure Call Standard for the ARM Architecture (AAPCS)* describes conventions for calling subroutines, even those written in assembly language. This standard, which is part of the *Application Binary Interface (ABI) for the ARM Architecture*, is not only used for C, but can be used for other languages as well. In the following diagrams, I will provide an example as to how to call an assembly language subroutine from a C program.

There are basically two techniques used to pass a variable's data in arguments to a subroutine:

- Pass by value: The value is passed in a register or on the stack to the subroutine, and the subroutine has no access to the source variable itself.
- Pass by reference: The memory address of a variable is passed, and the subroutine can actually update the variable in the calling routine's data area.

Figure E.4 contains a very short C main program that calls two assembly language routines that calculate the sum of an array of 32-bit integers:

- thesum: Subroutine that returns the sum to a reference argument
- fcnsum: Function that returns the sum as the return value of the function

```
1.      #include <stdio.h>
2.      #include <stdlib.h>
3.
4.      int main() {
5.          int count = 3, totalA, totalB, tstdat[] = {11, 45, 70};
6.          thesum (&totalA, tstdat, 3);
7.          totalB = fcnsum (tstdat, count);
8.          printf ("Sum from subroutine = %d\n", totalA);
9.          printf ("Sum from function = %d\n", totalB);
10.         return 0;
11.     }
```

Figure E.4: Main C program calling a subroutine and a function

Arguments	Pass by __	Location
thesum (&totalA, tstdat, 3);	**Subroutine**	
&totalA	Reference	[R0]
tstdat	Reference	[R1]
3	Value	R2
totalB = fcnsum (tstdat, count);	**Function**	
tstdat	Reference	[R0]
count	Value	R1
totalB	Value	R0

Table E.1: Subroutine and function arguments in C program

Generally speaking, the difference between a function and a subroutine is a function returns a value and a subroutine does not. This is very similar to the difference between a function and a task in Verilog.

As seen in this example, a subroutine can return one or more values through arguments "passed by reference." In situations where only one value is returned, it is advisable to only use a function because 1) it hides the location of the actual data, 2) it is more efficient, and 3) it is expected to be done (self documenting).

- Arguments are passed in registers R0 through R3. If there are more than four arguments, those are pushed onto the stack.
- The calling program does not expect the contents in registers R0 through R3 to be preserved.

Computer Architecture Tutorial Using an FPGA

- A function returns its value in register R0.
- The Link Register (LR) contains the return address.
- Arrays are passed by reference.
- Constants and single variables are passed by value.
- A single variable can be passed by reference if preceded by an ampersand.

1.	.global	thesum	@ Subroutine entry address to linker
2.	.global	fcnsum	@ Function entry address to linker
3.			
4. @		Subroutine thesum adds a variable number of integers.	
5. @		R0: Memory address of variable to receive the sum.	
6. @		R1: Memory address of array of integer values	
7. @		R2: Number of integers in the array	
8. @		LR: Contains the return address	
9. @		Registers R1 through R3 will be not saved.	
10.			
11. thesum:	ldr	R3,[R1],#4	@ Load first value.
12.	subs	R2,#1	@ Decrement number of integers.
13.	ble	retsub	@ Return with just one value.
14.	push	{R4}	@ R4 contents must be preserved.
15. thelp:	ldr	R4,[R1],#4	@ Load next interger in list.
16.	add	R3,R4	@ Add it to the running total.
17.	subs	R2,#1	@ Number of integers still to add.
18.	bne	thelp	@ Continue with next integer
19.	pop	{R4}	
20. retsub:	str	R3,[R0]	@ Return sum to calling program
21.	bx	LR	@ Return to calling program
22.			
23. @		Function fcnsum adds a variable number of integers.	
24. @		R0: Memory address of array of integer values	
25. @		R1: Number of integers in the array	
26. @		LR: Contains the return address	
27. @		R0: Return calculated sum to calling program.	
28. @		Registers R0 through R3 will be not saved.	
29.			
30. fcnsum:	ldr	R3,[R0],#4	@ Load first value.
31.	subs	R1,#1	@ Decrement number of integers.
32.	ble	retfcn	@ Return with just one value.
33. fcnlp:	ldr	R2,[R0],#4	@ Load next interger in list.
34.	add	R3,R2	@ Add it to the running total.
35.	subs	R1,#1	@ Number of integers still to add.
36.	bne	fcnlp	@ Continue with next integer
37. retfcn:	mov	R0,R3	@ Return sum to calling program
38.	bx	LR	@ Return to calling program
39.	.end		

Figure E.5: Assembly language subroutine and function called by C main program

Operating System Interface

One of the main responsibilities of an operating system, such as Windows and Linux, is to provide services for application programs. A large portion of these services involves reading and writing peripheral devices (display monitor, keyboard, mouse, network, etc.) and disk files (real spinning disks as well as solid-state memory devices). The calling program must provide the details of what is to be performed:

1. What is to be done (register R7)
2. Which device is to written or read (register R0)
3. Where the data buffer is in the program's memory (register R1)
4. How much data is to be written or read (register R2)

Debugger

The GNU debugger software ("gdb" command) helps programmers examine what's occurring inside a running machine code program by providing the following features:

1. Set breakpoints to pause program execution
2. Examine register contents
3. Change register contents
4. Examine memory contents
5. Single step through a program
6. Trace program execution

Design Patterns

A design pattern is not the program code itself, but rather a way to organize the functioning or structure of a program. The motives for using design patterns are to improve software design, reduce programming errors, and provide for much more effective long term maintenance.

Problems can be solved many different ways. A design pattern provides a general way of solving a problem that has been successful in the past. The classic Model View Controller pattern introduced with the Smalltalk object oriented language at Xerox's PARC during the 1970s is the classic example. It divides a program into three major components: Model (scientific, engineering, or business purpose of the program), View (user display), and Controller (user input interface and director).

The Adapter pattern is an example where a subroutine or group of subroutines is given a second "face." It's analogous to the Thumb instruction set providing a different approach to use the same registers, memory, and instructions as the ARM CPU 32-bit format.

Appendix F
Source Code Files

The Verilog source code files presented in this book build a working imitation of a 32-bit ARM computer. The files can be categorized by their purpose:

1. Demonstrate a computer component such as a shift register
2. Assemble ARM assembly language into ARM machine code
3. Assemble and execute an ARM machine code program

The component demonstrations are short and varied. They mostly appear in chapters 3 through 7 as working examples to introduce the Verilog language, the FPGA boards, and computer architectural building blocks. They can be identified by their top level Verilog module name being "TopLevel."

Working CPU models begin in Chapter 8 with source code having the top level module name of "CPU_UI" (CPU User Interface). The Verilog code in chapters 8 though 10 build a simple CPU for demonstration of basic CPU operation.

Beginning in Chapter 11, the full 32-bit ARM instruction format is used which contains the modules shown in Figure F.1. Each chapter begins with a file that only assembles ARM "assembly language statements" into ARM machine code. These files typically only contain modules CPU_UI, CPU, and ProgMod along with the macros, tasks, and functions that perform the assembly conversion.

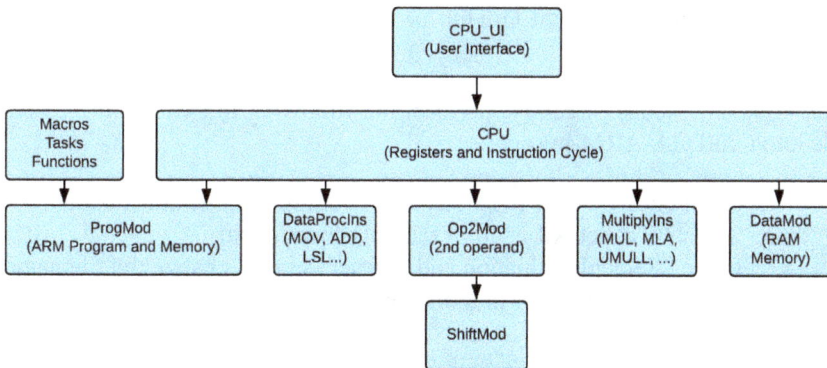

Figure F.1: Verilog modules for assembly and execution of ARM instructions

The remaining files in each chapter not only contain the assembler, but also the ARM imitation and ARM application programs. All of the modules shown in Figure F.1 are typically included in each of these files.

Each text file containing source code is complete. Every chapter in this book adds new features to the assembly and execution of ARM instructions, so the modules from one text file may not be compatibile with corresponding modules in another text file.

Source Code Download

This book contains almost 100 examples of Verilog coding. I have made them available on the Internet so they can be easily downloaded using the GitHub website. GitHub "is a code hosting platform for version control and collaboration." It is composed of multiple public and private "repositories" holding text, image, and video files.

Enter the following command in your Internet browser to initiate the load of all the listings in this book.

https://github.com/robertdunne/FPGA-ARM

I recommend that you download and unpack the source code files. If you are already familiar with and have experience with GitHub, then use a procedure with which you are most comfortable. Otherwise, please perform the following steps at the GitHub site:

1. Click on the green button labeled "Code" which will bring up a drop-down menu.
2. Select "Download Zip" from the drop-down menu which will download one file to your normal downloads directory.
3. You may now exit GitHub or close your browser since you will no longer need it.

From your downloads directory, perform the following to extract all the source code into C:\FPGA-ARM-master:

1. Right click on the FPGA-ARM-master.zip file just downloaded.
2. Select "Extract All..." from the pull-down menu.
3. In the "Select Destination and Extract" screen, change the file name to "C:\FPGA-ARM" or to a different directory name you chose for your work files.
4. Click on the "Extract" button.

The above procedure will generate all of the listing files as TXT files having file names corresponding to the captions under each listing in this book. Each will have to be copied and pasted as needed. The CSV files may be needed if you cannot locate your own for the FPGA boards. In addition, all GitHub repositories should have a README.md file containing pertinent information regarding the rest of the files.

Warning: The Verilog source code that appears in this book and is available for download is for learning computer architecture. No guarantee of its commercial utility is expressed or implied.

Listing	Description
3.1	Connect 10 switches to 10 LEDs
3.2	Example of nested modules in Verilog
4.1	Connect 7 switches to the 7 segments of HEX0
4.2	Binary on 7-segment displays
4.3	Combine *if* with *else*.
4.4	Base 4 display
4.5	Base 4 display using *case* statement
4.6	Base 8 display
4.7	Hexadecimal 7-segment display function
5.1	Logical AND gate with two inputs
5.2	Logical NOT with multiple outputs, but only one input
5.3	Logical AND gate
5.4	Logical AND gate
5.5	Structural model for elevator door example
5.6	Dataflow model for elevator door example
5.7	Behavioral model for elevator door example
5.8	Behavioral model with better design
5.9	Structural model for XOR from AND, OR, and NOT
5.10	Dataflow model for XOR from AND, OR, and NOT
5.11	Behavioral model for XOR from AND, OR, and NOT
5.12	Structural model for full adder
5.13	Dataflow model for full adder
5.14	Behavioral model for full adder
5.15	Verilog figures out the gates for full adder
5.16	4-bit adder using structural approach

Verilog code available for download and used as examples

Appendix G
ARM Instruction Format

Figure G.1 summarizes the machine code format of the ARM instructions demonstrated in this book. The following four instruction types are present:

1. Data processing: All arithmetic and logic instructions except multiplication.
2. Multiplication: All factors are in 32-bit registers; some products are 32 bits and others are 64 bits.
3. Branch: Change location for next instruction to execute
4. Load/store: Move a single byte or word between memory and a register.

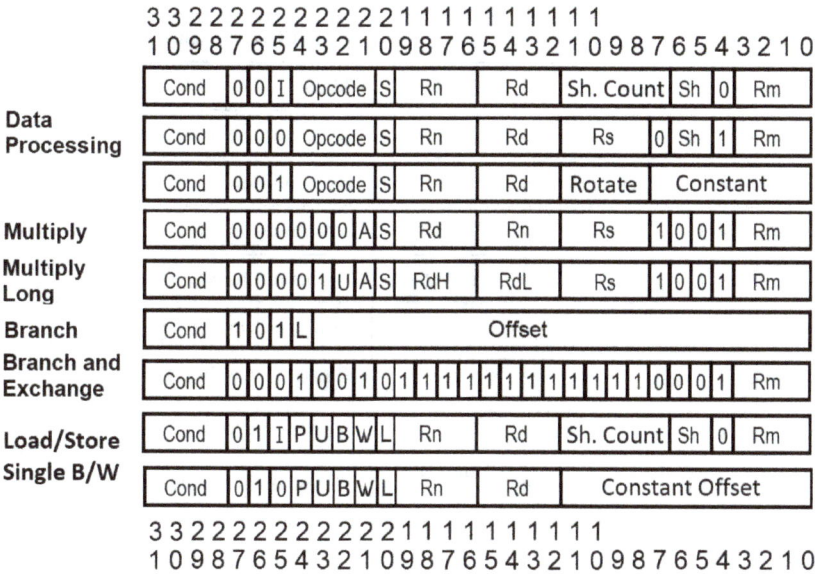

```
                3 3 2 2 2 2 2 2 2 2 2 2 1 1 1 1 1 1 1 1 1 1
                1 0 9 8 7 6 5 4 3 2 1 0 9 8 7 6 5 4 3 2 1 0 9 8 7 6 5 4 3 2 1 0

                  Cond  |0|0|I| Opcode |S| Rn  | Rd  | Sh. Count |Sh|0| Rm
Data
Processing        Cond  |0|0|0| Opcode |S| Rn  | Rd  | Rs      |0|Sh|1| Rm

                  Cond  |0|0|1| Opcode |S| Rn  | Rd  | Rotate | Constant

Multiply          Cond  |0|0|0|0|0|0|A|S| Rd  | Rn  | Rs    |1|0|0|1| Rm

Multiply
Long              Cond  |0|0|0|0|1|U|A|S| RdH | RdL | Rs    |1|0|0|1| Rm

Branch            Cond  |1|0|1|L|            Offset

Branch and
Exchange          Cond  |0|0|0|1|0|0|1|0|1|1|1|1|1|1|1|1|1|1|1|1|0|0|0|1| Rm

Load/Store        Cond  |0|1|I|P|U|B|W|L| Rn  | Rd  | Sh. Count |Sh|0| Rm
Single B/W
                  Cond  |0|1|0|P|U|B|W|L| Rn  | Rd  |   Constant Offset

                3 3 2 2 2 2 2 2 2 2 2 2 1 1 1 1 1 1 1 1 1 1
                1 0 9 8 7 6 5 4 3 2 1 0 9 8 7 6 5 4 3 2 1 0 9 8 7 6 5 4 3 2 1 0
```

Figure G.1: ARM machine codes used in this book

Conditional Execution Field

The high order four bits of every instruction determine whether the instruction will be executed or skipped. This four bit code in the cond field indicates what combination of the N (negative), Z (zero), C (carry), and V (overflow) status bits must be set or clear. These four status bits are contained in the 32-bit CPSR (Current Program Status Register) and are set by a previously executed data processing or multiplication instruction having the "S" bit set (bit 20 in the instruction).

N	**Negative**: Previous operation result was negative (i.e., bit 31 = 1)
Z	**Zero**: Previous operation result was zero (i.e., bits 31..0 = 0)
C	**Carry**: Previous operation resulted in a value that exceeded 32 bits.
V	**Overflow**: Previous operation resulted in a possible "sign" error.

Copy of Table 11.1: Status bits in the CPSR

Cond code	Assembly mnemonic	Necessary status bits	Meaning
0	EQ	Z	EQual (equals zero)
1	NE	!Z	Not Equal
2	CS or HS	C	Carry Set (unsigned Higher or Same)
3	CC or LO	!C	Carry Clear / unsigned LOwer
4	MI	N and !C	MInus
5	PL	!N	PLus (positive or zero)
6	VS	V	oVerflow Set
7	VC	!V	oVerflow Clear
8	HI	!C and !Z	HIgher (unsigned)
9	LS	!C or Z	Lower or Same (unsigned)
10	GE	N = V	Greater than or Equal (signed)
11	LT	N != V	Less Than (signed)
12	GT	!Z and (N = V)	Greater Than (signed)
13	LE	Z or (N != V)	Less than or Equal (signed)
14	AL	Always	Default (same as omitted)
15	Never	Reserved	Code 15 for future ARMs

Copy of Table 11.4: List of ARM assembly language condition codes

Data Processing Instructions

The Data Processing instructions include all of the arithmetic and logical operations except multiplication. There are basically three general formats, depending on the two "immediate" flag bits at bit positions 4 and 25.

1. **SUB R1, R2, #4, 2:** If bit 25 is set, then the second operand is a constant composed of an 8-bit value (bits 0 through 7) that is rotated to the right by two times the value in bits 8 through 11. If only one number is present in assembly language, then the rotate amount is assumed to be zero.
2. **SUB R1, R2, R3, LSR #17:** The second operand is a register that is shifted a constant amount (bit 4 = 0).
3. **SUB R1, R2, R3, LSR R6:** The second operand is a register that is shifted by the amount specified in another register (bit 4 = 1).

Algebraically, the general format of a subtraction instruction is $D = N - M \times 2^S$, where D and N represent register contents, and M and S can be either register contents or constants. This format is true of every one of the sixteen data processing instructions:

- **D**: Destination register to hold the 32-bit result
- **N**: First operand register: In subtraction, it is the minuend (quantity from which another quantity is to be subtracted)
- **M×2^S**: Second operand: In subtraction, it is the subtrahend (quantity to subtract). There are three possible formats for M and S:
    1. **M** and **S** are both constants: This is somewhat like scientific notation where M is a constant (0 through 255) and S is a shift count (0 through 30, even integers only).
    2. **M** is a register and **S** is a constant: The contents of register M are shifted (logical, algebraic, or circular) by a constant (range of 0 through 31).
    3. **M** and **S** are both in registers: The contents of register M are shifted by the value in register S.

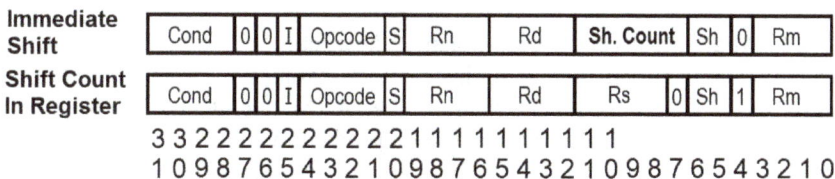

	Cond	0	0	I	Opcode	S	Rn	Rd	Sh. Count	Sh	0	Rm
Immediate Shift	Cond	0	0	I	Opcode	S	Rn	Rd	Sh. Count	Sh	0	Rm
Shift Count In Register	Cond	0	0	I	Opcode	S	Rn	Rd	Rs	0	Sh	1	Rm

```
3 3 2 2 2 2 2 2 2 2 2 2 1 1 1 1 1 1 1 1 1 1
1 0 9 8 7 6 5 4 3 2 1 0 9 8 7 6 5 4 3 2 1 0 9 8 7 6 5 4 3 2 1 0
```

Copy of Figure 12.18: Second operand contains a register with a shift.

Op Code Value	Op Code Name	Arithmetic or Logical Operation
0	AND	R <= R & op2;
1	EOR Exclusive OR	R <= R ^ op2;
2	SUB	R <= R - op2;
3	RSB Reverse Subtract	R <= op2 - R;
4	ADD	R <= R + op2;
5	ADC	R <= R + op2 + C;
6	SBC	R <= R - op2 + C - 1;
7	RSC	R <= op2 - R + C - 1;
8	TST	Set status for R & op2
9	TEQ	Set status for R ^ op2
10	CMP	Set status for R - op2
11	CMN	Set status for R + op2
12	ORR inclusive OR	R <= R	op2;
13	MOV Move (i.e., copy)	R <= op2;
14	BIC Bit Clear	R <= R & ~op2;
15	MVN Move NOT	R <= ~op2;

Copy of Table 11.6: All 16 ARM data processing instructions

Multiplication

In the ARM architecture, the multiplication instructions have a different structure than the other arithmetic and logic instructions. Multiplying two 32-bit numbers results in a 64-bit product. Within the ARM architecture, four of the multiplication instructions produce a 64-bit product, while two of the instructions only give the lower 32 bits (hoping the upper 32 bits are zeroes).

	Cond	0	0	I	Opcode	S	Rn	Rd	Rs	0	Sh	1	Rm
Data Processing	Cond	0	0	I	Opcode	S	Rn	Rd	Rs	0	Sh	1	Rm
Multiply	Cond	0 0 0 0 0 0	A	S	Rd	Rn	Rs	1 0 0 1	Rm				
Multiply Long	Cond	0 0 0 0 1	U	A	S	RdH	RdL	Rs	1 0 0 1	Rm			

```
3 3 2 2 2 2 2 2 2 2 2 2 1 1 1 1 1 1 1 1 1 1
1 0 9 8 7 6 5 4 3 2 1 0 9 8 7 6 5 4 3 2 1 0 9 8 7 6 5 4 3 2 1 0
```

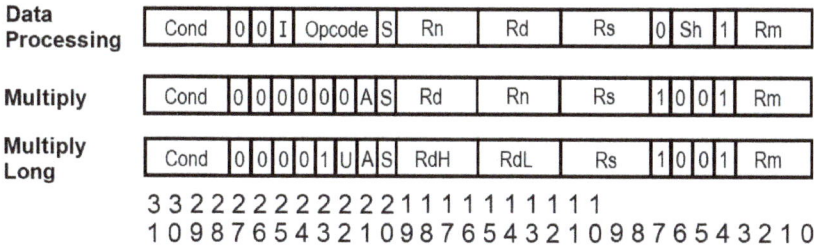

Copy of Figure 13.1: Compare bit positions in multiply instructions to data processing instructions

Three of the multiply instructions have the "A" bit set so they accumulate their resulting product to a running sum. If the "S" flag is set in the instruction, then the Negative and Zero status bits will be updated. The Carry and oVerflow CPSR status bits really don't have meaning for multiplication like they do for the data processing instructions.

- **MUL:** Rd <= (Rs × Rm) [31 : 0]
- **MLA:** Rd <= (Rs × Rm + Rn) [31 : 0]
- **SMULL:** {RdH, RdL} <= Rs × Rm
- **SMLAL:** {RdH, RdL} <= Rs × Rm + {RdH,RdL}
- **UMULL:** {RdH, RdL} <= Rs × Rm
- **UMLAL:** {RdH, RdL} <= Rs × Rm + {RdH,RdL}

Branch Instructions

The location of the next instruction to be executed can be set by branch instructions. The BL (Branch with Link, bit 24 = 1) saves a return address in the Link Register. Note: The B and BL instructions specify the "branch to" location by a relative <u>word</u> offset (signed 24-bit field). The BX (Branch and eXchange) instruction uses the absolute <u>byte</u> address in register Rm for the next instruction.

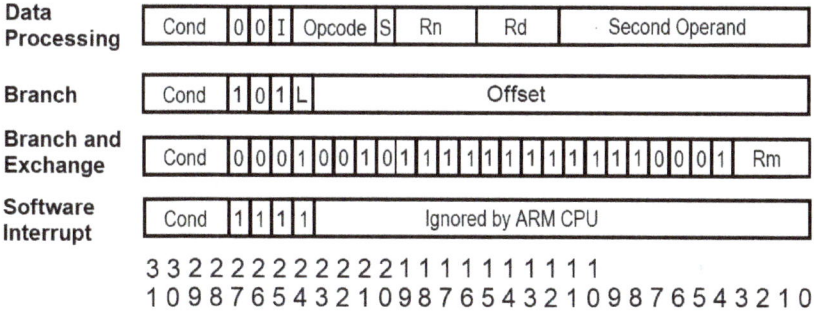

	Cond								
Data Processing	Cond	0 0	I	Opcode	S	Rn	Rd	Second Operand	
Branch	Cond	1 0 1	L	Offset					
Branch and Exchange	Cond	0 0 0 1 0 0 1 0 1 1 1 1 1 1 1 1 1 1 1 1 0 0 0 1	Rm						
Software Interrupt	Cond	1 1 1 1	Ignored by ARM CPU						

```
3 3 2 2 2 2 2 2 2 2 2 2 1 1 1 1 1 1 1 1 1 1
1 0 9 8 7 6 5 4 3 2 1 0 9 8 7 6 5 4 3 2 1 0 9 8 7 6 5 4 3 2 1 0
```

Copy of Figure 13.4: Compare branch to data processing and software interrupt instructions

Memory Load and Store Instructions

There are three addressing modes for the LDR and STR instructions which move data between a single register and a memory location:

- Indirect: The base register contains the complete memory address. Example: LDR R0,[R4].
- Pre-Indexed: The memory address is calculated by adding an offset to the contents of a base register. Example: LDR R0,[R4,#56].
- Post-Indexed: The base register contains the complete memory address, but then the base register contents will be updated after the address is used in the current instruction. Example: LDR R0,[R4],#56.

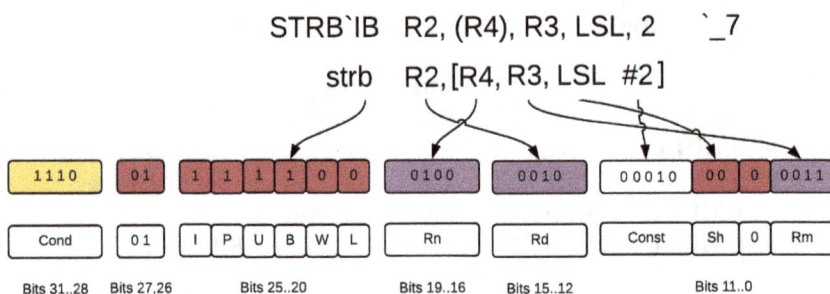

Copy of Figure 16.4: Store Byte pre-index mode instruction

- Cond (bits 31..28): Status indicating whether the instruction should be executed (1110 indicates always execute).
- 0 1 (bits 27,26): Opcode indicating that this is an LDR or STR.
- IPUBWL (bits 25..20): Bits indicating specifically what is to be done by the LDR/STR instruction.
    - I (bit 25): "1" indicates scaled Index (register with a shift), "0" indicates offset is immediate 12-bit constant.
    - P (bit 24): "1" indicates Pre-indexed, "0" for post-indexed.
    - U (bit 23): "1" indicates Up (add the offset), "0" indicates subtract the offset.
    - B (bit 22): "1" for Byte (STRB), "0" for word (STR).
    - W (bit 21): "1" indicates Write back. The exclamation point (!) to the assembler sets this bit.
    - L (bit 20): "1" indicates Load (LDR), "0" for store (STR).

- Rn (bits 19..16): Base register, R0 through R15
- Rd (bits 15..12): Register to contain result, R0 through R15
- Offset (bits 11..0): Immediate range of +4095 to -4095 (I-bit = 0) or index register scaled by a shift (I-bit = 1).

Answers to
Selected Questions

Questions marked with an asterisk (*) at the end of each chapter have their answers, or at least hints, provided below.

1.3 If a netlist is successfully downloaded, why might the FPGA not work properly as expected? In other words, why might the same source code work for one person, but not another (i.e., the LEDs do not light at all when expected)?

> a. The wrong target device may have been selected. Some FPGAs are close enough to accept the same netlist, but are definitely not the same (Figure 1.8).
> b. The wrong assignments file may have been imported. The hardware description is actually working, but signals are going to the wrong pins (Figure 1.20).

1.4 Name two reasons why the USB Blaster device driver is not available and does not appear on the download (i.e., program) page?

> a. The USB blaster was never downloaded to the computer being used (see Appendix A).
> b. The USB cable is connected to the wrong USB port on the FPGA board, and the operating system does not know it is there (Figure 1.26).

2.1 How can a NAND gate be converted to an inverter?

> Tie the two inputs together.

2.2 Why might you change the automated "instance" name (Figure 2.2) generated by the Quartus Prime compiler?

> a. For documentation purposes (match a drawing or other description).
> b. Sometimes, two gates get assigned the same instance name causing a fatal error, so one has to be changed to a different name.

2.4 Why is it that the data bus requires tri-state logic while the other two buses do not?

> The data bus can be written to by many devices: the CPU, memory chips, I/O devices, etc. The address and control buses are only written

by the CPU or its assistant such as a decoder.

2.8 How can an arch be a metaphor for describing a flip flop?

> In an arch, one side holds up the other. During construction, scaffolding is needed to hold up the arch, but after completion, the arch holds itself up. In an RS flip flop composed of two NAND gates, the output from each NAND gate "holds up" the other NAND gate, and is thereby stable.

3.2 Although we will generally ignore "blue" messages in the classroom, why should they all be checked in real production work?

> The warnings can indicate real coding errors like forgetting to connect a signal to an output port.

3.3 Why do you think we do not make all ports *inout* rather than specifically *input* or *output*?

> a. The *inout* capability should be reserved for true bidirectional lines that may need tri-state capability such as memory data lines.
> b. The Quartus Prime software can catch coding errors like having an input signal on the left side of an equals sign.

4.2 The Terasic boards light up their LED segments when they receive zeros, and the segments are off when they receive ones. About half the seven-segment displays on the market work this way, and the other half work the opposite. Search the Internet to find the names associated with the two types of displays. Which kind is used in the DE2-115 and DE10-Lite boards?

> The boards behave like they have "common anode" displays. If they were "common cathode" they would light when receiving a HIGH level (positive logic 1).

4.3 The seven segments of a seven-segment display are commonly assigned letter names. Search the Internet to find the names that would be assigned to segments used in this chapter.

> Bit 0 is is assigned to segment A at the top of the display. Bits 1 through 5 are assigned clockwise from there, and bit 6 is the middle of the display. In the examples switches 0 through 6 correspond to A through G, respectively.

4.4 "By hand, without a calculator or computer," convert the following numbers expressed in decimal to binary format. See Appendix C if you need some background in binary.

a. $21_{10} = 10101_2$
b. $63_{10} = 111111_2$
c. $16_{10} = 10000_2$
d. $129_{10} = 10000001_2$

4.5 "By hand, without a calculator or computer," convert the following numbers expressed in binary to decimal format. See Appendix C if you need some background in binary.

a. $1011_2 = 11_{10}$
b. $1100101_2 = 101_{10}$
c. $10110_2 = 22_{10}$
d. $100001_2 = 33_{10}$
e. $1111011_2 = 123_{10}$

5.4 Name a few benefits of the behavior coding style.

a. It's typically easier to write and maintain than structural or dataflow approaches.
b. Behavioral coding is somewhat self documenting. It closely represents the application description.
c. The Verilog compiler is usually much more knowledgeable of the internal features available on the target FPGA or CPLD. It can employ on-board multipliers and other special features to optimize performance.

6.3 How many input lines are needed for a decoder to select one of a possible 64 different outputs?

Six lines. See Table 6.1.

6.4 What is the difference between *tri* and *wire* variable types?

From an operational standpoint, it's only the spelling. However, from a documentation standpoint, a *tri* type variable would be used with a bus or other construct accepting tri-state values. With SystemVerilog, the *logic* type could also be used, but it would even be less helpful from a documentation perspective.

6.5 Convert the following two concatenations to 32-bit hexadecimal values: {8'b101, 16'hA47D, 4'd11, 4'o12}, {5'31, 9'hA4, 10'b11, 8'o11}

{8'b101, 16'hA47D, 4'd11, 4'o12} = {8'h05, 16'hA47D, 4'hB, 4'hA} = 32'h05A47DBA
{5'31, 9'hA4, 10'b11, 8'o11} = {5'b11111, 9'b10100100, 10'b11, 8'b001001} = 32'hFA900309

7.1 What is the largest unsigned number that can be placed in the operand field of the current 10-bit instruction format?

> The largest whole number in six bits is 63. See Appendix C for more details.

7.2 If the opcode field was five bits instead of four, how many opcodes would be possible?

> A 5-bit opcode field could support 32 different opcodes.

8.2 Each state of the instruction cycle (fetch, decode, and execute) is composed of its own circuit which is parallel to and mostly independent of the other states' circuits. What do the electronic circuits that perform the fetch and decode states do when the execute state is active?

> Basicly, nothing. They are just waiting for their turns. However, a technique known as "pipe lining" allows a CPU to look ahead and start processing the next instruction to be executed even before the current one is finished. That feature is not implemented in the CPU imitation in which we are currently working.

8.3 As described in this chapter, convert the instruction "ORR 13" to hexadecimal machine code.

> The machine code is 'hC0D unless the operand was assumed to be in hexadecimal in which case it would then be 'hC13. See Tables 7.2 and 8.3

8.4 As described in this chapter, convert the hexadecimal machine code 'h123 back into a text version of the instruction.

> The "human" text version is "EOR 'h23" or EOR 35 (decimal). See Tables 7.2 and 8.3

9.1 How does the Verilog macro capability (*define* statement) compare to that of a Verilog *function*?

> The macro capability within Verilog is basically a string substitution feature, and it transcends all modules after being defined. The Verilog *function* actually generates circuits. Also, the function is somewhat flexible in what it produces (*if* and *else* statements).

9.3 What is the scope (range of definition) of a *define* macro compared to that of the *parameter* statement?

> The macros built using the *define* statement are available in all modules following the macro definition. The *parameter* variables are

unique to the module in which they are created and not available to other modules.

10.1 Why is the ProgMem module "called" where it is instead of where it "logically" makes sense down in the fetch state?

> This is hardware, not software. A module cannot be "called" from behavioral coding within an *always* block. The ProgMem code is activated when its clock transitions.

10.4 What logical identity can be used to generate the AND operation from the other logical operations?

> Using DeMorgan's theorem, A & B = ~(~A | ~B), so the AND can be replaced by an OR with three NOTs.

11.1 Sequential circuits described in a Verilog always block should use non-blocking assignments (i.e., use <= instead of =). Why does line 108 of Listing 11.3 have "IP = 0" instead of "IP <= 0"?

> The integer IP is an "assembly time" variable that is neither blocked nor nonblocked. Inside the task on lines 94 and 96, the non blocked assignments using <= are present.

11.2. The factorial program in Listing 11.6 C won't work on an actual ARM processor, exactly as it is written. What is wrong with how it substitutes MOV PC instructions for real branch instructions?

> The value of the address labels, FLoop and MLoop, are different when a program runs than when it is assembled. This difference is primarily due to today's sophisticated operating systems which move programs to various absolute memory locations. This relocation will be discussed in upcoming chapters on memory and branching.

12.1 How can the decimal value 5120 be loaded into a 32-bit register with a single MOV instruction? Clue: 5120 = 4096+1024.

> MOV is one of the data processing instructions, all of which represent their immediate constants by an 8-bit base M rotated to the right by a 4-bit shift count S. The decimal value 4096 is 1×2^{12}, while 1024 is 1×2^{10}. Adding the two in binary is 0b1010000000000, which can be represented as 0b101 shifted left 10 bit position or rotated right 22 bit positions. Therefore the lower 12 bits of the MOV instruction which contains the shift count 11 and base 0b101 is 0xB05. Other possibilities that would work are 0xC14 and 0xD50.

13.1 "By hand, without a computer," convert the following decimal fractions into binary and provide the answers in hexadecimal.

a. $0.5_{10} = \frac{1}{2} = 0.1000_2 = 0.8_{16}$
b. $0.625_{10} = 0.1010_2 = 0.A_{16}$
c. $0.25_{10} = 0.0100_2 = 0.4_{16}$
d. $0.03125_{10} = 0.00001_2 = 0.08_{16}$
e. $0.0078125_{10} = 0.0000001_2 = 0.02_{16}$

13.2 "By hand, without a computer," convert the following binary fractions from hexadecimal back into real numbers in base 10.

a. $.C0000000_{16} = 0.11_2 = \frac{1}{2} + \frac{1}{4} = \frac{3}{4} = 0.75_{10}$
b. $.E0000000_{16} = 0.111_2 = 0.875_{10}$
c. $.10000000_{16} = 0.0001_2 = 0.0625_{10}$
d. $.50000000_{16} = 0.0101_2 = 0.3125_{10}$

13.4. How can you load $\frac{3}{4}$ into a register using a MOV instruction?

`MOV R1,3,2 // Load 8'hC0000000

13.5 Using a calculator for division by powers of 2, convert the following binary fractions from hexadecimal into the decimal fraction that each is "approaching."

a. $.33333333_{16} = 0.00110011001100110011001100110011_2 \Rightarrow 0.2_{10}$
b. $.66666666_{16} \Rightarrow 0.4_{10}$
c. $.CCCCCCCC_{16} \Rightarrow 0.8_{10}$
d. $.E6666666_{16} \Rightarrow 0.9_{10}$

13.6 Why will multiplying by 0.1 always result in a loss of precision in binary computers?

Base ten is not a multiple of base 2, like base 8 and base 16 are multiples. Some numbers like 0.1 cannot be represented exactly as a base two fraction for the same reason $1/3$ is $0.333333..._{10}$.

14.1 The addressing of computer memory has been a nightmare for decades. A large part of the problem has ironically been due to a good factor: the tremendous drop in memory prices. Please explain?

Large address spaces generally require a lot of bits in the instruction word, and CPU designers were reluctant to address much more memory than what seemed practical. In every CPU design, at least some of the instructions have to access memory. When the amount of available memory was small because it was extremely expensive, a reasonable maximum memory size was 65,536 (65K or 64K if you say K = 1024). This memory size requires a 16-bit address. Many mainframes, minicomputers, and microcomputers had a 65K addressing limit built into their instruction formats. As memory prices dropped much faster than CPU designs evolved, computer hardware systems designers and systems programers deployed a variety of memory management techniques to shoehorn in much larger memories into a restricted addressing space design.

14.2. Compare the merits of a byte-addressable computer architecture to one that is word addressable.

Word-addressable:

- Don't have to be concerned about big and little endian.
- Don't have to worry about alignment issues. Many CPUs will either degrade performance or not load/store "word" instructions on addresses that are not multiples of four bytes.

Byte-addressable:

- No complicated sub-word instruction options are needed. Many word-addressable CPUs had a special field within their instruction format to select a particular sixth or quarter word. Some didn't even have that, so bytes had to be masked and shifted into place.
- A variety of integer sizes are available: bytes, words, half-words, and double words.

15.1. Compare Verilog functions to functions in software? Likewise, how is a Verilog task different from a subroutine?

- A macro is called while the assembler is running, and a subroutine is called when the application program (being written) is running.
- A macro generates text lines that will later be "assembled," while a subroutine works with numbers and text of the running application.
- Each macro call makes the program physically larger and take up more memory, while subroutines generally reduce memory requirements by eliminating duplicate code.

15.4 Since recursive subroutines are less efficient, what is their advantage?

Many times the coding is shorter and easier to understand. This can result in quicker program development and better long term maintenance. However, the exit conditions must be clearly understood or else an infinite loop may result.

16.1 The second operand in the LDR and STR instructions is very similar to that of the data processing instructions. Both can be an immediate value or a register with a shift. What are two differences between the formats?

In the LDR/STR format, the 12 bit field is a straight 12 bits having a range of +-4095, but in the data processing instructions, it is an 8-bit base shifted by a 4-bit shift count. Secondly, the second operand can be taken from the contents of a register that is shifted. For data processing instructions, the shift count can be either a constant or the value contained in another register, but in the LDR/STR the shift count can only be a constant.

16.2. The LDR and STR instructions allow a negative offset. Is this negative direction set with two's complement or sign and magnitude format?

Technically, it's in the sign and magnitude format, but the sign and magnitude are not adjacent to each other. The "sign" is actually the U-bit in the LDR/STR instruction format, where U=1 means positive and U=0 means negative.

19.3. We switch between Thumb mode and ARM mode with the BX instruction. Although we could set the T-bit in the CPSR directly, why would this lead to problems if we did?

The quick and easy answer is that the ARM hardware reference manuals say don't do it, and a special way has been set up to perform the switch using the BX instruction. But also remember that the ARM has a "pipelined" architecture where by the time an instruction is actually executed, the following two instructions have already been fetched and are being prepared to execute. The 16-bit/32-bit instruction mode would be switched while instructions are in an intermediate state. The BX, like all branch instructions that execute, will clear the pipeline.

20.1. If the lane size for integer addition within NEON could be one bit wide, it would be exactly the same as which logical operation?

The exclusive OR operation. See the half-added in Figure 2.16.

20.2. When does a VADD.U8 produce a different result than a VADD.S8? How about the same question for S16 verses U16 or S32 verses U32 for any of the VADD, VSUB, etc. NEON instructions?

Never. Take a hex dump of the instructions produced by an assembler and see the "difference" in the code. For saturation arithmetic, VQADD, etc., the answer is quite different. Give it a try. You will like the results. Don't forget to try I8, I16, ect.

21.1 If R and V are two 16-bit integer vectors dimensioned as R[1000] and V

[1000,2]. If you needed to calculate R[i] = V[i,1] × V[i,2], which interleaved load and store instructions would you most likely use?

VLD2.16   {Q0-Q3}, [R1]! and VST2.16   {Q0-Q3}, [R1]!

21.2 If R and V are two 32-bit floating point vectors dimensioned as R[1000] and V [1000,3]. If you needed to calculate R[i] = V[i,1] × V[i,2] × V[i,3], which interleaved load and store instructions would you most likely use?

VLD3.32   {Q0 - Q2}, [R1]! and VST3.32   {Q0 - Q2}, [R1]!

22.1. What are the advantages of fixed point over floating point?

- Fixed point is exact. No error is present.
- Fixed point arithmetic is very fast.
- It is very easy and fast to convert between character representation and fixed point.

22.2 "By hand, without a computer," convert the following real numbers into single precision IEEE 754 floating point and provide the answers in hexadecimal.

a. 128.0 is 43000000 in floating point
b. 9.25 is 41140000 in floating point
c. -9.25 is C1140000 in floating point
d. 0.03125 is 3D000000 in floating point
e. 128.03125 is 43000800 in floating point
f. 0.0 is 00000000 in floating point
g. -0.0 is 80000000 in floating point

22.3 "By hand, without a computer," convert the following IEEE 754 floating point numbers from hexadecimal back into real numbers in base 10.

a. 42a80000 is 84.0
b. C1A80000 is -21.0
c. 424C8000 is 51.125
d. BF100000 is -0.5625
e. 3DCCCCCD is 0.1

Computer Architecture Tutorial Using an FPGA

Figures & Tables

A short description of each table and illustration appearing in the book follows:

Figure	Description
1.1	AND operation truth table
1.2	Random letters floating in soup or cereal
1.3	Gates "wired" to form netlist
1.4	Compiler converts digital design's source code into netlist for download to FPGA device.
1.5	Quartus Prime opening page
1.6	Use Project Wizard to specify target device and source language
1.7	Identify the project
1.8	Target device
1.9	Verify project location and target device
1.10	Specify source language
1.11	Specify schematic entry
1.12	Four commonly used tools for building digital schematics
1.13	Select an AND gate having two inputs.
1.14	Select input pin
1.15	Connect input pin to gate
1.16	Add output pin and provide pin names
1.17	Start compilation
1.18	File must be updated
1.19	Import list of pin name to pin locations
1.20	Select file containing the pin assignments
1.21	Pin location for each pin name
1.22	Switches and LED oin assignments on DE2-115 board
1.23	Programmer (download)
1.24	Click on the Schematic file name, then click "Start" to download

Figures & Tables

Computer Architecture Tutorial Using an FPGA

Illustrations appearing in this book

Figures & Tables   395

Tables appearing in this book

Glossary

This modified glossary summarizes ARM instructions and Verilog commands presented in this book. Each keyword (instruction or command) has a short description and two references to examples where it is used.

The first example is where the keyword first appears, while the second demonstrates a more complicated or typical use. Each reference contains a three number field indicating the chapter, listing number, and text line number where the keyword is used. For example, 3.5.27 indicates the keyword is on line 27 of Listing 5 in Chapter 3.

Keyword	Example 1	Example 2	Short Description
{ }	5.15.8	9.2.73	Braces used to concatenate bits
&	5.4.7	7.1.21	Bit by bit logical AND
& &	8.5.75	11.6.151	Conditional AND used in "if" statement
=	5.19.11	5.19.12	Blocking assignment in behavioral style
= =	4.2.9	10.2.109	Test for equality
< <	11.4.145	12.3.205	Compile-time logical shift left
<=	5.21.13	13.3.192	Non-blocking assignment in behavioral style
? :	6.1.5	15.3.203	Conditional operator: Condition ? Do if True : Do if False;
always	5.4.6	5.20.8	Beginning of block of behavioral coding
and	5.1.4	8.2.7	Logical AND gate (structural)
assign	3.2.11	5.3.4	Dataflow continuous signal path
begin	4.2.7	15.3.202	Start group of Verilog commands
case	4.5.3	7.1.20	Begin multiple conditions
define	9.2.38	11.4.69	Macro command specifying text substitution

else	4.3.5	15.1.145	Alternative circuit to if statement
end	4.2.10	15.3.217	End group of Verilog commands
endcase	4.5.8	7.1.31	End multiple conditions
endfunction	4.2.11	15.1.135	Close function description
endmodule	3.2.6	11.3.117	End of a Verilog circuit description
for	6.2.9	12.1.31	Loop to generate multiple Verilog commands
function	4.2.5	15.1.129	Produce one value or subcircuit at compile time
generate	6.2.8	12.5.35	Supports compile-time loops and conditions
genvar	6.2.4	12.5.34	Declare variable used within generate block
if	4.2.8	13.3.197	Single circuit for specific condition
input	3.2.2	9.1.39	Input port interface description
integer	6.5.5	12.1.29	Declare compile-time variable
module	3.2.1	11.3.68	Beginning of a Verilog circuit description
not	5.2.4	5.9.5	Logical NOT gate (structural)
or	5.5.6	5.9.9	Logical OR gate (structural)
output	3.2.3	9.1.41	Output port description
parameter	8.1.9	12.1.10	Substitute name for constant value (such as PI for 3.14)
reg	5.7.4	10.2.88	Signal lines and registers used in always block
task	10.1.89	11.3.91	Produce one or more compile-time values or subcircuits
wire	5.5.4	10.2.86	Structural continuous connection between components
xor	5.9.10	5.12.5	Logical Exclusive OR gate (structural)

Table Glossary.1: Verilog Keywords and commands

Opcode	Example 1	Example 2	Short Description
ADC	11.5.228	.	Add with carry
ADD	7.1.25	11.6.396	Addition
ALIGN	14.1.67	14.1.183	Align IP to multiple of bytes
AND	7.1.21	.	Logical AND (32 bits in parallel)
ASR	12.1.14	.	Algebraic Shift
B	15.1.62	15.4.749	Branch
BIC	7.1.28	11.5.391	Bit clear
BL	15.1.63	15.4.740	Branch with Link
BYTE	14.1.66	14.1.182	Initialize 8-bit byte in memory
BX	15.1.64	16.5.803	Branch and Exchange
CMN	11.5.234	.	Compare to Negative (status bits like ADD)
CMP	11.5.233	15.4.746	Compare two values (status bits like SUB)
EOR	7.1.22	11.5.393	Logical exclusive OR
LDR	14.1.63	16.4.803	Load full 32-bit R register
LDRB	14.1.64	14.1.178	Load lower 8 bits and zero fill upper 24 bits of register
LSL	12.1.12	13.5.597	Logical Shift Left
LSR	12.1.13	12.3.279	Logical Shift Right
MLA	13.3.115	.	Multiply, accumulate 32-bit product
MOV	7.1.27	11.6.390	Move (load register from register or constant)
MUL	13.3.114	15.4.746	Multiply giving 32-bit product
MVN	7.1.29	.	Move NOT (one's complement)
ORR	7.1.26	13.5.599	Logical inclusive OR
POP	16.3.62	16.5.808	Reload and remove register contents from the stack
PUSH	16.3.61	16.5.804	Append register contents to the stack
ROR	12.1.15	.	Rotate (circular) Right

Glossary

RSB	7.1.24	.	Reverse subtraction (minuend can be a register or constant)
RSC	11.5.230	.	Reverse subtraction with carry
SBC	11.5.229	.	Subtract with carry
SMLAL	13.1.181	.	Signed 64-bit product, accumulate
SMULL	13.3.116	.	Signed 64-bit product
STR	14.1.61	16.4.794	Store full 32-bit R register
STRB	14.1.62	.	Store low-order byte in register
SUB	7.1.23	13.5.600	Subtraction (subtrahend can be register or constant)
TEQ	11.5.232	.	Test Equal (status bits like EOR)
TST	11.5.231	.	Test Equal (status bits like AND)
UMLAL	13.3.119	.	Unsigned 64-bit product, accumulate
UMULL	13.3.118	13.4.605	Unsigned 64-bit product
WORD	14.1.65	14.1.184	Initialize 32-bit word in memory

Table Glossary 2: ARM instructions

Index